THE STRUCTURAL FOUNDATIONS OF QUANTUM GRAVITY

The Structural Foundations of Quantum Gravity

Edited by
DEAN RICKLES,
STEVEN FRENCH,
and
JUHA SAATSI

CLARENDON PRESS · OXFORD

OXFORD
UNIVERSITY PRESS

Great Clarendon Street, Oxford OX2 6DP

Oxford University Press is a department of the University of Oxford.
It furthers the University's objective of excellence in research, scholarship,
and education by publishing worldwide in

Oxford New York

Auckland Cape Town Dar es Salaam Hong Kong Karachi
Kuala Lumpur Madrid Melbourne Mexico City Nairobi
New Delhi Shanghai Taipei Toronto

With offices in

Argentina Austria Brazil Chile Czech Republic France Greece
Guatemala Hungary Italy Japan Poland Portugal Singapore
South Korea Switzerland Thailand Turkey Ukraine Vietnam

Oxford is a registered trade mark of Oxford University Press
in the UK and in certain other countries

Published in the United States
by Oxford University Press Inc., New York

First published 2006

British Library Cataloguing in Publication Data
Data available

Library of Congress Cataloging in Publication Data
Data available

Typeset by Laserwords Private Limited, Chennai, India
Printed in Great Britain
on acid-free paper by
Biddles Ltd, King's Lynn, Norfolk

ISBN 0-19-926969-6 978-0-19-926969-3

1 3 5 7 9 10 8 6 4 2

Preface

The papers in this collection are concerned with certain *foundational* aspects of quantum gravity. We have chosen to group them together under the banner of 'structuralism'. The ways in which the various papers fall under this banner, and the extent to which they do so, are somewhat diverse. Cao, for example views quantum gravity as lending support to a structural realist conception of ontology. Stachel likewise advocates a similar view, though the two authors' views are very dissimilar in their details. Dorato and Pauri's contribution, an attempt to motivate a position they call 'structural spacetime realism', bears many similarities to Stachel's position, though they steer clear of the *relationalist* label that Stachel is happy to wear. Pooley, on the other hand, cautions against the pulling of structural realist views from the formalism (specifically from diffeomorphism invariance). Instead he argues that there is a perfectly reasonable and preferable non-structural interpretation that can be given; namely, 'sophisticated substantivalism'. Interestingly, both Stachel and Pooley see the developments of quantum gravity (or diffeomorphism-invariant physics in general) as working against the view known as 'haecceitism'—namely the view that there can be possibilities that differ non-qualitatively (solely in virtue of which objects play which role in the qualitative structure). Stachel interprets this as motivating a relationalist position according to which objects, insofar as they exist at all, are simply conceived as deriving their existence from their place in a network of relations. Such a position gives a structural characterization of objects. But, as we mentioned, Pooley interprets matters differently, arguing that a substantivalist position can be defended within an anti-haecceitistic metaphysics too. Rickles argues that the ontological issue at the root of the substantivalism/relationalism debate cannot be settled by quantum gravity (or diffeomorphism-invariant physics) since an underdetermination takes hold in these contexts. However, he shows how a structuralist conception of the observables of these theories can help to avoid a serious problem in the context of the problems of time and change they face. The chapters by Smolin and Baez in very different ways argue that a metaphysics of relations is suggested by quantum gravity. Our contention is that *this* kind of relationalism is easily and best construed as a variety of structuralism. There is a common core to the views expressed in these papers, which can be characterized as the stance that relational structures are of equal or more fundamental ontological status than objects. The present volume seeks to extend what has been a fruitful dialogue between physicists and philosophers working on quantum gravity. Our goal is fairly modest: to suggest a possible, and fitting, interpretative and ontological perspective from which to view quantum gravity physics.

Acknowledgements

The following collection of papers began life as a workshop on structural aspects of quantum gravity held at the biennial PSA meeting in Milwaukee, 2002. The workshop was organized by Steven French and his two doctoral students Dean Rickles and Juha Saatsi. The participants there were: John Baez, Dean Rickles, Juha Saatsi, Lee Smolin, and John Stachel. The present volume seeks to extend the basic theme of that workshop with contributions from the aforementioned speakers (Juha Saatsi excepted) and other physicists and philosophers of physics. We thank the Philosophy of Science Association for providing us with the forum that sparked off this book. The editors would also like to thank the Department of Philosophy, University of Leeds for its support in providing travel funds to attend PSA 2002, at which the original symposium on the Structural Foundations of Quantum Gravity was presented. In addition, Dean Rickles would like to thank the International Collaboration on Complex Interventions and the Department of Philosophy at the University of Calgary for support during the final stages of this book.

The production team at Oxford University Press were excellent throughout: the referees appointed by the press bolstered our enthusiasm for this project and hunted out some blunders; the copy-editor Edwin Pritchard did a superb job resulting in numerous improvements to the original manuscript; Jenni Craig did a fine job in helping to push the book through to completion.

Finally, Steven would like to acknowledge, as always, the support of Dean and Morgan. Dean wishes to acknowledge Kirsty, Sophie, and Gaia for their patience and much needed grounding.

Dean Rickles, Steven French, Juha Saatsi

Contents

Notes on Contributors

John Baez (born 1961) received his BA degree in mathematics from the University of Princeton in 1982, and his Ph.D. in mathematics from the Massachusetts Institute of Technology in 1986. After his postdoctoral work at Yale University he went to the University of California, Riverside, in 1989, where he now teaches in the Mathematics Department. He has always been interested in mathematical physics—at first quantum field theory, but these days, quantum gravity and the theory of n-categories. These two themes get combined in 'spin foam models' of quantum gravity. He spends a lot of time explaining these things in his column 'This Week's Finds in Mathematical Physics', which can be found at his website, **http://math.ucr.edu/home/baez.**

Tian Yu Cao received his Ph.D. degree from University of Cambridge in 1987, and now is Associate Professor of Philosophy in the Department of Philosophy, Boston University. His research interests include the application of structural realism to conceptual history and foundational issues of physical theories. He published a book *Conceptual Developments of 20th Century Field Theories* (Cambridge University Press, 1997) and edited a volume *Conceptual Foundation of Quantum Field Theory* (CUP, 1999). He is working on a book dealing with the genesis of QCD and string from the perspective of structural realism.

Mauro Dorato teaches Philosophy of Science at the Department of Philosophy and the Department of Physics of the University of Rome III, Italy. He has a degree in Philosophy and in Mathematics from the University of Rome 'La Sapienza', and a Ph.D. in Philosophy from the Johns Hopkins University, Baltimore. He has written on the relationship between the physics and the metaphysics of time (*Time and Reality*, Bologna: Clueb, 1995), on the history and philosophy of scientific laws (*The Software of the Universe*, Aldershot: Ashgate, 2005), and on the relationship between physics and metaphysics.

Steven French is Professor of Philosophy of Science at the University of Leeds. He has a degree in physics and studied for his Ph.D. on the philosophical foundations of quantum mechanics under the supervision of Professor Michael Redhead. His research covers issues of particle identity and individuality in the foundations of quantum physics, as well as the development of structuralist approaches to a range of issues in the philosophy of science. His co-authored book, *Science and Partial Truth: A Unitary Account of Models and Scientific Reasoning*, with Newton da Costa, was published by Oxford University Press in 2003 and his recent work, *Identity in Physics: A Historical, Philosophical and Formal Account*, with Decio Krause, will also be published by OUP in 2006.

Massimo Pauri is full Professor of Relativity and Quantum Theory at the Physics Department of the University of Parma, Italy. He has a degree in Theoretical Physics from the University of Milan. He has written papers in various fields of theoretical physics. He has also contributed to various books of philosophy of physics on the issue of time, on the relations between the physical description of the world and free will, as well as on foundational issues of general relativity and quantum theory. He is titular member of the International Academy of Philosophy of Sciences (since 1992) and Fellow of the Pittsburgh Center for Philosophy of Science (since 1994).

Oliver Pooley read Physics and Philosophy at Balliol College, and took Part III of the Maths Tripos at St John's College, Cambridge, before returning to Oxford to do graduate work in Philosophy. Before taking up his current positions as Fellow and Tutor in Philosophy at Oriel, he held a British Academy Postdoctoral Fellowship and college lectureship at Exeter College, Oxford. Most of his research is in the philosophy of physics, where he is especially interested in topics that overlap with metaphysics and the philosophy of language. He is completing a book called *The Reality of Spacetime* (under contract with Oxford University Press).

Dean Rickles is a postdoctoral fellow in Philosophy of Science at the University of Calgary. He obtained his Ph.D. in History and Philosophy of Science in 2004 from the University of Leeds, under the supervision of Steven French. His research interests lie within and around the philosophy of physics, philosophy of science, and metaphysics. He is editing a textbook on the philosophy of physics for Ashgate (to appear as *The Ashgate Companion to the New Philosophy of Physics*), and is currently writing two other books: *Points, Particles, and Permutations* and *The Vices and Virtues of Randomization.*

Juha Saatsi is a postdoctoral research fellow in Philosophy at the University of Leeds. He studied theoretical physics in Finland before coming to Britain to study mathematics and philosophy. He received his Ph.D. in Philosophy at the University of Leeds in 2006.

Lee Smolin was born in New York City and educated at Hampshire College and Harvard University. He is currently a long-term researcher at the Perimeter Institute for Theoretical Physics and an adjunct professor of physics at the University of Waterloo. He was formerly a professor at Yale, Syracuse, and Penn State Universities and held postdoctoral positions at the Institute for Advanced Study, Princeton, The Institute for Theoretical Physics, Santa Barbara, and the Enrico Fermi Institute, the University of Chicago. He has been a visiting professor at Imperial College London and has held various visiting positions at Oxford and Cambridge Universities and the Universities of Rome and Trento, and SISSA, in Italy. He is the author of *The Life of The Cosmos* (London: Weidenfeld & Nicolson, 1997) and *Three Roads to Quantum Gravity* (London: Weidenfeld & Nicolson, 2000).

John Stachel is Professor Emeritus of Physics and Director of the Center for Einstein Studies at Boston University. He is the founding editor of Princeton University Press's *The Collected Papers of Albert Einstein*, co-editor of the series Einstein Studies, and author of *Einstein from 'B' to 'Z'* (Boston: Birkhäuser, 2002) and over 100 papers on theoretical physics, and history and philosophy of science.

1

Quantum Gravity Meets Structuralism: Interweaving Relations in the Foundations of Physics

Dean Rickles and Steven French

In this introductory chapter we aim to provide some of the technical and philosophical background to the issues discussed in this volume. We hope that, together with the other chapters, it will motivate the view that 'going structural' is well supported by this most pressing area of physics.

1.1 QUANTUM GRAVITY: BACKGROUND, CONCEPTS, AND METHODS

The physics of gravity is inextricably connected to the geometry of space and time. In Einstein's theory of general relativity—the best theory of *classical* gravity that we have—the geometry (curvature) of spacetime, as encoded in the metric tensor $g^{\mu\nu}$, is *identified* with the gravitational field. But the metric field is also responsible for the characteristic structures of space and time too (causal structure, notions of distance, and so on). Hence, the metric plays a dual role in general relativity: it serves to generate both the gravitational field structures and the chronometric, spatio-temporal structures (cf. Stachel 1993). In the context of general relativistic physics, of course, the metric—and, therefore, the geometry of space—is *dynamical*: the metric on spacetime is not *fixed* across the physically admissible models of the theory (as it is in, for example, Newtonian and specially relativistic theories). The geometry of spacetime is affected by matter in such a way that different distributions of matter yield different geometries—the coupling and the dynamics is described by Einstein's field equation. In other words, general relativity does not depend on the fixed metrical structure of spacetime; rather, the metric itself, and hence the geometry, comes only once a matter distribution has been specified (and the dynamical equation has been solved). Classically, this feature, called *background independence*,[1] is rather

[1] Background independence is, more properly, defined as the freedom from 'background structures', where a background structure is some element of the theory that is fixed across the

remarkable, but it is, at least, fairly easy to make physical and conceptual sense of. We can draw, for example, the following features from background independence in general relativity: the geometry of spacetime satisfies equations of motion; its curvature produces the gravitational force; since it produces a force it has energy momentum; and so on.

However, in the quantum theory of gravity, the spacetime metric will most probably have to be quantized, so that we will have to consider it a quantum theory of spacetime (or quantum *geometry*), and will need to make sense of superpositions of macroscopically distinct spacetime geometries.[2] This is *not* so easy to make *any* kind of sense of; prima facie quantum gravity faces all of the technical and interpretative problems of quantum theory, and then some. However, there are real interpretative problems even at the classical level; most notably, the 'hole argument' (Earman and Norton 1987). The hole argument has played an important role in the small portion of work that exists on the interpretation of quantum gravity, at least in its canonical (i.e. Hamiltonian) guise. Let us begin by saying a little about this argument, its root and its significance—for it is central to the claims being made about the 'structural' and 'relational' aspects, and features quite heavily in what follows. We then discuss the impact of quantum gravity (including a few words about the relevance of the hole argument in this context), and say something about the various methods and concepts employed in the field of quantum gravity. The emphasis throughout will be on the notion of background independence (and the related notions of 'background structure' and 'background dependence').

1.1.1 The Hole Argument and Spacetime Ontology

The hole argument is most easily couched in terms of models $\langle \mathcal{M}, \mathcal{D} \rangle$ (where $\mathcal{D}$ is a set of dynamical fields on $\mathcal{M}$—any further background fields are, of course, absent in general relativity[3]). Let us restrict ourselves, purely for simplicity, to the vacuum case and therefore assume that $\mathcal{D} = g$, so that the (Lorentzian) metric is the only dynamical field on $\mathcal{M}$. The models $\langle \mathcal{M}, g \rangle$ then minimally correspond to

models of a theory—this 'fixity' can be cashed out in a variety of ways, a common one of which is the idea that a structure is fixed if it is not varied in the action of the theory. However, background independence, when used in the context of quantum gravity, is usually meant in a restricted sense, covering the freedom from a background metric alone. The other side of the coin is, of course, background *dependence*, which is simply a dependence on background structures. See §1.1.7 and the contributions of Baez and Smolin (in this volume) for more details. See also Butterfield and Isham (1999) for a very nice disentangling of the various notions of 'fixity' in this context.

[2] Though several authors have argued that gravity might in fact act so as to collapse wave-functions that would result in such geometric superpositions—see Károlyházy et al. (1986) and Penrose (1986).

[3] It is important to note that the manifold and the topological and differential structure of the manifold *are* background structures in the theory—we can, in other words, 'factor' the manifold into these other structures and when we do we find that they appear in the theory as background structures. Though one may choose different manifolds and topologies (spaces of dimension other than four, for example) once chosen they remain fixed in place, and are not sensitive to the dynamical goings on in that space. Again, see the contributions of Baez and Smolin for more on this issue.

a 'bare' manifold possessing only topological and differential structure along with geometrical structure determined *dynamically* (i.e. post-solution) by g in accordance with the vacuum Einstein equation:

$$G_{\mu\nu} \equiv R_{\mu\nu} - \frac{1}{2} R^{\alpha}_{\alpha} g_{\mu\nu} = 0. \tag{1.1}$$

The crucial property of Einstein's equations, as regards the hole argument, is that they are *generally covariant*: if $\langle \mathcal{M}, g \rangle$ satisfies (i.e. solves) the Einstein equation then so does the diffeomorphic copy $\langle \mathcal{M}, \phi^* g \rangle$, $\forall \phi \in \mathsf{Diff}(\mathcal{M})$.[4] The 'carried along' field $\phi^* g$ will generally be different in the sense that, given a global chart on $\mathcal{M}$ with coordinates $\{x^i\}$, $\phi^* g(x) \neq g(x)$. When this happens we have the beginnings of a hole argument: there will be many metrics that solve the equations that will give (locally—i.e. at a specific point or within some region) different results. Hence, choose a region of the manifold, $\mathcal{H} \subset \mathcal{M}$ (the hole), and suppose that we can solve completely for all points outside the hole in the region $\overline{\mathcal{H}} = \mathcal{M} - \mathcal{H}$ (i.e. we know $g(x)$, $\forall x \in \overline{\mathcal{H}}$). Now let $\phi_{\mathcal{H}}$ be a diffeomorphism that acts as the identity on $\overline{\mathcal{H}}$ but not within $\mathcal{H}$: then the field equations do *not* uniquely determine $g(x)$ for $x \in \mathcal{H}$; for both $g(x)$ and $\phi^*_{\mathcal{H}} g(x)$ are solutions (thanks to general covariance), and yet $\phi^*_{\mathcal{H}} g(x) \neq g(x)$ for at least one point within the hole. If we put the hole to the future of some initial slice then this signals a violation of determinism: the Einstein equation cannot *uniquely* determine the spread of the metric field over the points of the spacetime manifold.

Earman and Norton argued that this form of indeterminism places the manifold substantivalist[5] in serious trouble: if the points of spacetime are real existents, and are independent of any dynamical goings on at or around them, then he will surely have to view the two solutions above as representing *physically distinct* possible

[4] Recall that, for a given manifold $\mathcal{M}$, a diffeomorphism ϕ is a smooth (i.e. C^∞), invertible mapping $\phi : \mathcal{M} \to \mathcal{M}$. Diffeomorphisms also act on field structures on $\mathcal{M}$ by 'carrying them along' to new diffeomorphic field structures. Thus, given a field structure $d \in \mathcal{D}$ (e.g. some tensor field) on $\mathcal{M}$, the action of a diffeomorphism ϕ produces a new field structure $\tilde{d} = \phi^* d$ called the *carry along* of d by ϕ. $\mathsf{Diff}(\mathcal{M})$ is simply the group of all such diffeomorphisms on $\mathcal{M}$—this group is usually understood to be the *gauge group* of general relativity.

[5] Following Sklar, let us define substantivalism as that view that takes 'spacetime to be an entity over and above the material inhabitants of the spacetime ... that could exist even were there no material inhabitants of the spacetime' (1985: 8). Relationalism is just the denial of this: what the substantivalist calls 'spacetime' is 'nothing but a misleading way of representing the fact that there is ordinary matter and that there are spatiotemporal relations among material happenings' (ibid. 10). There are two important things to note about these definitions: (1) there is assumed a straightforward distinction between 'matter' and 'space(time)'; (2) the distinction between the positions is grounded in a basic ontological priority claim involving matter or space—in the case of general relativity these distinctions become rather fragile, and this fragility leads to a remarkable degree of resemblance between relationalist and substantivalist interpretations of spacetime in general relativity (compare, for example, Hoefer's (1996) 'metric field substantivalism', Stachel's (1993) 'relationalism,' and Saunders's (2003) 'non-eliminative relationalism'). However, according to Earman and Norton's best choice for the substantivalist (viz. 'manifold substantivalism') it is the manifold of points, along with their topological and differential properties and relations, that best represents spacetime conceived as a substantival entity. At least the distinction between matter and space(time) is rather more robust on this view.

states of affairs (distinct possible worlds, if you prefer); but if that is the case, then the indeterminism that is exposed by the hole argument is genuine and physical. Earman and Norton conclude from this that we should reject substantivalism, for metaphysical positions are not the kinds of thing that should be leading us into such difficulties—only reasons of physics should be doing that, they say. But substantivalists have not been deterred, adopting ever more subtle forms of the basic position than that of Earman and Norton's naive manifold realism. In our view, however, the best way to respect the hole argument, while remaining a realist about spacetime, is to adopt a structuralist position. Many of the more recent positions that call themselves 'substantivalist' (especially the 'sophisticated' ones: e.g. Hoefer 1996; Pooley, this volume) and 'relationalist' (particularly, the physicist inspired ones: e.g., Stachel 1993, this volume; Saunders 2003; Smolin, this volume; and Rovelli 2004) turn out to be of just this kind. The basic idea, to be developed in the following subsections, is that the fundamental ontology of the theory is given by relational structures rather than individual objects; inasmuch as objects exist at all, they derive their properties and individuality from the relational network in which they are embedded. Before we move on to consider quantum gravity, and the structural stance in more generality, let us first pause to consider the nature of the hole argument in a little more depth.

1.1.2 Gauging the Hole Argument

In order to fully appreciate the inner workings of the hole argument (*qua* problem of determinism, at any rate), it is better to shift to the canonical (constrained Hamiltonian) formulation of general relativity, and thus construct its phase space Γ (parts of which are 'unphysical'—see below).[6] Adopting this stance does two things for us: (1) it allows us to make sense of general relativity as a theory about the *dynamics of space*; (2) it allows us to make sense of the way in which general relativity is a *gauge theory*.[7]

First we take spacetime to be a four-dimensional manifold $\mathcal{M}$ diffeomorphic to $S \times \mathbb{R}$—with S a (compact, orientable) 3-manifold taken to represent 'space' and where $\mathbb{R}$ is taken to represent 'time'. We choose S so that it is spacelike with respect to g and so that it is a Cauchy surface—let's now call this surface Σ. Let t be the function on $\mathcal{M}$ associated with the foliation by Σ and whose level surfaces are the leaves of the foliation. Thus far we have simply defined the background structure of

[6] Here, we focus on the *geometrodynamical* formulation according to which the configuration variable is the 3-metric q on a hypersurface. There are—as the chapters by Cao, Stachel, and Smolin will highlight—alternative 'polarizations' of general relativity's phase space (one might use, for example, a *connection* on a hypersurface as the configuration variable), but the differences are largely irrelevant for our purposes, and would overly complicate our account (see Rickles 2005a for a discussion of the hole argument transplanted into these different contexts). The chapter by Rickles in this volume offers a more detailed and general introduction to the canonical formalism.

[7] Recall that a gauge theory is one whose *physical* content is captured by those dynamical variables, the observables, that are invariant under the action of the (gauge) symmetry group, i.e. those unchanged by gauge transformations (these are those transformations generated by the first class constraints—see below, and see Dirac 1964, for the classic exposition).

the theory. A phase space Γ is then constructed using this background by taking the basic dynamical variables of the theory to be the 3-metric q_{ab} on Σ (playing the role of canonical 'position' variable) and p^{ab} (playing the role of canonical 'momentum' variable conjugate to q_{ab})[8]—both are induced by the $3 + 1$ 'split' together with g. Thus, an instantaneous state of the gravitational field is given by pairs $(q, p) \in \Gamma$. However, not any old pairs will do—i.e. not all points in Γ are physically 'kosher'. The reason for this has to do with the four-dimensional diffeomorphism invariance of the covariant theory which is 'translated' into a pair of constraints on the initial data (Σ, q, p) so that, in order to count as *physically admissible* (i.e. dynamically possible), they must satisfy both the *diffeomorphism* (or *vector*) constraint and the *Hamiltonian* (or *scalar* constraint). We can express these formally—with ${}^3\mathsf{R}$ being the Ricci curvature scalar of q on Σ—in the geometrodynamical formulation as follows:

$$\mathcal{D}_a(q, p) = -2q_{ac}\nabla_b p^{bc} = 0 \tag{1.2}$$

$$\mathcal{H}_\perp(q, p) = \det(q)^{-1/2}\left[q_{ac}q_{bd} - \frac{1}{2}q_{ab}q_{cd}\right]p^{ab}p^{cd} - \det(q)^{1/2}\,{}^3\mathsf{R} = 0 \tag{1.3}$$

This analysis brings out some of the gauge-theoretic aspects of general relativity—though, it has to be said, these aspects are at their most transparent in the connection formulation. In phase space terms we see that the full phase space Γ does not correctly represent the physically possible worlds of general relativity, for not all points will satisfy the constraints. However, the points that *do* satisfy the constraints form a submanifold $\mathcal{C} \subset \Gamma$ known as the *constraint surface*. The crucial feature of this setup, vis-à-vis the hole argument, is that the constraint surface is partitioned into gauge orbits whose elements (phase points) correspond to those states related by the symmetries (i.e. the diffeomorphisms) generated by the constraints—see the essays by Dorato and Pauri and Rickles (in this volume) for more details.

The connection to the hole argument is now obvious: the spacetime diffeomorphisms utilized therein correspond to 'unphysical' gauge motions generated by the constraints. The relevant constraint for the hole argument is $\mathcal{D}_a$ since this generates *spatial* diffeomorphisms of Σ.[9] (This distinguishing of the constraints may seem a little unnatural, since the hole argument calls upon the *full* group of spacetime

[8] The momentum variable is related to the extrinsic curvature K^{ab} of Σ by $p^{ab} \equiv \det(q)^{1/2}(K^{ab} - K^c_c q^{ab})$, where K_{ab} describes the embedding of Σ in $\langle\mathcal{M}, g\rangle$.

[9] However, an analogous problem also holds for the Hamiltonian constraint, though its treatment and interpretation leads to even thornier issues connected with time and change (known as the *problem of the frozen formalism* in the classical theory and the *problem of time* in the quantum theory). Very roughly, the problem is that if we interpret the Hamiltonian constraint as both the generator of time evolution (as is standard) *and* a generator of gauge transformations (and being a first class constraint, following Dirac 1964, we should indeed view it as such—but see Kuchař (1992) for a view to the contrary), then it seems as if there is no change, for time evolution corresponds to an unphysical gauge motion. The quantum version of the problem simply follows from the fact that if the classical Hamiltonian is zero, then the Schrödinger equation for relevant wave functions Ψ (e.g. 'the wave function of the universe') will be $i\frac{\partial\Psi}{\partial t} = \hat{H}\Psi = 0$, and we will be without quantum dynamics. See Belot (1996), Belot and Earman (1999, 2001) and Rickles (this volume) for more details.

diffeomorphisms. However, in the canonical formalism, we can envisage the normal deformations generated by the Hamiltonian constraint to be zero and yet still generate hole argument situations, and likewise, in setting the tangential deformations generated by the diffeomorphism constraint to zero, we can still generate 'problem of time' situations—whether anything of real significance rests on this fact we leave to the reader to decide.) The gauge motions—the transformations generated by the constraints—act on all points of Γ *including* those points lying within $\mathcal{C}$. In fact, the constraints have the effect of shifting phase points along orbits of the gauge group. Distinct points lying on the same gauge orbit are physically indistinguishable, representing equivalent (with respect to the 'genuine' observables) descriptions of the same physical state. Hence, even after disposing of the physically impossible states (by focusing on the constraint surface), there is something of an overabundance of physically possible states; this surplus structure is known as 'gauge freedom', and it is this that is responsible for the indeterminism that the hole argument exposes so well.

Now, the view that the constraints generate gauge motions—so that the diffeomorphisms utilized in the hole argument are gauge—leads to a natural resolution (that we might refer to as 'the physicist's resolution'[10]) of the hole argument: the indeterminism is simply unphysical, it is gauge. All that the hole argument shows us is that there are no observables of the form '$F(\mathbf{x})$': since points of space or spacetime are not diffeomorphism invariant, neither are quantities defined with respect to them. Hence, the value of the metric field at a certain *independently specified* spacetime point is not admissible; that we can talk about such a thing at all is merely the result of the surplus degrees of freedom (the gauge freedom) in the mathematical framework we use to formulate the theory. Thus, the observables of the theory should not distinguish between gauge-equivalent states (i.e. states lying within the same gauge orbit); rather, they should be constant along gauge orbits (so that their Poisson bracket with the constraints vanish) and dependence should be at the level of entire gauge orbits.[11] There are a number of ways of cashing this out, both formally and technically. For example, we might see it as motivating (or, perhaps, as being underwritten by) an anti-haecceitistic metaphysics, according to which there is no physical difference (i.e. a difference between possible worlds) without a qualitative difference (this is the line of Stachel and Pooley in this volume). The natural formal setup for making sense of the 'eradication' of the unphysical gauge degrees of freedom is to construct the *reduced* phase space $\overline{\Gamma}$ (roughly, $\overline{\Gamma} = \mathcal{C}/\mathsf{Diff}(\Sigma)$, where $\mathcal{C} \subset \Gamma$ and $\mathsf{Diff}(\Sigma)$ include the diffeomorphisms tangent and normal to the

[10] See, for example, Wald (1984: 259–60) and Hawking and Ellis (1973: 227–8) for a pair of classic statements of this viewpoint.

[11] Note, as hinted at in the previous footnote, that Kuchař argues that the constraints of general relativity should be distinguished: the diffeomorphism constraint, generating spatial diffeomorphisms, should be viewed as a gauge transformation, so that observables should be insensitive to their action, but the Hamiltonian constraint is a different matter for it generates changes in the variables from one hypersurface to another. This is, of course, related to the problem of the frozen formalism: if the Hamiltonian constraint is taken to generate gauge transformations, then observables must be constants of the motion, which, Kuchař maintains, is absurd. Rickles reviews the interpretative options in his contribution to this volume.

hypersurface Σ) with phase points given by equivalence classes of models under all the diffeomorphisms. Earman and Norton (and, more recently, Belot and Earman 1999, 2001) took this space to be out of bounds for substantivalists; after all adopting it is, more or less, tantamount to implementing Leibniz equivalence (i.e. the idea that diffeomorphic models represent one and the same physically possible world). But they view the adoption of Leibniz equivalence (or, equivalently, commutation of physical quantities with all of the constraints) as underwriting relationalist (or, at least, *anti*-substantivalist) positions. We don't wish to enter this debate here; it is well trodden and many of the chapters in this volume cover the central issues. What we want to suggest is that structuralist views sit nicely in this space, and avoid the (often seemingly *verbal*) dispute between relationalists and sophisticated substantivalists (i.e. substantivalists who endorse Leibniz equivalence). Indeed, as mentioned previously, we say that these latter positions sit very happily under the more general banner of 'structuralism' (see §1.2). Let us now turn to the subject of quantum gravity, and consider the bearing of background independence in this context. We shall then connect this to structuralism.

1.1.3 Enter Quantum Gravity

The problem of quantum gravity involves finding a way of describing the gravitational field in those high-energy, small-scale regimes in which its quantum mechanical features cannot be swept under the carpet. However, *quantum gravity*, as a label, does not yet denote any existing theory; rather, there are a number of distinct research programmes in competition for that title. There are certain minimal constraints that these approaches must satisfy to qualify, or at least be in the running. How and to what extent these constraints are met, and indeed *what* the precise constraints are is a matter of debate between the various camps. Minimally, though, it seems that what is required is a quantum theory that has general relativity as a classical limit, so that the success of general relativity can be explained from the perspective of the new theory. We might understand this in terms of a *synthesis* (or unification) of quantum field theory and general relativity (say, a generally relativistic quantum field theory); but even if we can make sense of such a notion, it is not clear that synthesis or unification is a *general* requirement.[12] For one, it has never been decisively demonstrated that there is an a priori conflict between the *formalisms* of classical general relativity and quantum field theory.[13] It should be noted, also, that there is no clear empirical problem that requires quantum gravity for its resolution, nor is there, at present, any way to probe quantum gravitational sectors empirically (though, recently, there has

[12] This 'synthetic' view seems to be the one adopted by those working on loop quantum gravity, and the canonical approaches more generally—this picture seems to correspond to that favoured by Cao (2001, and in this volume). The string theorist, by contrast, appear to follow a more 'accommodationist' line: quantum gravity is contained in the general framework of the theory in virtue of there being a massless spin-2 particle (the graviton) in the string spectrum.

[13] On the other hand, there does appear to be a 'conceptual mismatch' at the level of the views of spacetime that each calls upon: general relativity is not set against a background spacetime, whereas all quantum theories constructed so far, have been—see p. 13.

been some progress in this latter respect: see Amelino-Camelia 1999). In light of this, let us begin by assessing the possible reasons for wishing to construct a quantum theory of gravity—this brief detour will act as a primer on the kinds of issue and areas that a theory of quantum gravity might be expected to deal with. We shall then sketch in *very* broad brushstrokes the kinds of methods that have been employed to implement a theory of quantum gravity, and then finally indicate the way in which structuralism enters the picture by appealing to background independence. The following sections will then attempt to consolidate this suggested 'structuralist turn' in quantum gravity by situating it within wider historical and philosophical issues pertaining to structuralism.

1.1.3.1 Why Bother?

One of the first questions one faces when thinking about quantum gravity is why one should bother constructing such a theory at all. There are, after all, no phenomena that are out of the reach of the theories we have at out disposal already. Why should we require a revolution when there is nothing to *revolt* against? It is true that many times in the history of physics, when a revolution has occurred, it has occurred because of some *lack* with the theories then current. Either there was an inconsistency in the theory, or else the theory could not deal with some new (or old) piece of observational data. However, conflict with the observed data is not necessary for a revolution; in the next subsections we present several alternative reasons for requiring another revolution in physics.

Dimensions of Quantum Gravity

Max Planck demonstrated over a century ago that the three fundamental constants of nature—c (speed of light *in vacuo*), G (the gravitational constant), and $\hbar$ (Planck's constant: the quantum of action)—can be uniquely combined in such a way so as to produce 'natural' units of *length*, *time*, and *mass*. We get:

$$l_p = \left(\frac{\hbar G}{c^3}\right)^{\frac{1}{2}} \approx 1.62 \times 10^{-33} cm. \tag{1.4}$$

$$t_p = \frac{l_p}{c} = \left(\frac{\hbar G}{c^5}\right)^{\frac{1}{2}} \approx 5.40 \times 10^{-44} s. \tag{1.5}$$

$$m_p = \frac{\hbar}{l_p c} = \left(\frac{\hbar c}{G}\right)^{\frac{1}{2}} \approx 2.17 \times 10^{-5} g. \tag{1.6}$$

At these scales, in the 'Planck regime', general relativity and quantum field theory stop working: singularities, and other craziness emerge that lie outside of their domain of applicability. It is here that quantum gravity is expected to reign triumphant, and provide an adequate framework—'formally' adequate, in the sense of providing a consistent mathematical theory (perhaps by eliminating the 'craziness'); and 'experimentally' adequate in the sense of offering up confirming instances of data. Of course, this is not an *empirical* problem with general relativity and quantum field theory simply because these dimensions are 'out of reach' as far as empirical accessibility goes—though we might class it as a 'potentially empirical problem'.[14] Rather, the lack is purely conceptual; it is a problem with the frameworks of quantum theory and general relativity that lies beyond what is empirically accessible. In addressing the issue of what happens in this realm we are ineluctably led to consider what happens at scales when one or the other theory becomes relevant for the other, so that both theories have to be considered acting together. When we do this, then certain other, even deeper, conceptual problems surface, problems to do with the radically divergent conceptual schemes the theories employ.

The Principle of Unification

A great many revolutionary advances in physics have come about by means of a synthesis between two theories that were thought to be disparate. For example, special relativity was conceived by trying to hold the principle of Galilean relativity and Maxwell's theory of electromagnetism together. The problem is that Maxwell's theory is not Galilean invariant. By unifying the two, Einstein realized that electromagnetic phenomena must look the same in uniformly moving reference frames. Quantum field theory was the result of concerted efforts to bring together special relativity and quantum mechanics. General relativity was conceived in an attempt to unify Newton's theory of gravity with the principle of locality of special relativity. In each case there was supposed to be some fundamentally conflicting pair of theories or pieces of data that were both taken 'seriously' for the purposes of unification. The end product is a theory of the piece of data that respects both in some ways, and departs in other ways. If this is the dialectic of progress in physics, then we should expect a theory of quantum gravity to emerge from the unification of quantum field theory and general relativity, the latest in a series of conflicting pairs that physics has presented us with.

However, the concept of unification is not straightforward, and admits certain ambiguities in the context of physical theories. Unification can mean any number of distinct, though often related, concepts. We can range a number of such concepts in order of 'strength' as follows: (1) reductionism, (2) synthesis, or (3) compatibility (encompassing 'accommodation'). It isn't clear that all revolutions in physics occur at the same strength, or even that they involve any kind of unification. Above, we have been speaking of unification as synthesis, whereby two incompatible theories are 'merged', in some sense, into one that takes important features from both, and discards other features. In the case of general relativity, the key distinguishing feature that is expected (by many) to be retained is the diffeomorphism invariance of the theory, and the background independence that it implies.[15] Quantum field

[14] We should add, however, that the field of 'quantum gravity phenomenology' has gone some way towards demonstrating that features from the Planck scale might be accessible through certain potentially observable effects, such as the violation of Lorentz invariance—see Amelino-Camelia (1999) for a clear review.

[15] The argument, in a nutshell, is that (1) diffeomorphism invariance means that the physical quantities of the theory are insensitive to diffeomorphisms; (2) diffeomorphisms act (inversely) on the dynamical fields of the theory. Therefore, (3) there cannot be a background metric, for if there were then the diffeomorphisms *would* make a difference to the physical quantities.

theory will, most likely, retain the probabilistic structure as encoded in the operator algebraic representation of observables, and the representation of states as elements of Hilbert space (or, possibly, linear functionals over the operator algebra). Thus, what is required is a background-independent quantum field theory, or, equivalently, a quantum theory on a differentiable manifold.[16] In his contribution John Baez presents one such possibility in the form of topological quantum field theory. His idea involves the tools of category theory which he uses to demonstrate certain deep analogies (at the category theoretic level) between quantum theory and general relativity. Such a theory would certainly be structuralist. The reason for this is to do with the absence of a background metric with which to ground absolute locations in spacetime. This leads to the view that spacetime localization is relativized to something other than spacetime points—this may be physical objects or fields, or, if we view the identification of the metric with the gravitational field as marking an ontological identity, then we are able to localize with respect to the metric field itself (or, more properly, the points defined by the metric field).[17]

Coping with Singularities

One area where there does appear to be some breakdown in current physics, albeit in a (currently) non-empirical sense, is black hole physics. We know from general relativity, and the singularity theorems of Penrose and Hawking, that very many admissible initial data sets, for gravity plus matter, will result, under the evolution described by Einstein equations, in a gravitational collapse so extreme that a singularity will be produced. A singularity, you will recall, is (roughly) a region of spacetime at which the gravitational curvature becomes infinitely large; physically this may correspond to, for example, a material body collapsing to a point.[18] Now, a physically reasonable normative requirement on our theories is that infinite quantities should be avoided or, more strongly, that infinities do not correspond to anything physical. Whether one views this constraint as reasonable

[16] It is for this reason that the hole argument becomes a pressing issue in quantum gravity (conceived of in terms of a background independent quantum field theory). For one must make certain non-trivial choices regarding how one deals with the symmetries utilized in that argument. In capsule form these choices concern the question of whether or not we should quantize *with* or *without* the symmetries generated by the constraints—or, in more technical terms, whether we should use the machinery of non-constrained or constrained quantization. The choice is non-trivial because there will be degrees of freedom being quantized in the unconstrained approach that are not contained in the constrained approach; these can have potentially physical consequences (cf. Gotay 1984).

[17] This isn't quite right. In the case of the topological quantum field theories that Baez discusses, there are no local degrees of freedom at all. This is a problem because general relativity *does* have local degrees of freedom, only they are determined dynamically by solving the Einstein equation. However, the topological quantum field theories Baez presents nonetheless display structuralist tendencies on account of the weight they put on relations as opposed to objects.

[18] This is a rough characterization, and the details are much more complex. Less roughly we can define a singular spacetime to be one containing a geodesic of finite total *affine* length, such that a scalar invariant considered along it becomes infinite (see Earman 1995)—though there are problems even with this 'received' definition: see, for example, Geroch (1968).

or not, however, it seems clear that such singularities would pose a severe problem for our current physical theories. The reason has to do with the simple dimensional argument presented above: when the spacetime curvature is of the order of the Planck length the quantum fluctuations of the spacetime metric would no longer admit a representation by means of a smooth (pseudo-) Riemannian manifold. Thus, some new physics is inevitably required to deal with the gravitational field in such circumstances (cf. Penrose 1978).

Of course, another area in which infinities raise a problem is in quantum field theory. As in classical electrodynamics, in quantum electrodynamics there is the problem of electrons interacting with their own field. To resolve this problem, manifested as divergent integrals, one looks to renormalization theory and, more recently, renormalization group theory, to cope with or make sense of the difficulties. However, when we run general relativity through this sausage machine, we find that more infinities are produced: the theory is non-renormalizable.[19] These ineradicable infinities might be taken as signalling the need for a shift to a new or modified theory; and, indeed, many have taken the lesson of non-renormalizability to be as signalling a shift to a non-perturbative approach.[20]

It might be the case that these two types of infinities can be dealt with together as a package. Indeed, this seems to be the case in some approaches to quantum gravity; certainly it is in string theory and loop quantum gravity (the two main lines of attack): stringy dynamics has the effect of 'delocalizing' interactions (i.e. 'smearing' them out, away from points), so that the point interactions responsible for the ultraviolet divergencies are outlawed. Likewise the singularities of spacetime are avoided in string theory, since there gravity simply corresponds to a certain vibrational mode of the string, and in loop quantum gravity the micro-structure of space is discrete (since the geometry, and therefore geometrical quantities that depend on the metric, are quantized). In this way, the success of renormalization procedures is made a little clearer from a physical point of view. Thus, we might say, roughly, that the loop gravity and string theory programmes avoid the singularities by adding non-locality at the level of 'space' and 'objects' respectively.

Cosmological Quantum Theory

If quantum theory is about observations, then we need observed things and observers. What about the universe as a whole? Bell calls this 'an embarrassing concept' (1981: 622). Why? Because we do not see the universe in a superposition of states. Why not if there is no observer to observe it? Shouldn't the universe be in a grand superposition of, for example, macroscopically distinct volume states? In this case we cannot call

[19] See Deser and van Nieuwenhuizen (1974) for a classic discussion. However, the pudding is in the proof, on which see Goroff and Sagnotti (1986).

[20] However, we need not desert a theory just because it is non-renormalizable. The theory of the renormalization group (as devised by Wilson and co.—see Binney et al. 1992), and the programme of effective field theories show us how we might view general relativity as an *effective* field theory that is nonetheless capable of making physical predictions (cf. Donoghue 1994, 1996). See Burgess (2004) for a very readable account of this viewpoint. Castellani (2002) offers a nice elementary survey of effective field theories and their philosophical implications.

on environmental effects (and thus modifying the state by including these variables), since there is no environment. But with no measuring 'agency' the universe should be in such a state if quantum theory is universally valid. In typically colourful style, Bell writes that:

> It would seem that [quantum] theory is exclusively concerned with 'results of measurement' and has nothing to say about anything else. When the 'system' in question is the whole world where is the 'measurer' to be found? Inside, rather than outside, presumably. What exactly qualifies some subsystems to play this role? Was the wave function waiting to jump for thousands of millions of years until a single-celled living creature appeared? Or did it have to wait a little longer for some highly qualified measurer—with a Ph.D? If the theory is to apply to anything but idealized laboratory operations, are we not obliged to admit that more or less 'measurement-like' processes are going on more or less all the time more or less everywhere? Is there ever then a moment when there is no jumping and the Schrödinger equation applies? (1981: 611)

This brings into sharp focus the problems that quantum cosmology poses to the interpretation of quantum theory. If it is measurement that prevents the wave function from applying to everything at all times, then what measures the whole universe? The universe is, after all, a valid object of study in general relativity, therefore we should it expect to remain a valid object of study in quantum general relativity. We might expect, then, that no measurement takes place, and that the evolution is plain linear Schrödinger-style evolution—of course, the problem of time, mentioned earlier, becomes highly relevant here. Given this, how can we make sense of the universe's being in a superposition of (e.g. geometrical) states? There are several possibilities: some go down the Everettian route, and some go down the de Broglie–Bohm route. Neither route leads to jumps or collapses.

One thing seems certain, however, and that is that the traditional Copenhagen interpretation is put under pressure in this context. Recall that according to this 'orthodox' interpretation of quantum theory, any measurement interaction requires an observer that is *external* to the system that is being measured and is classical. Yet if we want quantum theory to be *universal* (i.e. independent of scale and applicable to all dynamical systems) then this category of interpretation faces a very simple problem. Quantum theory being universally valid means that it applies to systems of any size. The universe as a whole can be viewed as a physical system. Indeed, in cosmology this is a perfectly reasonable object of study. Yet there is, by definition, no observer outside the universe. Moreover, physics on cosmological scales, and so the universe as a whole system, is the domain of general relativity. It seems as though we have wandered into the territory of quantum gravity, and it seems that the Copenhagen interpretation is at a loss to deal with it.

Though many view the problems of quantum cosmology as strictly independent from the problem of quantum gravity, there are some who see the two problems as entangled (e.g. Smolin 1991, 2003). Our view is that there is a definite asymmetry here: a theory of quantum gravity should certainly give us an account of quantum cosmology, but the converse of this need not be true; for example, there are proposals to make sense of quantum cosmology that, strictly speaking, lie outside quantum gravity proper—e.g. consistent histories and Hartle's 'spacetime quantum mechanics' (1995).[21]

Problems with the Semiclassical Theory

There is a fairly simple argument that demonstrates that a semiclassical theory of quantum gravity (that is, a coupling of a classical gravitational field with quantum matter) results in superluminal signalling (Eppley and Hannah 1977). Suppose we have two spacelike separated observers, at sites A and B, and that they are making continuous measurements on the gravitational field. Suppose now that in between A and B we perform a beam-splitting experiment on a photon (let us suppose, for the sake of simplicity, that the photon has a mass of one unit—one pound, say). The experiment results in a probability distribution according to which it is at site A with a probability of a half, and at site B with a probability of a half. Since the gravitational field is classical, this will manifest itself as a warping of spacetime equivalent to half a pound at A and half a pound at B. Now suppose that observer A makes a measurement to determine the position of the photon. If the photon is at A, then the wave function collapses in such a way as to produce a warping equivalent to a one-pound mass, and the gravitational field at B will diminish by an amount equivalent to half a pound. Otherwise, the field at A will be diminished by half a pound of curvature and will increase to a pound of curvature at B. This happens instantaneously. Thus, observer A could use this setup to send a message to B; he could make a measurement to send the message, say, 'Yes, the bomb has been launched,' and not make any measurement to say 'No, the bomb has not been launched.' All B has to do is continue to measure the state of the gravitational field, and watch out for the increase or decrease. This suggests that we have to consider the gravitational field as quantized too, in order to avoid conflict with the principle of locality in general relativity.[22]

Time and Space in GR and QFT

One of the most obvious areas where philosophers can apply themselves is to the radically divergent concepts of space and time that are employed in quantum field theory (QFT) and general relativity (GR). In a nutshell the theories are incompatible because they employ incompatible ideas of space and time: quantum field theory (those forms we have at present) is necessarily background dependent (in order that the states, operators, and even the fundamental axioms can be defined); general relativity is background independent. Many of the problems and issues raised in the contributions in this book can be traced back to this single source. Here we do no more than merely hint at the scope of the problems.

The central point of difference between the two types of theory concerns, of course, the treatment of the metric (or connection) on the spacetime manifold. In existing quantum field theories (and quantum theories more generally), space and

[21] See also Gell-Mann and Hartle (1990) for an attempt to formulate a (non-quantum gravitational) quantum theory capable of dealing with the universe as a whole.

[22] We don't discuss this further, though there has been some recent philosophical work on this argument: see Mattingly (2006); Callender and Huggett (2001); and Wüttrich (2004).

time possess a *fixed* metric and connection structure. That is to say, the metric is imposed prior to solving any equations of motion for the other fields, is not allowed to vary in the action, and so is not affected by the behaviour of the quantum fields defined with respect to it. We hinted above at the fact that the metric is crucial in quantum field theory for the mathematical and conceptual foundations of the theory. For example, it is an axiom of the theory that for any pair of spacelike separated (relativistic) quantum fields (i.e. field operators with support in regions of spacetime that lie at spacelike distances from each other) $\hat{\Phi}(X^i)$ and $\hat{\Phi}(Y^j)$

$$[\hat{\Phi}(X^i), \hat{\Phi}(Y^j)] = 0. \tag{1.7}$$

This is known as the 'microcausality condition', and it encapsulates the specially relativistic basis of the theory. The fundamental conceptual conflict (vis-à-vis the nature of spacetime) between quantum field theory and general relativity can be captured if we consider the quantum gravitational analogue of the microcausality condition. Recall that in a quantum theory of gravity the spacetime metric will be an operator. Yet the metric field is responsible for chronogeometrical structure in addition to gravitational field structure, which implies that it is responsible for the causal structure too (microcausal structure included). In other words, since the causal structure is dependent on the metric and the causal structure determines whether two events are spacelike or not, and given that the metric is prone to quantum fluctuations, it follows that the notion of spacelikeness, and therefore microcausality itself, becomes subject to quantum fluctuations: one of the fundamental axioms of quantum field theory is thus rendered meaningless (cf. Wald 1984: 381–2). We need, then, a new conception of spacetime that goes beyond the conceptions that we find in quantum field theory and general relativity, and that will take something special indeed: prima facie, we need to either reject background-independence (scrap general relativity) or else find a way to set up a background independent quantum field theory (scrap quantum field theory as it is understood at present).

The question of why we should attempt to construct a quantum theory of gravity duly dealt with (albeit in a very cursory manner), the next question we face is *how* to go about it. The fact that there are many unconnected approaches makes philosophizing about quantum gravity a difficult matter. Of course, in an article of this nature we can barely pay lip service to the welter of methods that have been devised to resolve the problem of quantum gravity.

1.1.4 Categorizing the Manifold Methods

There are so many distinct approaches to quantum gravity that the task of categorizing them is rendered surprisingly difficult. Indeed, there is quite a diverse range of suggestions within the quite substantial literature on the subject. Let us begin by outlining several of these, before we lay down our preferred version.

- Relativity vs Particle Physics based: the various methods are divided according to the principle that there are methods that favour general relativity over quantum field theory, and those that reverse the preference.

- Additional Structure based: the methods are distinguished by various novel elements that are added to the foundations of one or another ingredient theory until quantum gravity is accounted for. Examples might be supersymmetry, extra dimensions of spacetime, and so on.
- Covariant vs Canonical based: methods are distinguished by the method of quantization used.
- Perturbative vs Non-perturbative: methods should be distinguished according to whether they use perturbative or non-perturbative technology.

The latter two taxonomies are often run together, with covariant and perturbative methods set together against canonical and non-perturbative methods. Certainly loop quantum gravity is both canonical and non-perturbative and the first revolution (pre-1995) picture of string theory was covariant and perturbative. However, there is no necessary connection between these distinctions. For example, recent work on string theory, though remaining manifestly covariant, aims to be thoroughly non-perturbative. However, the kinds of connections that *do* hold between these two distinctions certainly deserves some serious attention from physicists and philosophers of physics. Focusing on these two taxonomies independently of one another, they face the same problem; namely that there are methods that simply fall outside their scope.

The first taxonomy appears to be largely based in dogma and prejudice, rather than on any deep underlying divisions concerning the subject matter of the approaches thus divided. The relativists follow their geometric training whilst the particle physicists follow their analytic training. Fortunately too there are signs that this once 'great divide' is eroding, with the appearance of certain researchers who have feet in multiple camps (e.g. Baez 1994 and Smolin 2000). The second taxonomy hardly constitutes a taxonomy at all, since there would most likely be but a single theory or approach to each category. The third taxonomy makes a useful cut between the approaches, but the division is rather weak since one can associate (via a Legendre transformation) a canonical approach to each covariant approach, and vice versa—furthermore, these formulations should be equivalent. Moreover, again, there are approaches that fall outside the remit of this taxonomy—we are thinking of those approaches, such as causal set theory, which do not work by *quantizing* a theory at all. Our preferred way is to focus on background structure and dependence or independence on it and from it. This is related to the fourth taxonomy on our list, since the perturbative approaches tend to be those that make use of a fixed, flat background spacetime, while the non-perturbative ones do not follow this procedure. It strikes us that this division reflects the most fundamental and central division that separates the distinct approaches to quantum gravity. However, it seems that the balance between the two sides of this carving of the approaches is becoming increasingly lopsided, with the majority of physicists acknowledging the importance of having a background-independent theory—the background-dependent approaches appear to be slowly dying off. Let us consider this distinction a little further by exposing the pitfalls of background-dependent methods, and the virtues of the background-independent methods.[23]

1.1.5 What's Wrong with Background-Dependent Methods?

Among the first serious attempts to produce a quantum theory of gravity were background-dependent, covariant perturbation quantizations. This method was generally adopted within the particle physics community. The idea was, as Ashtekar so nicely puts it (1988: 1), to do unto the gravitational field as was done to the electromagnetic field: quantize the gravitational field to get a particle (the *graviton*) that mediates the interaction. However, just as photons require background metrical structure, so does the graviton.[24] One begins the analysis in terms of *weak* gravitational waves moving about in Minkowski spacetime. This is accomplished by splitting the spacetime metric $g_{\mu\nu}$ into a *background* part and a *perturbation*; the background part corresponds to flat Minkowski spacetime, with metric $\eta_{\mu\nu} = \mathrm{diag}(-1, 1, 1, 1)$, and the perturbation term $p_{\mu\nu}$, measures the 'deviation' from the flat (classical) background. Thus, one has

$$g_{\mu\nu} = \eta_{\mu\nu} + p_{\mu\nu}. \tag{1.8}$$

This procedure is done to make the quantization job easier; one has all of the machinery of a fixed spacetime so that, for example, microcausality conditions are defined with respect to this rather than the full metric. The helicity states[25] of the gravitational waves on the background become the quantum states of the graviton. Utilizing the representations of the Poincaré group, one is able to define the graviton as a spin-2 particle. We know, also, that this particle must be massless because the gravitational interaction works long range, and the slightest mass would contradict results concerning the deflection of light.

Weinberg (1995; see also 1979), in his definitive discussion on covariant quantum gravity, showed that, in the vacuum case, one can *derive* the equivalence principle from the Lorentz invariance of the spin-2 quantum field theory of the graviton. Thus, it is sometimes claimed (mostly by string theorists) that the spin-2 theory is equivalent to general relativity and follows from the quantum theory. The upshot of this is that any theory with gravitons is a theory that can accommodate general relativity (in some appropriate limit). This analysis forms the basis of string theory's claim that it is a candidate theory of quantum gravity: since there is a string vibration mode corresponding to a massless spin-2 particle, there is an account of general relativity (see Kiefer 2004: 34).

[23] For accounts of some of the other lines of research in quantum gravity, see the chapters by Stachel and by Smolin in this volume.

[24] Indeed, so does *any* particle at all. What's more, the background must be flat in order to help oneself to the Poincaré symmetry and thus define a preferred vacuum state, from which one derives the particle content of the theory. Thus, dynamical curved spacetimes are especially problematic from this perspective. See Wald (1994) for the reasons why this is so.

[25] If $\phi \rightarrow e^{ih\theta}$ is a transformation of a plane wave under the the action of a rotation about the direction of propagation, then h is the helicity of the wave.

But this analysis has proceeded from substantive assumptions, that we have available a flat background, and that we proceed using a linear approximation (so that the physical interpretation is one of a few gravitons propagating on Minkowski spacetime). The concept of the graviton, and this way of doing quantum gravity, is an approximation, albeit a pragmatic one that is, perhaps, required to do 'real' physics. Worse, the attempt to use perturbative methods leads to a non-renormalizable theory; any attempt to eradicate the divergences that result from probing the local fields at arbitrarily small distances fails, simply producing yet more divergences. Weinberg knows all of this, of course, but he refrains from ruling out the perturbative approach *tout court*. As we mentioned above, one may view the theory as *effective*; for sufficiently small energies, the theory may still produce testable physical predictions. However, be that as it may, the non-renormalizability of quantum general relativity *is* unequivocal with respect to the 'fundamental' status of the theory: it cannot be fundamental, for this would require consideration of Planckian physics that lies outside of the simple linear approximation. But we should have expected this, says Rovelli, for 'GR has changed the notions of space and time too radically to docilely agree with flat space quantum field theory' (2004: 4).

String theory is, however, one way—by far the most heavily researched—of sticking to the perturbative, covariant, background-dependent methodology of quantum field theory while avoiding the divergences. Another response to the non-renormalizability was to consider 'corrections' to the theory in the form of additional particles with quantum loop amplitudes that serve to cancel out the divergences associated with the gravitons. This is the way of 'supergravity' theories.

As philosophers, what conclusions might we draw from this? The general understanding is that the problems that the old covariant perturbation approaches face stem, at least in large part, from the background dependence that is imposed. The existence of a background, continuous spacetime implies, *ceteris paribus*, that the local fields have no limit of resolution; one can probe them to whatever distances and energies one likes. The metric remains fixed and classical. Divergences follow, as we mentioned above. The answer to the puzzle seems to be that the limitless resolution be limited in some way. One very ingenious way was to add dimensions to the fundamental objects of the theory so that interactions are 'delocalized' away from spacetime points. This is conservative as regards spacetime, since one can retain the fixed, classical background: the revision is applied to the 'material' side of the ontology. This is the way of string theory and M-theory, of course. Alternatively we can delocalize the points themselves, perhaps by making their coordinates 'non-commuting' q-numbers. We might also attempt to make the theory background independent, so that no fixed metric appears in the definitions of the states and observables of the theory; the metric will be a dynamical entity and become an operator in the quantum theory. This way the points, inasmuch as they exist at all, are dynamically individuated by the metric field, and so spacetime geometry itself is quantized—in the loop quantum gravity approach the geometry of space is found to be discrete (in that the geometrical operators on a spatial slice when quantized have discrete spectra).

1.1.6 What's Right about Background-Independent Methods?

Perturbative background-dependent methods attempt to stick as much as possible to the old ways of quantizing fields. Faced with a non-linear field, one treats the non-linearities as perturbations about some linear equation. Likewise, in perturbative quantizations of gravity the trick was to view the variety of curved metrics as perturbations about a fixed background metric. This background supplies all of the machinery of standard quantum field theory, representation theory, and the renormalization techniques. However, it faces a problem: recall from §1.1.3.1 that we should expect quantum gravitational effects to become significant at the Planck scale. Hence, as one approaches higher and higher energies (smaller and smaller length scales) the metrical fluctuations should become ever more non-negligible. It becomes harder and harder to sustain the perturbative idea of treating the various metrics of the various solutions as small perturbations about a flat, fixed background. This idea prompted the search for non-perturbative methods of quantization.

Canonical quantization methods follow this non-perturbative line (quantizing the full metric), and attempt to do physics in a background-independent manner.[26] Originally, the path involved using a configuration space of Riemannan metrics on a three-dimensional hypersurface, so that general relativity was rendered a dynamical theory of the geometry of space—the approach was called 'geometrodynamics' (see Arnowitt et al. 1962). However, that approach faced many problems. A recent modification, ushered in by Ashtekar's change of variables, uses connections on a principle $SL(2, \mathbb{C})$ bundle over a three-dimensional hypersurface—see the articles in this volume by Dorato and Pauri, Rickles, and Smolin for more details on the canonical approach.[27]

Much conceptually interesting material in the canonical approach comes from the way in which the spacetime diffeomorphisms—conceptually interesting in their own right, of course—are implemented by means of constraints. In the quantum theory, these constraints must be enforced too, so that wave functions are invariant under spacetime diffeomorphisms by being annihilated by the constraints. In the 'new

[26] In this case, non-perturbative methods and background independence seem to be two sides of the same coin. Indeed, in the context of quantum gravity they are often discussed as if they were synonyms (see Smolin, this volume, p. 209). The precise relations that hold between these two concepts, in this restricted context, merit further investigation. As a first step consider the following link between background independence and non-perturbative methods: (1) in non-perturbative methods the full metric is quantized; (2) the metric represents space(time); (3) the metric is dynamical. These three factors seem to give us no choice: quantizing a dynamical metric, without making a perturbative split, *enforces* background independence. String theory might seem like a counter-example to this; however, note that still very little is known about the non-perturbative extension to string theory (viz. M-Theory), and the more that is discovered about it, the more it seems like background independence will be one of its features.

[27] In fact, Smolin was one of the physicists involved in the creation of the approach called 'loop quantum gravity' (arguably the only serious rival to strings). Also, John Baez (likewise a contributor to this volume) has done much to make the mathematical foundations of loop quantum gravity more solid. Carlo Rovelli, the other creator of loop gravity, has written an excellent textbook (Rovelli 2004) on the theory that does much to expose its philosophical implications.

variables' approach, mentioned above, there is an additional constraint that comes from the connection, namely the Gauss law constraint. This expresses invariance under gauge transformations: again the physical wave functions in the quantum theory must be invariant with respect to these too. The Ashtekar variables had the effect of transforming the phase space of general relativity into a copy of the phase space of a Yang–Mills theory.[28] This in turn allowed for the application of mathematical techniques that had proven fruitful in the Yang–Mills context; of particular importance was the loop representation that it afforded. This representation (roughly an infinite Fourier transformation from the connection variables) admits natural solutions to the Gauss-law constraint (i.e. it is gauge invariant), and solutions can be found to the other constraints by considering equivalence classes of loops (under spatial diffeomorphisms), or knots, and intersections of knots. Quantizing the theory led to the application of *spin networks*, introduced in the 1970s by Roger Penrose. In the context of loop quantum gravity it is found that the spin networks form a basis for the quantum states. Penrose's original idea was to dispense with the continuous spacetime manifold, and replace it with a combinatorial structure. He writes, in typically visionary form, that

> A reformulation is suggested in which quantities normally requiring continuous coordinates for their description are eliminated from primary consideration. In particular, space and time have therefore to be eliminated, and what might be called a form of Mach's principle must be invoked: a relationship of an object to some background space should not be considered—only relationships of objects to each other can have significance. (1971: 151)

Following Penrose's line, the claim of many of those working on loop quantum gravity is that spin networks point towards a relational conception of space. Why? The reason, so far as we can see, is connected to the hole argument. The claim is that spin networks represent quantum space (i.e. a quantized version of the spatial part of the gravitational field). However, in order to accomplish this, the states must be diffeomorphism invariant. Yet spin networks are defined on a (compact three-dimensional) manifold, just like the metric was in the classical case. Hitting a spin network with a diffeomorphism shifts it around the manifold. Thus, we need to impose the constraints (i.e. we need to solve the quantum Einstein equations). This is achieved by taking the equivalence class of spin networks under these diffeomorphisms, giving us a diffeomorphism invariant *s-knot* (for 'spin'-knot) or 'abstract' spin network. The idea is that the *s*-knot is 'smeared out' over the manifold; it is not a localized entity—so hitting an *s*-knot with a diffeomorphism does nothing, we simply get the same state back. However, any other fields must then be localized with respect to these *s*-knots; the *s*-knots represent space and define location. Since the *s*-knots are *dynamical* entities—being, roughly, a quantum analogue of the classical metric field—it seems as though localization has been relativized: localization is relational. However, the ontological conclusion regarding the relationalist conception

[28] There is nonetheless, of course, a crucial difference between the two theories: Yang–Mills theories are formulated with respect to a metric manifold (i.e. they are background dependent) whereas general relativity is not.

of space seems to be drawn from nothing more than the fact that Leibniz equivalence has been imposed—i.e. by solving the diffeomorphism constraint in the move to *s*-knots. Relational localization cannot itself deliver relationalism about space(time), since, on the understanding that the 3-metric and *s*-knot state represent classical and quantum space, the localization is relativized to space! But this then simply begs the question about the ontological nature of space. Thus, this is a non sequitur as has been shown by many substantivalists who also adopt Leibniz equivalence, and as Pooley nicely charts in his contribution to this volume. It is, we say, much better understood as underwriting a structuralist stance. What is objective is the structure that is formed by abstracting the invariant core from the symmetries of the individual localized spin networks. What we get is a delocalized structure which can be understood as encoding relational features; relations between fields.[29]

1.1.7 Background Independence and Structuralism

The chapters that follow this essay are, more or less, united in their focus on background-independent approaches to the problem of quantum gravity. Thus, in general, string theory, and other background-dependent approaches, are mentioned only as examples of how *not* to go about constructing a theory of quantum gravity. One of the main points we wish to make here is that background independence and structuralism are well-matched bedfellows; better matched, in fact, than are the traditional positions of substantivalism and relationalism. Let us spend some time developing this line of thought before turning our attentions to structuralism (and structural realism).

To understand background independence, we need to introduce the notion of *background structure*. This will, no doubt, be already quite familiar to philosophers, though most likely under the somewhat scholastic *sobriquet* of 'absolute object'.[30] The technical use of this term, in the context of spacetime theories and in the sense we intend it here, originated with Anderson (1967), where he used it to refer to those objects that are dynamically decoupled in one direction from the other objects in the theoretical ontology. This means that they can affect the behaviour of other objects—i.e. play a role in determining the kinematical and dynamical properties and relations of a set of fields, for example—without being likewise affected. The idea of absolute object, though intuitively easily graspable, is a notoriously slippery customer. Friedman (1983: 62–70), for example, discerns three distinct senses that can be marshalled under its banner:

- The first arises in the context of the debate about the ontological status of spacetime structures (Friedman 1983: 62–3); that is, the debate between absolute (or, more

[29] See Rickles (2005a, 2005b) for a detailed philosophical examination of some of these themes.

[30] We should perhaps point out that Smolin, in a relevant early paper, refers to background structures as 'ideal elements' and characterizes them as 'contingent, in the sense that they may be altered without altering the basic character of the theory, play a role in the dynamical equations of the theory, and are not themselves determined by solving any dynamical equations of the theory' (1991: 231). However, he has since converted to the present terminology, and sticks with it for his contribution to this volume.

properly, substantivalist) and relational spacetime. The issue here concerns the *range* of the ontologies employed: the relationalist will wish to reduce spacetime structures to relations between physical objects, so that the ontology is coextensive with the set of physical events (or, if he is a little cleverer, the set of *possible* physical events); the substantivalist, on the other hand, will claim a larger ontological domain containing an independent manifold of possibly unoccupied spacetime points. The distinction between this sense of absolutism and relationalism is just the distinction outlined in Earman's 'R2'; namely, that holding between those views that take spatio-temporal properties and relations to be 'parasitic on relations among a substratum of space points that underlie bodies or space-time points that underlie events' (1989: 12) and those that do not. We are, in effect, back with Sklar's notion of substantivalism; hence, this first sense of absolute has been reassigned to the notion of *substantival* spacetime.

- The second sense concerns the dependence or independence of *quantities* from frames of reference or coordinate systems (Friedman 1983: 63). The absolute objects are taken to be those quantities that are thus independent. An example is simultaneity. In a spacetime with structure $\mathbb{E}^3 \times \mathbb{R}$ the notion of simultaneity can be defined independently of reference frames and coordinate charts; it is, therefore, an absolute quantity. However, shift to a relativistic picture and the notion is relativized to a reference frame. This is, for sure, connected to absolute objects; as Earman (1989: 12) points out in his characterization of 'traditional relationalism': '[R1] All motion is the relative motion of bodies, and consequently, space-time does not have, and cannot have, structures that support absolute quantities of motion.' But although this sense depends upon absolute objects, it does not characterize them. Rather, the absolute quantities are grounded in the absoluteness of the background framework given by the type of spacetime structures employed.
- The third sense is Anderson's, which we mentioned above. It is this sense that we are interested in. Friedman defines an absolute object in this category as a 'geometrical structure ... that affects the material contents of space-time (through laws of motion, for example) but is not affected in turn' (Friedman 1983: 64).

However, we don't intend to add anything new to the clarification of the concept of absolute object. We simply wish to line up what physicists call 'background structures' with the third of Friedman's 'senses', and with what Anderson means by absolute object.[31] There is a sense, then, in which the introduction of the background-independent/dependent distinction 'cleanses' the concept of absolute object of one of its ambiguities, but an interpretative problem nonetheless remains: what is the conceptual significance of background independence?

Smolin (1998: 2–3) characterizes background-independence/dependence in the context of quantum gravity as follows:

The background dependent approaches are those in which the definitions of the states, operators and inner product of the theory require the specification of the classical metric

[31] For a very careful disentangling of the various senses of 'absolute object', see Rynasiewicz (2000).

geometry. The quantum theory then describes quanta moving on this background. The theory may allow the description of quanta fluctuating around a large class of backgrounds, but nevertheless, some classical background must be specified before any physical situation can be described or any calculation can be done. All weak coupling perturbative approaches are background dependent, as are a number of non-perturbative developments. ... The background independent approaches are those in which no classical metric appears in the definition of the states, operators and inner product of the theory.... [T]he metric and connection enter the theory only as operators, and no classical metric appears in the definition of the state space, dynamics or gauge symmetries.

Now, as we mentioned, the received view amongst physicists is that background independence implies relationalism about space(time). Smolin is quite explicit about this in his contribution, writing

Thus, we often take background independent and relational as synonymous. The debate between philosophers that used to be phrased in terms of absolute vrs relational theories of space and time is continued in a debate between physicists who argue about background dependent vrs background independent theories. (p. 22)

Rovelli sketches the supposed implication—on the understanding that (active) diffeomorphism invariance implements background independence in general relativity—as follows:

[Diffeomorphism invariance] implies that spacetime localization is relational, for the following reason. If (ψ, X_n) is a solution of the equations of motion, then so is $(\phi(\psi), \phi(X_n))$ [where ϕ is a diffeomorphism]. But ϕ might be the identity for all coordinate times t before a given t_0 and differ from the identity for some $t > t_0$. The value of a field at a given point in $\mathcal{M}$, or the position of a particle in $\mathcal{M}$, changes under the active diffeomorphism ϕ. If they were observable, determinism would be lost, because equal initial data could evolve in physically distinguishable ways respecting the equations of motion. Therefore classical determinism forces us to interpret the invariance under $\mathrm{Diff}_{\mathcal{M}}$ as a gauge invariance: we must assume that diffeomorphic configurations are physically indistinguishable. (1999: 3)

Hence, the 'physical' aspects of a system are not given by specifying a single field configuration, but instead by the 'equivalence class of field configurations ... related by diffeomorphisms' (ibid.). The observables of such a system are then given by diffeomorphism invariant quantities. Such specifications of states and observables are clearly independent of any background metric: only gauge-invariant quantities are to enter into such specification, and any reference to a background metric (via, for example, fixed coordinates or functions on $\mathcal{M}$) yields non-gauge-invariant quantities. Thus, diffeomorphisms change the localization of fields on $\mathcal{M}$; this is represented in the Hamiltonian scenario by the action of the constraints. However, the localization is a gauge freedom, so any states or observables involving localization to points will not be physical. Smolin sees a direct connection between taking the equivalence class of metrics, which pushes towards a relational view of localization, and relationalism about spacetime:

The basic postulate, which makes GR a relational theory is [that a] physical spacetime is defined to correspond, not to a single (M, g_{ab}, f), but to an equivalence class of manifolds, metrics and fields under all actions of *Diff*(M). (This volume, p. 206)

Rovelli makes the case in more detail as follows:

[t]he point is that only physically meaningful definition of location within GR is relational. GR describes the world as a set of interacting fields including $g_{\mu\nu}(x)$, and possibly other objects, and motion can be defined only by positions and displacements of these dynamical objects relative to each other. ... All this is coded in the active diffeomorphism invariance ... of GR. Because active diff invariance is gauge, the physical content of GR is expressed only by those quantities, derived from the basic dynamical variables, which are fully independent from the points of the manifold. ... [Diff invariance] gets rid of the manifold. (Rovelli 2001: 108)

What status are we to attribute to the manifold once we remove dependency upon its coordinates, and smear out reference to points with the action of Diff($\mathcal{M}$)? Rovelli suggests that the manifold is an 'auxiliary mathematical device for describing spatiotemporal relations between dynamical objects' (2001: 4). Of course, spacetime coordinates enter into many areas of physics, especially mechanics and field theories, i.e. as positions of objects (particles, strings, etc.) or as the argument of a local field operator. Many physicists believe that general relativity rules out such *absolute* local quantities; there are local degrees of freedom, but the locality is grounded dynamically. This is, again, seen to follow from the practice of taking an equivalence class of manifolds and metrics under diffeomorphisms as the correct description of a spacetime in general relativity. Smolin claims that a consequence of this view is that

there are no points in a physical spacetime ... [since] a point is not a diffeomorphism invariant entity, for diffeomorphisms move the points around. There are hence no observables of the form of the value of some field as a given point of a manifold, x. (2000: 5)

The latter point, that there are no local (i.e. localized to a *particular* spacetime point) observables in general relativity, is perfectly true of course—we think this is the real 'lesson' of the hole argument, as we mentioned above. However, Smolin gets things the wrong way around. It is physical quantities that must be diffeomorphism invariant, and this does indeed supply the result that there are no observables localized to points of the manifold. But it is a big step from here to relationalism and the absence of points.

This 'relationalism from relational localization' move is, then, fairly common,[32] but it is, for the reasons we have given, also a non sequitur: substantivalism is perfectly compatible with the view that observables of general relativity are relational, and it is compatible with the shift to equivalence classes—the sophisticated substantivalists

32 The view of the physicists is a far cry from the 'received view' amongst philosophers, which is that general relativity supports spacetime substantivalism. Underlying this belief is the availability of 'empty space' solutions—i.e. those consisting of a differentiable manifold and metric tensor and without any matter fields. But this is as problematic as the move to relationalism: the empty space solutions might just as well be taken to describe a physically real field; as Stachel points out '[a]n empty spacetime could also be called a pure gravitational field, and it seems to me that the gravitational field is just as real as any other' (1993: 144). These features have been used by both sides of the debate to claim victory of the other. The substantivalists have sought to pull the gravitational field to their 'spacetime' side, and the relationalists have sought to pull it to their 'material' side. This tug of war has gone on for some time, and we think that time has come to accept that neither side is given more or less support than the other: we need to look beyond the physics to support these positions, or else look to an alternative view.

have demonstrated this (see Pooley, this volume). Once again (as with the hole argument), we have an interpretative underdetermination: both substantivalists and relationalists can lay claim to this setup. We view this underdetermination as a problem similar to the interpretative underdetermination that plagues quantum statistical mechanics, where there are conceptually incompatible interpretations of quantum particles that are nonetheless both compatible with the quantum formalism.[33] Our response is to evade the underdetermination by adopting a structuralist metaphysics: forget points and forget individual material fields, the structure as characterized by the equivalence class of metrics is where our ontological commitments should lie.

To wrap up this section, we shall indicate how what is called relationalism can be understood as a form of structuralism. Let us begin with a statement of a variety of mathematical structuralism—that of Resnik:

> In mathematics, I claim, we do not have objects with an 'internal' composition arranged in structures, we have only structures. The objects of mathematics, that is, the entities which our mathematical constants and quantifiers denote, are structureless points or positions in structures. As positions in structures, they have no identity outside of a structure. Furthermore, the various results of mathematics which seem to show that mathematical objects such as the numbers do have internal structures, e.g., their identification with sets, are in fact interstructural relationships. (Resnik 1981: 530)

Thus, mathematical objects, for Resnik, have their identities fixed only through their relationships to each other.[34] The overall structure determines the objects' identities. This basic feature—the identities of things being derived from a relational structure—is what characterizes a structuralist position. Now consider the following passage from Smolin:

> Observables associated with classical general relativity with cosmological boundary conditions measure relations between physical fields. Points have no intrinsic meaning and are only identified through the coincidence of field values. The diffeomorphism invariance of the classical theory is thus an expression that that theory is background independent (up to the specification of the topology of the manifold.) (1998: 10)

Recall, as we said above, that the idea that objects are 'identified' and 'have meaning' in virtue of relations to some other things is part and parcel of structuralism. Moreover, the way in which this is cashed out through background independence is firmly within the structuralist camp. There is something akin here—and in the

[33] See French and Rickles (2003) for a review of the ins and outs of this debate, and a discussion of the connections to the interpretation of spacetime theories vis-à-vis relationalism *vs* substantivalism. Pooley (this volume) strongly disagrees with us that there is an analogy to be had here—we reserve the right to save our response for another occasion!

[34] See Resnik (2000) for a well-developed account of his position. See Parsons (1990) for a critical analysis of this position in which he argues that it cannot in fact be extended to the most elementary objects of mathematics (we'd like to thank one of the referees for pointing this out to us). More recently, Busch (2003) and Psillos (forthcoming) have drawn on the comparison with mathematical structuralism to develop criticisms of the form of (physical) structuralism advocated here. For a response see French (2006).

passages of Smolin and Rovelli given previously—for example, to Mundy's (1992) notion of 'spacetime structuralism', according to which spacetime theories should be recast in non-coordinate geometry terms, using relational predicates, and then shifting to the isomorphism class as the object that encodes the various *equivalent* coordinate spacetimes. Mundy then argues that 'points, like numbers, are structural roles in isomorphism classes of models of certain theories' (p. 523), and with this we are back to Resnik's comments and the central core of structuralism. Baez's discussion of topological quantum field theory (and quantum field theory and general relativity from the point of view of category theory) fits well with this perspective too, since category theory places the weight on morphisms (generalized functions) over objects: the objects are defined by the relations they bear to other objects. In the next section we present some key themes from the history and philosophy of structuralism, so that these connections will be all the more transparent.

1.2 STRUCTURALISM AND STRUCTURAL REALISM

1.2.1 Motivating Structuralism

Recent years have seen the beginnings of appropriate philosophical investigation of quantum gravity. It is notable—and, to some, surprising—that many physicists have welcomed this interest from philosophers (e.g. Rovelli 1997: 182; Baez 2001: 177—see also Rickles 2005b), and one can find philosophers speaking about quantum gravity at physicists' conferences and publishing in physics journals, and vice versa (cf. Callender and Huggett 2001: 1). Our aim in this chapter has been to indicate how this dialogue might be further pursued from a structural perspective; we now propose to place these developments in their wider historical context.

1.2.2 What is Structuralism?

Defining structuralism is itself a philosophical issue and one of the points we want to press is precisely that it should not be conceived of as a monolithic philosophical position but as a heterogeneous movement composed of a number of intertwined strands. In perhaps its broadest characterization, as already used in this chapter, structuralism can be understood as urging a shift in one's ontology, away from objects, as traditionally conceived, and towards structures, typically conceived of in terms of relations. Crudely put, on the traditional conception, objects ontologically underpin the relevant structures, in the sense that they are the relata for the relations which hold between them. Structuralism shifts the focus onto the relational structures themselves and away from the objects, which must then be reconceived, in some sense, from the structure. The extent of this reconceptualization will then depend on both the form of structuralism adopted and the view of objects from which one begins. A 'weak' form of structuralism might adopt a weak form of reconceptualization and leave the objects as ontologically underpinning the structure, but insist that epistemologically they are 'hidden' in some sense. On this *epistemic* form of structuralism the claim is that we have epistemic access only to the structure, not what might lie behind it; therefore, that is what we should be concerned with in our interpretations of our physical theories. A 'stronger' form of structuralism might urge a more radical reconceptualization of objects, such that they come to be understood as mere 'nodes' or 'intersections' in the structure. More generally, objects might be understood as being *secondary* to the structure; the relations are then to be regarded as having ontological primacy over the objects. Alternatively, one might eschew talk of 'primacy' and adopt a view that is committed to both categories but privileges neither over the other (see e.g. Rickles, this volume).

1.2.3 ... and Where Does it Come from?

Of course, how dramatic a shift in focus this amounts to will depend on the view of objects one starts with. Historically, many structuralists took a 'substantivalist' view of objects, in the sense that they were conceived of in terms of some form of Lockean substratum underlying the properties they possess. In these terms, structuralism has been seen as a move towards the 'liberation' of physical ontology from the substance paradigm. This was certainly the stance adopted by Cassirer and Eddington, for example, whose broadly structuralist responses to both General Relativity and Quantum Mechanics have unfortunately been overshadowed by the work of Russell in modern structuralists' own retrospective narratives of the movement's origins.

Both Cassirer and Eddington included a fundamental subjective element in their positions, which perhaps explains their relative neglect in today's more realist context. Cassirer, in particular, was, famously, a neo-Kantian who insisted that, far from being ruled out by the developments of early twentieth-century physics, Kant's philosophy, properly understood, offered the most appropriate framework for accommodating such developments (for an excellent introduction to Cassirer's ideas, see Friedman 2004). The structuralist element of his philosophy was grounded in his reflections on the nature of space and was hugely influenced by Klein's Erlanger programme. This offered a structural conception of geometrical objects which shifts the focus from individual geometrical figures, grasped intuitively, to the relevant geometrical transformations and the associated laws. This shift underpinned Cassirer's insistence on 'the priority of the concept of law over the concept of object'. From this perspective, 'objects' dissolve into a 'web of relations', held together by certain symmetry principles which represent that which is invariant in the web of relations itself.

Cassirer famously applied this structuralist framework to the foundations of relativity theory and argued that the unity of the concept of object, which is apparently lost through the relativistic transformations, is effectively reinstated in structuralist terms via the 'lawful unity' of inertial systems offered by the Lorentz transformations. The shift from a substantivalist conception of objects to a structuralist one is furthered by the General Theory of Relativity and what we are left with is an understanding of the objects of a theory as defined by those transformations which leave the relevant physical magnitudes invariant. General covariance then functions as a principle of objectivity which offers a 'deanthropomorphized' conception of a physical object

(Ryckman 1999). Thus Cassirer saw General Relativity as the natural conclusion of the structuralist tendency:

With the demand that laws of nature be generally covariant, physics has completed the transposition of the substantial into the functional—it is no longer the existence of particular entities, definite permanencies propagating in space and time, that form 'the ultimate stratum of objectivity' but rather 'the invariance of relations between magnitudes'. (Ibid. 606, citing Cassirer 1957: 467).

When it comes to quantum mechanics, there is a similar shift from things-as-substances to relations as the ground of objectivity in science; or as Cassirer put it, '[w]e are concerned not so much with the existence of things as with the objective validity of relations; and all our knowledge of atoms can be led back to, and depends on, this validity' (Cassirer 1937: 143). In classical mechanics objectivity rests on the spatio-temporal persistence of individual objects and here, '"[o]bjective" denotes a being which can be recognized as the same in spite of all changes in its individual determinations, and this recognition is possible only if we posit a spatial substratum' (ibid. 177). It is not only the notion of spatio-temporal persistence that quantum mechanics threatens (under the standard interpretation) but the individuality of the particle itself. What is an electron then, Cassirer asks? Not, he answers, an individual object (ibid. 180), as such a conception appears to be undermined by quantum statistics (ibid. 184). At best, quantum particles 'are describable as "points of intersection" of certain relations' (ibid.). From this structuralist perspective, the entity 'constitutes no longer the self-evident starting point but the final goal and end of the considerations: the *terminus a quo* has become a *terminus ad quem*' (ibid. 131).

Eddington's importance in the history of General Relativity is well known, of course and in both his 'popular' and 'professional' works he presented structuralism as offering the most appropriate way of understanding the foundations of the theory:

The investigation of the external world in physics is a quest for structure rather than substance. A structure can best be represented as a complex of relations and relata; and in conformity with this we endeavour to reduce the phenomena to their expressions in terms of the relations which we call intervals and the relata which we call events. (Eddington 1923: 41).

Beginning with point events, the aggregate of which constitute 'the World' and which is postulated to be four-dimensional, the interval can then be defined, as a quantitative relation, and the operation of comparing intervals eventually yields—via a fair bit of jiggery-pokery—the field equations. Eddington insisted that these should be read from left to right, not as laws of the World relating the continuum of points events and matter, but as mathematical identifications denoting 'definite and absolute' conditions of the world (Kilmister 1994: 44–6). Hence, 'Matter does not cause an unevenness in the gravitational field; the unevenness is matter' (Eddington 1923: 152). By matter here, Eddington means matter as substance and thus this construction is seen as eliminating substance from our ontology in favour of relational structures, which were taken to be of a kind defined and investigated by group theory (see Eddington 1936: ch. XII and 1939: , ch. IX).

From this perspective, substantivalist and relationist metaphysics, as traditionally conceived, are nothing more than embellishments to the 'bare structural description' which the structuralist focuses on. Thus, taking the example of uniform spherical space, all that we know about such a space, Eddington argued, is that it has the structure of the rotation group. 'When we introduce spherical space into physics we refer to something—we know not what—which has this structure' (1939: 146). Similarly, Euclidean space and Riemannian space are referred to as something with a specifiable group structure. The usual attempts to describe space in terms of more or less familiar metaphysical categories are an 'unauthorized addition' to physical knowledge. Here again the structuralism is underpinned by a shift away from entities, in this case spacetime points and Eddington insisted that 'Space is not a lot of points close together; it is a lot of distances interlocked' (1923: 10).

Furthermore, as in the case of Cassirer, Eddington took the implications of quantum statistics for particle individuality as opening the door to a structuralist accommodation of quantum physics. And again, it is a substantival conception of object that must be abandoned, in favour of a group-theoretic understanding (for more on Eddington's structuralist conception of quantum particles, see French 2003). What we obtain, then, is a structuralist view of all of science:

> Physical science consists of purely structural knowledge, so that we know only the structure of the universe which it describes. This is not a conjecture as to the nature of physical knowledge; it is precisely what physical knowledge as formulated in present-day theory states itself to be. In fundamental investigations the conception of group-structure appears quite explicitly as the starting point; and nowhere in the subsequent development do we admit material not derived from group-structure. (Eddington 1939: 142–3).

Eddington's later work, particularly as presented in his *Fundamental Theory*, represents an attempt to articulate a unified theory of physics—that is, in part, a theory of Quantum Gravity—within such a structuralist perspective. That it remains barely comprehensible, if at all, should not detract from the heroic effort involved![35]

Of course, to modern eyes, the concern with substance might seem somewhat idiosyncratic. And if one were to initially regard an object, not as a substance possessing properties, but as nothing more than a bundle of such properties and relations (perhaps united by some kind of primitive 'compresence' relation) to begin with, then the structuralist shift may not seem quite so radical after all.[36] Furthermore, it has come to be appreciated that one can in fact maintain a view of objects as individuals in the context of quantum statistics, where this individuality can be understood as grounded in either some form of substantival metaphysics, or a broadly Scholastic notion of haecceity or 'primitive thisness' (French 1989; French and Krause 2006). However, the bundle theory, just mentioned, appears not to fare that well, since it requires the acceptance of some form of Leibniz's Principle of the Identity of Indiscernibles—so that no two 'bundles' can be exactly alike—and this appears to be ruled out by quantum mechanics (French and Redhead 1988). Nevertheless, Saunders has recently elaborated a kind of 'modernized' form

[35] For an almost equally heroic effort to render it comprehensible and relate it to modern concerns, see Durham (2005).

[36] This point is made in French (2001) and also in Pooley's contribution to this volume.

of the Principle which is compatible with quantum theory (Saunders 2003; for comments, see French and Rickles 2003). Interestingly, this form grants relations an individuating role and it can thus be regarded as yielding a form of structuralism, in that the very individuality of the object is grounded in the latter's relations with other objects. Similarly, but less plausibly, perhaps, Stachel suggests, in his contribution to this collection, that haecceity or 'primitive thisness' can have a relational basis too; one might wonder how the haecceity can still be regarded as 'primitive' under such a conception.

It can also be argued that even if one were to accept the traditional implication of quantum statistics with regard to individuality, one does not have to give up a metaphysics of objects entirely since the supposed non-individuality can be captured via some non-standard formal framework which accommodates a conception of objects still, albeit of a strange kind (see Krause 1992; French and Krause 2006). Nevertheless, that both these metaphysical packages—'quantum objects-as-non-individuals' and 'quantum objects-as-individuals'—are effectively supported by the physics provides an alternative motivation for structuralism. Put simply the idea is that what we have here is a form of 'metaphysical underdetermination' in which the metaphysical interpretation—in this case of quantum objects as either individuals or non-individuals—is underdetermined by the physics itself. This can be taken to raise a fundamental problem, in that we can no longer ascertain which metaphysics of objects—at the most basic level of their individuality—is implied by the physics. This problem can then be resolved, or 'sidestepped', by reconceptualizing the notion of object in structuralist terms, for it only afflicts object-based ontologies (be they individuals based or non-individuals based). Pooley (this volume) has questioned the strength of this motivation, on the grounds that the underdetermination only exists in 'logical space'. Here he seems to be following Redhead and Teller, who have argued that the non-individuals package meshes better with quantum field theory and hence we have grounds for choosing that horn of the apparent dilemma, so the underdetermination evaporates.[37]

Of course, these alternative metaphysical packages were articulated in the context of 'first quantized' quantum mechanics and it should, perhaps, come as no surprise that the force of an apparent underdetermination weakens once one broadens the theoretical context. But note, it would be a mistake to view the move to quantum field-theory as truly *resolving* the underdetermination since in the field-theoretic context we do not, strictly speaking, have objects at all but only field excitations. It is rather a case of a particular underdetermination which exists in one theoretical context, not featuring in another, and this should not be unexpected. Nevertheless, new forms of underdetermination might arise in these new contexts and indeed, in the field-theoretic context, Redhead has located the structuralist stance as laying between the two questions 'what is a field?' and 'what are the equations which govern its behaviour?' (Redhead 1995: 18). The standard answers to the first—that a field is some kind of substance or merely a set of properties instantiated at spacetime

[37] For discussion of this line of argument, see French and Krause (2006).

points or regions—are not exhausted by the answer to the second. Harking back to the history again, Cassirer, for example, rejected the substantival account for philosophical reasons and insisted that a field is not a 'thing' but rather a 'system of effects' (1937: 178). Those who prefer their structuralism less in thrall to an already given philosophical position might want to articulate another form of metaphysical underdetermination—this time between fields as substances and fields as instantiated properties. Even if one were to follow Cassirer and choose the latter horn of the underdetermination, this would still leave the nature of spacetime as a potentially non-structural element of one's ontology.

Whether similar motivations can be articulated in the context of the foundations of spacetime theory is a further, interesting question. Certainly it is debatable whether the traditional dichotomy between substantival and relationalist views of spacetime can be understood as a form of metaphysical underdetermination in the above sense (Pooley, this volume, argues not). Recent interest in 'spacetime structuralism' has been motivated, in large measure, by the hole argument, which, as we have seen, presents an apparent dilemma of either giving up manifold substantivalism or accepting a form of indeterminism. In an attempt to avoid having to succumb to a relationalist position, various structuralist alternatives have been articulated—though, as we mentioned previously, these positions often parade under a label other than structuralism (Dorato 2000 is an exception). Still, concerns over the individuation of spacetime points may still drive one to a form of spacetime structuralism (see, for example, Stein 1967). We recall Eddington's understanding of space as 'not a lot of points close together; it is a lot of distances interlocked' (1923: 10) and more recently, Dorato has asserted that 'To say that spacetime exists just means that the physical world exemplifies, or instantiates, a web of spatiotemporal relations that are described mathematically' (2000: 7).

This suggests that spacetime has an objective existence that is not grounded in some form of substantivalism, but then Dorato appears to agree with Cao (1997) that the existence of spatio-temporal relations must be underpinned by the existence of the gravitational field, understood as a 'concrete' and hence, presumably, substantive, entity. As far as Cao is concerned a field is a 'hypothetical entity', employed as the basis for generating the field equations which describe the structural aspects of these entities and from which particles emerge as 'observable manifestations' (2003). But this just pushes the question back: what is this hypothetical entity, metaphysically speaking? The structuralist's answer is that the field is just the structure, the whole structure, and nothing but the structure (French and Ladyman 2003).

1.2.4 Structural Realism

Interest in the structuralist programme has recently been reawakened in the context of the realism–antirealism debate in the philosophy of science. Psillos's characterization of different forms of structuralism in terms of the 'upwards' and 'downwards' epistemic paths represents a useful way of framing the recent discussions in a way which connects current positions to their predecessors (2001). Broadly speaking, when we follow the 'upwards' path we begin with supposedly secure knowledge and then infer what we can know on that basis. Thus Russell, for example, began with 'percepts', which represent our experiences and which we know via direct acquaintance, and then used his causal theory of perception to infer that all that we can know of the external world on that basis is its structure (Russell 1927). Thus he writes,

When we are dealing with inferred entities, as to which we know nothing beyond structure, we may be said to know the equations, but not what they mean: so long as they lead to the same results as regards percepts, all interpretations are equally legitimate. (p. 287)

Now this view famously came under attack from the mathematician Newman (1928), who argued that if we know only the structure of the world, then we actually know very little indeed. The argument is apparently straightforward: given any 'aggregate' of relata A, a system of relations can be found having any assigned structure compatible with the cardinality of A; hence, the statement 'there exists a system of relations, defined over A, which has the assigned structure' yields information only about the *cardinality* of A. In other words, to say we know the structure of the world is to say nothing more than that we know the cardinality of the world. Russell himself appears to have been convinced by Newman's conclusion and in the context of our history above, it is worth noting that Braithwaite also deployed it against Eddington (Braithwaite 1940), writing that

his [Newman's] strictures are applicable to Eddington's group-structure. If Newman's conclusive criticism had received proper attention from philosophers, less nonsense would have been written during the last twelve years on the epistemological virtue of pure structure. (Braithwaite 1940: 463)

Unlike Russell, however, Eddington was less impressed, arguing that Newman's conclusion depends on a mathematical distinction between *elements* of a set and the relevant *relations*, but that from the group-theoretical perspective on which his form of structuralism is founded, no such distinction is possible: 'The element is what it is because of its *relation* to the group structure' (Eddington 1941: 269; his emphasis). In particular, he contrasts Russell's 'vague' conception of structure as a pattern of entities—or, perhaps, a pattern of relations—with his group-theoretic understanding of structure as a pattern of 'interweaving', or a 'pattern of interrelatedness of relations'. As an example, he presents the algebra of operators representing rotations acting on elements, for which the 'pattern of interrelatedness' is manifested in the associated multiplication table and, he insists, the information encoded in such a table is by no means trivial in the way Newman indicated (for further discussion, see French 2003).

Nevertheless, the Newman argument continues to be presented by critics of structuralism,[38] possibly because these critics see structuralism as following the 'upwards' path in general and as beholden to Russell in particular. However, Russell's account emerged at a specific time, historically (1926–7) and although it contains

[38] See, for example, Demopoulos and Friedman (1995); Psillos (1999); Ketland (2004); for a response, see Melia and Saatsi (forthcoming).

a good representation of the then current understanding of spacetime theory, the implications of the new quantum mechanics were only dimly appreciated. Indeed, given Cassirer's and Eddington's concern to develop a form of structuralism that could accommodate these implications, one might suggest that it is to these authors, rather than Russell (and Newman) that both structuralists and their critics should look.

In its modern form, a structuralist accommodation of modern physics can be characterized in terms of what Psillos calls the 'downward path'. Here one begins with the full, theoretical edifice, as it were, and then undertakes a strategic, epistemic retreat according to what one learns from reflection on both the progress of science and its metaphysical implications (or lack thereof). Thus Worrall has famously presented a form of structuralism as a response to Laudan's 'Pessimistic Meta-Induction' (Worrall 1996). Put rather crudely this asserts that the history of science is, to a significant extent, a history of changing ontologies—as one moves from the particle theory of light to the wave theory to Maxwell's theory and so on, to take one example—and given this, one has good reason not to be a realist with regard to the ontology of our current best theories. Worrall's response, again put rather simply, is to note that the same history suggests that important structural elements of theories are preserved through these changes. By 'ontology' here is meant the theoretical representation of scientific entities, such as light, electrons, etc. The relevant structures, on the other hand, are represented for Worrall by the appropriate mathematical equations—Snell's Laws are incorporated into Maxwell's Equations and so on. Thus, whereas the ontological component of a theory may be subjected to a pessimistic meta-induction, as far as the structural component is concerned things look quite optimistic.

This gives rise to a form of 'Structural Realism' (SR) which holds that one can, and should, adopt a realist attitude towards the well-confirmed structural aspects of theories (see also Redhead 1995). As Ladyman has pointed out (1998), this should be regarded as an epistemic form of SR since it holds that all that we know are the structures, while the objects themselves remain epistemologically inaccessible. Again there is a historical aspect to these developments, since in defending this position Worrall draws on those famous passages from *Science and Hypothesis* where Poincaré writes that theoretical terms 'are merely names of the images we substituted for the real objects which Nature will hide forever from our eyes. The true relations between these real objects are the only reality we can ever obtain' (1905: 162).[39]

Setting aside these historical issues again, Worrall noted that this form of structuralism might be capable of accommodating quantum physics, although he did not develop this aspect of his account. However, in retaining the idea of epistemologically inaccessible objects, hidden behind the structures as it were, Worrall appears to run up against the very implications that Cassirer and Eddington took to underpin their

[39] Elsewhere Poincaré presents group theory as the most appropriate representation of these 'true relations' and displays certain Kantian inclinations which hardly commend him to the realist. Such inclinations appear again and again through the history of structuralism and the issue arises as to whether, in drawing on this history for her understanding of 'structure', the structural realist can neatly peel them off from the rest of structuralist programme.

forms of structuralism, namely that quantum particles are not individuals in some sense. Here, in the context of defending realism, the above underdetermination between objects-as-individuals and objects-as-non-individuals has a particular bite: van Fraassen has argued that realism should be understood as requiring a commitment to a metaphysical interpretation, at this most basic level (van Fraassen 1991). However, this underdetermination indicates that no such interpretation can be given founded on the physics itself. The realist is thus faced with a problem.

Ladyman's 'ontic' form of SR (Ladyman 1998) can be seen as responding to this concern (as well as to the pessimistic meta-induction) by effectively eliminating the objects completely, leaving only the structures. Again, put simply, the idea is that it is not just that all that we know are the structures but that all that there *is* are the structures. It is important to emphasize (because some critics appear incapable of grasping this) that although Ladyman's view bears some resemblance to earlier forms of structuralism—in taking the ontology of the world to be structural most crucially—the underlying argument is quite different (it is not that quantum mechanics implies that quantum objects are in some sense non-individuals but that on the basis of the physics alone we cannot say whether the particles are individuals or not and hence if we want a realism compatible with our current best theories, we had better adopt a different ontology). The elaboration and development of this position has raised a number of interesting issues, to do with the representation and metaphysics of structure, the conceivability of structures without any underlying objects, the identity conditions for such structures, and so on, some of which, at least, have been addressed elsewhere (Ladyman 1998; French 2001; French and Ladyman 2003). It is important to realize that in eliminating objects from the realist's ontology, the structuralist is not advocating the view that physicists cannot talk, whether theoretically or informally or whatever, of 'electrons', 'quarks', etc., but rather is insisting that from a metaphysical perspective these entities must be reconceptualized in structural terms. Furthermore, this form of 'ontic structuralism' can be extended to quantum field theory along the lines already sketched above (French and Ladyman 2003).

Of course we are not suggesting that to understand the foundations of quantum gravity one must be a structural *realist*. One could be a structural empiricist and adopt van Fraassen's modal stance towards interpretations of theory and insist that what these interpretations tell us is how the world could be (see Bueno 2000, for steps leading in this direction). In the case of structuralism, what we are asserting, according to this empiricist stance, is that the world could be, metaphysically, structural. In either case—that of the structural realist or that of the empiricist—what is important for our purposes is that we are provided with the resources for giving an ontological account of the foundations of quantum gravity. These resources will include the representational, whether they be group theoretic, as Eddington advocated, or category theoretic, as Baez and others have suggested, and the metaphysical, as in the claim, defended, again, by Eddington and, more recently, Rickles, that relations and the objects which act as their relata come together in a structuralist package, as it were; or the view, proposed, in various forms, by Saunders, Stachel, and others, that it is relations 'all the way down', even to the level of the individuality of the objects, so the latter emerge as mere 'nodes' in the structure, or intersections of relations, as Cassirer thought. And of course there is still a great deal of work to be done in articulating those resources fully and properly and there are a number of criticisms that must be faced (see, most notably, Chakravartty 1998 and Psillos, forthcoming) but we hope, at the very least, that the essays contained in this collection will lead to a greater appreciation of both the virtues of this approach and the obstacles still to be overcome.

REFERENCES

Alexander, H. G. (ed.) (1957) *The Leibniz–Clarke Correspondence.* Manchester: University of Manchester Press.

Amelino-Camelia, G. (1999) "Are we at the Dawn of Quantum Gravity Phenomenology?" In J. Kowalski-Glikman (ed.), *Towards Quantum Gravity.* Heidelberg: Springer-Verlag, 2000.

Anderson, J. L. (1967) *Principles of Relativity Physics.* New York: Academic Press.

Arnowitt, R., S. Deser, and C. W. Misner (1962) "The Dynamics of General Relativity". In L. Witten (ed.), *Gravitation: An Introduction to Current Research.* London: John Wiley & Sons (pp. 227–65).

Ashtekar, A. (1988) "Introduction". In A. Ashtekar and J. Stachel (eds.), *Conceptual Problems of Quantum Gravity: Proceedings of the 1988 Osgood Hill Conference.* New York: Birkhäuser, 1991.

—— and R. Tate (1996) *Lectures on Non-Perturbative Canonical Gravity.* Singapore: World Scientific Publishing.

Baez, J. (1994) "Strings, Loops, and Quantum Gravity". In J. Baez (ed.), *Knots and Quantum Gravity.* Oxford: Oxford University Press (pp. 133–68).

—— (2001) "Higher-Dimensional Algebra and Planck-Scale Physics". In C. Callender and N. Huggett (eds.), *Physics Meets Philosophy at the Planck Length.* Cambridge: Cambridge University Press (pp. 177–95).

Bell, J. S. (1981) 'Quantum Mechanics for Cosmologists'. In C. J. Isham, R. Penrose, and D. W. Sciama (eds.), *Quantum Gravity 2: A Second Oxford Symposium.* Oxford: Clarendon Press, (p. 611).

Belot, G. (1996) "Why General Relativity *Does* Need an Interpretation". *Philosophy of Science,* 63: 80–8.

—— and J. Earman (1999) "From Metaphysics to Physics". In J. N. Butterfield and C. Pagonis (eds.), *From Physics to Philosophy.* Cambridge: Cambridge University Press (pp. 166–86).

—— (2001) "Pre-Socratic Quantum Gravity". In C. Callender and N. Huggett (eds.), *Physics Meets Philosophy at the Planck Scale.* Cambridge: Cambridge University Press (pp. 213–55).

Bergmann, P. G., and A. Komar (1960) "Poisson Brackets between Locally Defined Observables in General Relativity". *Physical Review Letters,* 4: 432–3.

Binney, J. J., N. J. Dowrick, A. J. Fisher, and M. E. J. Newman (1992) *The Theory of Critical Phenomena: An Introduction to the Renormalization Group.* Oxford: Clarendon Press.

Braithwaite, E. (1940) "Critical Notice of *Philosophy of Physical Science*". *Mind,* 49: 455–66.

Brighouse, C. (1994) "Spacetime and Holes". *Proceedings of the Philosophy of Science Association,* 1: 117–25.

Brügmann, B. (1994) "Loop Representations". In J. Ehlers and H. Friedrich (eds.), *Canonical Gravity: From Classical to Quantum.* Berlin: Springer-Verlag (pp. 213–53).

Bueno, O. (2000) "What is Structural Empiricism? Scientific Change in an Empiricist Setting". *Erkenntnis,* 50: 59–85.

Burgess, C. P. (2004) "Quantum Gravity in Everyday Life: General Relativity as an Effective Field Theory". *Library of Living Reviews.* **http://relativity.livingreviews.org/Articles/lrr-2004-5.**

Busch, J. (2003) "What Structures Could Not Be". *International Studies in the Philosophy of Science,* 17: 211–25.

Butterfield, J. N. (1989) "The Hole Truth". *British Journal for the Philosophy of Science,* 40: 1–28.

—— and C. J. Isham (1999) "On the Emergence of Time in Quantum Gravity". In J. Butterfield (ed.), *The Arguments of Time.* Oxford: Oxford University Press (pp. 111–68).

Callender, C., and N. Huggett (2001) "Why Quantize Gravity (or Any Other Field for that Matter)?" *Philosophy of Science,* 68 (Proceedings): S382–94.

Cao, T. Y. (1997) *Conceptual Development of 20th Century Field Theories.* Cambridge: Cambridge University Press.

—— (2001) 'Prerequisites for a Consistent Framework of Quantum Gravity', *Studies in the History and Philosophy of Modern Physics,* 32(2): 181–204.

—— (2003) "Can We Dissolve Physical Entities into Mathematical Structures?" *Synthese,* 136: 57–71.

Cassirer, E. (1937/1956) *Determinism and Indeterminism in Modern Physics.* New Haven: Yale University, 1956. (Translation of *Determinismus und Indeterminismus in der modern Physik,* Goteborg: Elanders Boktryckeri Aktiebolag, 1937.)

—— (1957) *The Philosophy of Symbolic Forms,* III: *The Phenomenology of Knowledge.* New Haven: Yale University Press.

Castellani, E. (2002) "Reductionism, Emergence, and Effective Field Theories". *Studies in History and Philosophy of Modern Physics,* 33: 251–67.

Chakravartty A. (1998) "Semirealism". *Studies in History and Philosophy of Modern Science,* 29: 391–408.

Demopoulos, W., and M. Friedman (1985) "Critical Notice: Bertrand Russell's The Analysis of Matter: Its Historical Context and Contemporary Interest". *Philosophy of Science,* 52: 621–39.

Deser, S., and P. van Nieuwenhuizen (1974) "Nonrenormalizability of the Quantized Dirac-Einstein System". *Physical Review D,* 10: 411–20.

Dirac, P. A. M. (1964) *Lectures on Quantum Mechanics.* New York: Belfer Graduate School of Science Monographs Series.

Donoghue, J. F. (1994) "General Relativity as an Effective Field Theory: The Leading Quantum Corrections". *Physical Review D,* 50: 3874.

—— (1996) "The Quantum Theory of General Relativity at Low Energies". *Helvetica Physica Acta,* 69: 269–75.

Dorato, M. (2000) "Substantivalism, Relationism and Structural Spacetime Realism". *Foundations of Physics,* 30(10): 1605–28.

Duff, M. J. (1978) "Covariant Quantization". In C. J. Isham and R. Penrose (eds.), *Quantum Gravity: An Oxford Symposium.* Oxford: Clarendon Press.

Durham, I. (2005) 'Sir Arthur Eddington and the Foundations of Modern Physics'. Unpublished Ph.D. Thesis, University of St Andrews.

Earman, J. (1989) *World Enough and Space-Time: Absolute Versus Relational Theories of Space and Time.* Cambridge, MA: MIT Press.

—— (1995) *Bangs, Whimpers, and Shrieks.* Cambridge: Cambridge University Press.

—— and J. Norton (1987) "What Price Substantivalism? The Hole Story". *British Journal for the Philosophy of Science,* 38: 515–25.

Eddington, A. (1923) *Mathematical Theory of Relativity*. Cambridge: Cambridge University Press.

____(1936) *Relativity Theory of Protons and Electrons*. Cambridge: Cambridge University Press.

____(1939) *The Philosophy of Physical Science*. Cambridge: Cambridge University Press.

____(1941) "Discussion: Group Structure in Physical Science". *Mind*, 50: 268–79.

____(1946) *Fundamental Theory*. Cambridge: Cambridge University Press.

Eppley, K., and E. Hannah (1977) "The Necessity of Quantizing the Gravitational Field". *Foundations of Physics*, 7: 51–68.

Fairbairn, W., and C. Rovelli (2004) "Separable Hilbert Space in Loop Quantum Gravity". *Journal of Mathematical Physics*, 45: 2802–14.

French, S. (1989) "Identity and Individuality in Classical and Quantum Physics". *Australasian Journal of Philosophy*, 67: 432–46.

____(1998) "On the Withering away of Physical Objects". In E. Castellani (ed.), *Interpreting Bodies: Classical and Quantum Objects in Modern Physics*. Princeton: Princeton University Press (pp. 93–113).

____(2001) "Symmetry, Structure and the Constitution of Objects". *Pittsburgh Archive for the Philosophy of Science*: **http://philsci-archive.pitt.edu/documents/disk0/00/00/03/27/index.html.**

____(2003) "Scribbling on the Blank Sheet: Eddington's Structuralist Conception of Objects". *Studies in History and Philosophy of Modern Physics*, 34: 227–59.

____(2006) 'Structure as a Weapon of the Realist', *Proceedings of the Aristotelian Society*, 106: 167–85.

____and D. Krause (2006) *Identity and Individuality in Quantum Physics: A Historical, Philosophical and Logical Analysis*.

____and J. Ladyman (2003) "Remodelling Structural Realism: Quantum Physics and the Metaphysics of Structure". *Synthese*, 136: 31–56.

____and M. L. G. Redhead (1988) "Quantum Physics and the Identity of Indiscernibles". *British Journal for the Philosophy of Science*, 39: 233–46.

____and D. Rickles (2003) "Understanding Permutation Symmetry". In K. Brading and E. Castellani, *Symmetries in Physics: New Reflections*. Cambridge: Cambridge University Press (pp. 212–38).

Friedman, M. (1983) *Foundations of Space-Time Theories*. Princeton: Princeton University Press.

____(2004) "Ernst Cassirer". *Stanford Encyclopedia of Philosophy*: **http://plato.stanford.edu/entries/cassirer/.**

Gambini, R., and J. Pullin (1996) *Loops, Gauge Fields, Knots, and Quantum Gravity*. Cambridge: Cambridge University Press.

Gell-Mann, M., and J. Hartle (1990) "Quantum Mechanics in the Light of Quantum Cosmology". In W. H. Zurek (ed.), *Complexity, Entropy, and the Physics of Information, Santa Fe Institute Studies in the Sciences of Complexity VIII*. Reading: Addison-Wesley.

Geroch, R. (1968) "What is a Singularity in General Relativity?" *Annals of Physics*, 48: 526–40.

Goroff, M. H., and A. Sagnotti (1986) "The Ultraviolet Behavior of Einstein Gravity". *Nuclear Physics*, B 266: 709.

Gotay, M. (1984) "Constraints, Reduction, and Quantization". *J. Math. Phys.* 27(8): 2051–66.

Hartle, J. (1995) "Spacetime Quantum Mechanics and the Quantum Mechanics of Spacetime". In B. Julia and J. Zinn-Justin (eds.), *Gravitation and Quantizations: Proceedings of*

the 1992 Les Houches Summer School. Les Houches Summer School Proceedings Vol. LVII. Amsterdam: North Holland.
Hawking, S. W., and G. F. R. Ellis (1973) *The Large Scale Structure of Space-Time.* Cambridge: Cambridge University Press.
Hoefer, C. (1996) "The Metaphysics of Spacetime Substantivalism". *Journal of Philosophy,* 93: 5–27.
Károlyházy, F., A. Frenkel, and B. Lukács (1986) "Gravity and State Vector Reduction". In R. Penrose and C. J. Isham (eds.), *Quantum Concepts in Space and Time.* Oxford: Clarendon Press (pp. 109–28).
Ketland, J. (2004) "Empirical Adequacy and Ramsification". *British Journal for the Philosophy of Science,* 55(2): 287–300.
Kiefer, C. (2004) *Quantum Gravity.* Oxford: Clarendon Press.
Kilmister, C. (1994) *Eddington's Search for a Fundamental Theory.* Cambridge: Cambridge University Press.
Komar, A. (1955) "Degenerate Scalar Invariants and the Groups of Motion of a Riemann Space". *Proceedings of the National Academy of Science,* 41: 758–62.
Krause, D. (1992) "On a Quasi-set Theory". *Notre Dame Journal of Formal Logic,* 33: 402–11.
Kuchař, K. (1992) "Time and Interpretations of Quantum Gravity". In G. Kunstatter, D. E. Vincent, and J. G. Williams (eds.), *Proceedings of the 4th Canadian Conference on General Relativity and Relativistic Astrophysics.* Singapore: World Scientific (pp. 211–314).
Ladyman, J. (1998) "What is Structural Realism?" *Studies in History and Philosophy of Science,* 29: 409–24.
Maidens, A. (1993) 'The Hole Argument: Substantivalism and Determinism in General Relativity'. Unpublished doctoral thesis, University of Cambridge.
Mattingly, J. (2006) 'Is Quantum Gravity Necessary?' In Jean Eisenstaedt and Anne Kox (eds.), *The Universe of General Relativity* (Einstein Studies, vol. 11). Boston: Birkhäuser.
Maudlin, T. (1988) "The Essence of Space-time". *Proceedings of the Philosophy of Science Association, 1988,* 2: 82–91.
Melia, J., and J. Saatsi (forthcoming) 'Ramseyfication and Theoretical Content'.
Mundy, B. (1992) 'Space-Time and Isomorphism'. In D. Hull, M. Forbes, and K. Okruhlik (eds.), *Proceedings of the 1992 Biennial Meeting of the Philosophy of Science Association,* vol. i, East Lansing, Mich.: Philosophy of Science Association (pp. 515–27).
Newman, M. H. A. (1928) "Mr Russell's Causal Theory of Perception". *Mind,* 37: 137–48.
Parsons, C. (1990) "The Structuralist View of Mathematical Objects". *Synthese,* 84: 303–46.
Penrose, R. (1971) "Angular Momentum: An Approach to Combinatorial Space Time". In T. Bastin (ed.), *Quantum Theory and Beyond.* Cambridge: Cambridge University Press (pp. 151–80).
—— (1978) "Singularities of Spacetime". In N. R. Lebovitz, W. H. Reid, and P. O. Vandervoort (eds.), *Theoretical Principles in Astrophysics and Relativity.* Chicago: University of Chicago Press (pp. 217–43).
—— (1986) "Gravity and State Vector Reduction". In R. Penrose and C. J. Isham (eds.), *Quantum Concepts in Space and Time.* Oxford: Clarendon Press (pp. 129–46).
Poincaré, H. (1905) *Science and Hypothesis.* New York: Dover.
Pooley, O. (2002) 'The Reality of Spacetime'. Unpublished doctoral thesis, Oxford University.
Psillos, S. (1999) *Scientific Realism: How Science Tracks Truth.* London: Routledge.
—— (2001) 'Is Structural Realism Possible?' *Philosophy of Science* (Supplement) 68(3): S13–S24.
—— (forthcoming) "The Structure, The Whole Structure and Nothing but the Structure". *Philosophy of Science (Proceedings).*

Redhead, M. L. G. (1995) *From Physics to Metaphysics.* Cambridge: Cambridge University Press.

—— and P. Teller (1992) "Particle Labels and the Theory of Indistinguishable Particles in Quantum Mechanics". *British Journal for the Philosophy of Science*, 43: 201–18.

Resnik, M. (1981) "Mathematics as a Science of Patterns: Ontology and Reference". *Noûs*, 15: 529–50.

—— (2000) *Mathematics as a Science of Patterns.* Oxford: Clarendon Press.

Rickles, D. P. (2005a) "A New Spin on the Hole Argument". *Studies in the History and Philosophy of Modern Physics*, 36: 415–34.

—— (2005b) "Interpreting Quantum Gravity". *Studies in the History and Philosophy of Modern Physics*, 36: 691–715.

Rovelli, C. (1997) "Halfway through the Woods: Contemporary Research on Space and Time". In J. Earman and J. Norton (eds.), *The Cosmos of Science.* Pittsburgh: University of Pittsburgh Press (pp. 180–223).

—— (1999) "Loop Quantum Gravity". *Library of Living Reviews:* **http://relativity.livingreviews.org/Articles/lrr-1998-1/.**

—— (2001) "Quantum Spacetime: What do we Know?" In C. Callender and N. Huggett (eds.), *Physics Meets Philosophy at the Planck Scale.* Cambridge: Cambridge University Press (pp. 101–22).

—— (2003) "A Dialogue on Quantum Gravity". *International Journal of Modern Physics*, D12: 1509–28.

—— (2004) *Quantum Gravity.* Cambridge: Cambridge University Press.

Russell, B. (1927) *The Analysis of Matter.* London: Kegan Paul, Trench, Trubner and Co.

Ryckman, T. (1999) "Einstein, Cassirer, and General Covariance—Then and Now". *Science in Context*, 12: 585–619.

Rynasiewicz, R. (2000) "On the Distinction between Absolute and Relative Motion". *Philosophy of Science*, 67(1): 70–93.

Saunders, S. (2003) "Indiscernibles, General Covariance, and Other Symmetries: The Case for Non-eliminativist Relationalism". In A. Ashtekar, D. Howard, J. Renn, S. Sarkar, and A. Shimony (eds.), *Revisiting the Foundations of Relativistic Physics: Festschrift in Honour of John Stachel.* Dordrecht: Kluwer (pp. 151–73).

Sklar, L. (1974) *Space, Time, and Spacetime.* Berkeley: University of California Press.

—— (1985) *Philosophy and Spacetime Physics.* Berkeley, CA: University of California Press.

Smolin, L. (1991) "Space and Time in the Quantum Universe". In A. Ashtekar and J. Stachel (eds.), *Conceptual Problems in Quantum Gravity.* New York: Birkhäuser (pp. 228–91).

—— (1997). "The Future of Spin Networks". ArXiv:gr-qc/9702030.

—— (1998) "Towards a Background Independent Approach to $\mathcal{M}$-Theory". ArXiv:hep-th/9808192.

—— (2000) *Three Roads to Quantum Gravity: A New Understanding of Space, Time and the Universe.* London: Weidenfeld & Nicolson.

—— (2003) "Time, Structure and Evolution in Cosmology". In A. Ashtekar, D. Howard, J. Renn, S. Sarkar, and A. Shimony (eds.), *Revisiting the Foundations of Relativistic Physics: Festschrift in Honour of John Stachel.* Dordrecht: Kluwer (pp. 221–74).

—— (2004b) "An Invitation to Loop Quantum Gravity". ArXiv:hep-th/0408048. To appear in *Reviews of Modern Physics.*

Stachel, J. (1980) "Einstein's Search for General Covariance, 1912–1915". In D. Howard and J. Stachel (eds.), *Einstein and the History of General Relativity. Einstein Studies*, vol. i. Boston: Birkhäuser, 1989 (pp. 63–100).

____(1993) 'The Meaning of General Covariance'. In J. Earman et al. (eds.), *Philosophical Problems of the Internal and External Worlds: Essays on the Philosophy of Adolf Grünbaum.* Pittsburgh: University of Pittsburgh Press/Konstanz: Universitaetsverlag Konstanz.

Stein, H. (1967) "Newtonian Space-Time". *The Texas Quarterly*, 10: 174–200. Reprinted in R. Palter (ed.), *The Annus Mirabilis of Sir Isaac Newton.* Cambridge, Mass: MIT Press, 1970 (pp. 258–84).

Thiemann, T. (2001) "Introduction to Modern Canonical Quantum General Relativity". ArXiv:gr-qc/0110034.

____(2003) "Lectures on Loop Quantum Gravity". In D. Giulini, C. Kiefer, and C. Lämmerzahl (eds.), *Quantum Gravity: From Theory to Experimental Search.* Berlin: Springer-Verlag (pp. 41–135).

Van Fraassen, B. (1991) *Quantum Mechanics: An Empiricist View.* Oxford: Oxford University Press.

Wald, R. (1984) *General Relativity.* Chicago: University of Chicago Press.

____(1994) *Quantum Field Theory in Curved Spacetime and Black Hole Thermodynamics.* Chicago: University of Chicago Press.

Weinberg, S. (1979) "Phenomenological Lagrangians". *Physica*, A 96: 327.

____(1995) *The Quantum Theory of Fields*, I: *Foundations.* Cambridge: Cambridge University Press.

Weingard, R. (1988) "A Philosopher Looks at String Theory". *Proceedings of the Philosophy of Science Association 1988*, 2: 95–106.

Worrall, J. (1996) "Structural Realism: The Best of Both Worlds?" *Dialectica*, 43 (1989), 99–124. Reprinted in D. Papineau (ed.), *The Philosophy of Science.* Oxford: Oxford University Press (pp. 139–65).

Wüttrich, C. (2004) "To Quantize or Not to Quantize? Fact and Folklore in Quantum Gravity". Forthcoming in *Philosophy of Science.*

Zweibach, B. (2004) *A First Course in String Theory.* Cambridge: Cambridge University Press.

2

Structural Realism and Quantum Gravity

Tian Yu Cao

Technically, it is very difficult to construct a tenable quantum theory of gravity. As a philosopher, however, my major concern is with having a consistent strategy to guide technical moves. If we look at quantum gravity this way, we immediately face a question of theoretical constraints imposed by general relativity and quantum field theory, which are the two most successful theories in fundamental physics: one deals with gravity in a classical field-theoretical framework, the other deals with quantum fields. Since quantum gravity means a quantum theory of the gravitational field,[1] what should we do so that we can secure a chance of success if we cannot meet these constraints in their original forms, and thus have to go beyond the two theories?

As I have argued elsewhere (Cao 1999, 2001), the trouble is that it is impossible to meet the constraints imposed by these two theories in a single theory without radically revising each of them. Briefly, quantum field theory requires a Minkowskian spacetime as a fixed background,[2] which is rejected by general relativity; and the latter requires a continuous manifold that cannot stand violent quantum fluctuations.[3]

[1] Some might argue that many implementations of quantum gravity are not quantization of the gravitational field (string theory being the most obvious, of course). But, in terms of the conceptual framework, string theory is only a variation of quantum field theory. For more on this see (Cao 1999); string theorist Joseph Polchinski has also agreed with this judgement (private communication).

[2] A Minkowskian background is sufficient for formulating quantum field theory. But it is not necessary. The Equivalence Principle allows one to extend quantum field theory to non-dynamical curved background manifolds. The extension is unambiguous for the Dirac action and for spin-one gauge actions, and for scalars becomes unambiguous when one specifies that massless scalar field theories should be conformal invariant. In cases where gravity is important (but geometries are still static) while quantum gravity is not, such as Hawking radiation (as a limiting case where the initial and final geometries are static [a star before gravitational collapse or a black hole after], concrete results can be obtained, although in general non-static cases nothing is unproblematic). For initial ideas about quantum field theory in curved spacetime, see Isham et al. (1975) and Hawking and Israel (1979); for more recent developments, see Wald (1994) and Kay (1996); for the breakthrough in formulating a spectral condition in a curved background manifold where there is no Poincaré symmetry, see Brunetti et al. (1996). I am grateful to Stephen Adler for suggesting a more precise description of the role played by the Minkowskian background in formulating quantum field theory, and for his giving me the references to the extensions that have been achieved since the mid-1970s (see Cao 2001). I am also grateful to Dean Rickles (private communication) for reminding me of this subtle point.

[3] In the substantivalist view of spacetime, the substantival manifold itself would be subject to quantum fluctuations, as John Wheeler once argued (1973). In the relationalist view of spacetime,

What is the way out? Before any attempt is made to address this crucial issue, we have to take a closer look at the notions of physical reality offered by general relativity and quantum field theory respectively. This question is interesting in its own right, in addition to its relevance for our construction of a tenable quantum theory of gravity, because confusions in this regard have to be cleared before we can have a correct understanding of general relativity and quantum field theory. Forget quantum gravity for a moment.

The claim I wish to make in this article is that structural realism is a framework in which the aforementioned confusions can be cleared and constraints met satisfactorily, and thus a strategy in guiding technical moves for constructing a consistent quantum theory of gravity can be suggested. But what is structural realism in the first place?

2.1 STRUCTURAL REALISM

There are various versions of structural realism these days.[4] The basic ideas of my own version that are most pertinent to the arguments in this chapter can be briefly summarized as follows.

1. The physical world consists of entities that are all structured and/or involved in larger structures.[5]
2. There are two types of structures. A structure of first type, call it componential structure, is formed by elements through a structuring agency, and thus the elements enjoy ontological priority over the structure as a whole. In contrast, a structure of second type, call it holistic structure, enjoys ontological priority over its elements, meaning that the elements, either as unstructured raw stuff or as place-holders, derive their individuality from the places they occupy and the functions they play in the structure.[6]
3. The difference in the ontological status of structures (versus that of their elements) in the two types has its root in different allocation of causal power. In the first

the relevant fluctuations refer to the fluctuations of physical properties of the gravitational field that is constitutive of the spacetime structures.

[4] Cao (1985, 1997, 2003a, 2003b); Chakravartty (1998); Ladyman (1998) and references therein; French and Ladyman (2003).

[5] John Stachel (2002) also holds this view.

[6] An example of the first type is the atomic structure of hydrogen. It is the electron's negative electric charge and the proton's positive electric charge that make the formation of a hydrogen atom possible through their electromagnetic interactions; while the hydrogen atom has no causal power over the existence and identity of the electron and proton. This kind of ontological priority of elements over the structure of which they are components is reflected in the fact that the existence of electrons and protons does not depend on the existence of the hydrogen atom, while the latter clearly depends on the former (see Strawson 1959). Among the examples of the second type stand prominently fields and spacetime structures. A field is a structure with an infinite number of components (its values at spacetime points). Although its components are individually detectable through the probing of test bodies put at proper locations, none of them, as manifestations of the field in its interaction with the test body at the location the component sits, has self-subsistent existence aside from being a place-holder in the field configuration. More discussions on spacetime structures will be given in §2.2.

type, it is the causally effective properties of elements that make it possible for a structure to be formed through the causal interactions of elements (the structuring of elements). In the second type, the elements are causally idle with regard to the structure of which they are components; what is causally effective is the structure, which is thus constitutive of the individuality of its components.

The most important implication of point (1) is that entities of any kind can be approached through their internal and external structural properties and relations that are epistemically accessible to us. In fact, our conception of the reality of any unobservable entity can only be constructed through such a structural approach.

In the process of construction, the structural knowledge just mentioned can be divided into two categories: mathematical and physical statements. A mathematical structure (e.g. connection) involved in mathematical statements (e.g. those in general relativity) may represent a physical structure (e.g. an inertio-gravitational field) which itself is also a physical entity. But this is not generally true. In most cases, what a mathematical structure represents is a relationship between physical entities (e.g. a field equation involving a coupling term between two fields), or the relational aspect of a physical entity (e.g. a metric representing the relational aspects of the connection through the compatibility conditions),[7] or even a general cognitive structure (e.g. a manifold representing the epistemic necessity for having a parameter spacetime to start with).[8] More on the latter two points in §2.2.

A physical structure involved in physical statements may be substantial, meaning it has energy and momentum,[9] and thus be a physical entity itself; or merely relational, such as the spacetime structure, which itself cannot be regarded as a physical entity. A substantial structure can be componential or holistic, but a purely relational structure can only be holistic. It is worth stressing that in line with point (1), any purely relational structure must be ontologically supported or constituted by substantial physical structures, and thus be a representation of the relational aspect of the latter. In my view, no free floating purely relational structure can exist without any ontological underpinning.

It should be stressed that the meaning of ontological priority is different in the two types of structures. In a componential structure, the ontological priority of elements over the structure formed by the elements means that the structure derives

[7] These relational aspects may (as in the case of metric) or may not (as in the case of automorphism group) form a relational physical structure. See the next paragraph.

[8] Cao (2001, 2003c); Ashtekar said: 'I fully agree with your view on unstructured background. In fact in the late eighties, I had a discussion with Julian Barbour in Syracuse where I was trying to make the same points to him but in more technical terms. Classically, he wanted to say that [in writing] the constraint equations in the Hamiltonian formulation ... points of manifolds have no reality at all. But I pointed out to him that one can not even write the constraint equations if one does not have a manifold as such without identification between points. The analogy you make with quanta to drive home the idea that there is a distinction between "lack of identity" and "lack of reality" is very nice' (2003). Since 'a general cognitive structure' can only be defined historically, the notion Ashtekar used, 'a manifold as such' is somewhat too strong a notion.

[9] It is not easy to directly define the energy and momentum possessed by a gravitational field, although the field is not unrelated to energy and momentum. For this reason, perhaps a new criterion for being substantial that is wider than having energy and momentum is needed.

its very existence from the existence of its elements. In a holistic structure, however, the ontological priority of a structure over its components only means that it is the individuality of the components that is constituted by the structure, not that their existence is derived from the structure of which they are components. Ontologically, in any structure of the second type, the components are always embedded in the structure and thus being individuated, their existence or reality and their individuality cannot be separated. The abstract way of talking about unstructured stuff and place-holders, or about something that exists but lacks individuality, makes sense only in the realm of epistemology,[10] when we try to approach the individuality (and thus the full reality) of components in a holistic structure through its constituting agent, the structure itself or the structural features of the components dictated by the structure. It should be clear now that structural realism in ontology has underlain and justified constructive realism in epistemology,[11] and thus is directly relevant to the construction of a tenable theory of quantum gravity, as we will see in §2.3.

2.2 PHYSICAL REALITY OF SPACETIME AND QUANTUM FIELDS

The current accepted interpretation of general relativity is that its chrono-geometric structures describe phenomenal spatio-temporal relations in the physical world. The phenomenological reality of spacetime, however, should not be misconceived as a substantival reality because it has no existence of its own, but only expresses the spatio-temporal relations among physical entities.

The foundation of this relationalist interpretation of spacetime is the so-called hole argument (Stachel 1980, 1986). The upshot of the argument is this. In a generally covariant theory such as general relativity, homogeneous indistinguishable points of the manifold on which the dynamic equations are defined have no physical reality because they have no individuality, otherwise the causality principle would be compromised. Only when a manifold is equipped with a metric, a solution to the dynamic equation, and its points are enmeshed into the metric structure and thus function as place-holders, can the manifold be used to define spatio-temporal relations. But even in this case, the absolute positions in spacetime, due to the lack

[10] What about field quanta? In my understanding of quantum field theory, field quanta, as registers of a field's interactions with other entities, do not exist prior to the interactions, and once a field quantum is registered in an interaction, it has already externally acquired its individuality through the interaction. This remark, concerning only the individuality of a structure's components, is not meant to be a general claim that there is no physical entity without individuality. Quantum particles in non-relativistic quantum mechanics seem to offer a ready counter-example. But it is not clear to me the extent to which these quantum particles can be viewed as particles existing in their own right, or otherwise only as a special case of being field quanta. More thorough investigations have to be done before any general claim can be made.

[11] According to constructive realism, reality is structurally constructed step by step. It should be stressed that the process of construction is only an epistemic process, not a Platonist ontic process of imposing structures 'upon an otherwise unstructured reality (formless and passive matter) from without', which was rightly rejected by John Stachel (2002).

of reality of the manifold points, remain undefinable. Thus, the absolute view of spacetime, in the sense that there is a preferred reference frame, is replaced by a relative view that no such frame can be found or defined. Furthermore, since the metric, with which the defining of spatio-temporal relations becomes possible, is dynamical, the fixed view of spacetime is replaced by a dynamical view. Taking spacetime as something substantival is thus regarded by some philosophers (see e.g. Dorato 2000) as having mistakenly reified the relational structure of spacetime (which is represented by the mathematical structure of a manifold equipped with metrics) and taken it as a structure possessing its own existence and causal power. Since the ontologically independent existence of the manifold points (interpreted as spacetime points) is definitely rejected by the hole argument, the substantivalist view seems to be definitely replaced by the relationalist view. It seems that the negative side of the argument is convincing. But what about its positive side?

Unfortunately, in its positive side, the relationalist view is deeply flawed. The old version of Machian type, according to which the spatio-temporal relations are determined by material bodies, is untenable simply because of the existence of vacuum solutions. The newer Grünbaumian version (see Grünbaum 1977) according to which these relations are constituted by rods and clocks, is also untenable because this external view is in contradiction with (i) the intrinsic view initiated by Gauss and Riemann, (ii) Einstein's view that they are constituted by the gravitational field, and (iii) Einstein's view that the separation of rods and clocks from all other physical entities is inconsistent.[12]

The combination of the prevailing relationalist rhetoric and the lack of concrete relationalist understanding of spacetime results in a strange phenomenon. Many self-claimed relationalists appear to substitute dynamical talk for relationalist talk, in particular in the context of dealing with pure gravity without any other physical entities being involved. Rather than directly arguing in support of relationalism, proponents defend a dynamical understanding and then claim that since no non-dynamical fixed background spacetime is acceptable, spacetime is purely relational. Of course, dynamical spacetime can still be a substantivalist one if it can exist in its own right, such as the case in Wheeler's geometrodynamics (1962).[13]

A proper understanding of spacetime, which would preserve all the merits of the relationalist view without its flaws, is a structural-constitutive-constructive (SCC) one. The major ideas of the SCC view can be briefly summarized as follows.

1. It shares the relationalist view that the manifold points or manifold itself have no direct spatio-temporal meaning, and that the phenomenological reality of spacetime is constituted by the metric or Riemann tensor, which endow the points with individuality either through imposing non-reflexive metrical relationships upon manifold points (see Saunders 2003), or characterizing the points with

[12] For more details, cf. §5.1 of Cao (1997).

[13] See also Stein (1967), in which Howard Stein explicitly argues against the view that 'dynamical' means 'relational'. I am grateful to Dean Rickles for bringing this reference to my attention.

four or more invariants of the Riemann tensor.[14] For this reason, the points of the manifold or the manifold itself enjoy no ontological priority over the chrono-geometrical structures.

However, different from the relationalist view, the SCC view maintains that a minimally structured manifold (with only a global topological structure) is real and is the starting point for our further construction of the reality of spacetime for two reasons. First, without initially assuming the existence of the manifold as a parameter spacetime, no further investigations of spacetime structures would be possible.[15] Here the Kantian argument for the epistemic necessity of spacetime looms large. Second, the very dimensionality of the manifold represents the most general feature of spacetime. But the acknowledgement of the primitive (in the epistemically constructive sense) reality of such a minimally structured manifold should not be taken as an attempt to revive the substantivalist view, because here the spacetime is not to be taken as the totality of (ontologically basic) events represented by the manifold points; rather, the events or points are deeply structural in the sense that their individuality is constituted by complicated structures (or themselves are characterized by structural features dictated by the larger structure in which they are embedded).

2. It shares the relationalist view that spacetime is not fixed but dynamical, not substantival but relational. However, it maintains that mere dynamicity is not enough to reject the substantivalism, thus more convincing arguments are needed. Furthermore, the roots of dynamic interactions lie in the causal power of physical properties possessed by substantial physical entities. Thus being dynamical presumes being substantial. But the chrono-geometrical structures that constitute the individuality of manifold points are not substantial but purely relational. So substantial entities have to be found that are supposed to be constitutive of the chrono-geometrical structures. A solution suggested by the SCC view to this dual task is this. The gravitational field, as a substantial physical entity represented by the connection field, is the required dynamical entity, whose relational aspects are represented by the metric tensor field, which represents the chrono-geometrical structure, through the compatibility condition.[16] Therefore, although the spatio-temporal relations are constituted by the chromo-geometrical structure (the metric), the latter itself is constituted, or ontologically supported, by the inertio-gravitational field (the connection). Once the relationship between the

[14] This is associated with Bergmann (1957) and Komar (1958), see also Stachel (1993). I am grateful to John Stachel and Dean Rickles for bringing these references to my attention.

[15] More arguments in this regard can be found in Cao (2001).

[16] In the geodesic equation, if we take the components of the connection with respect to any basis as being numerically equal to the Christoffel symbols of a metric tensor, then the metric tensor thus obtained is compatible to the connection. Observationally, this means that particles' trajectories, as the result of their interactions with the gravitational field represented by the connection, are geodesics with respect to the metric that is compatible with the connection, or only the metric that is compatible with the connection can describe the observable spatio-temporal behaviour of physical entities, and thus the claim that a compatible metric represents the relational aspect of the connection is justified.

metric and the connection is thus clarified, the mystery about the dynamicity of the metric (as a purely relational structure) is dispelled: it is only an epiphenomenon of the dynamical behaviour of the connection, a substantial physical structure or entity.

3. It takes the metric and connection as holistic structures that enjoy ontological priority over their components. For the metric, manifold points are only place-holders for the spatio-temporal relations it stipulates; for the connection field, it is both a holistic structure, stipulating the possible behaviours of test bodies if they are put somewhere in the field and interacting with parts of the field (as place-holders) there, and a substantial entity.
4. For the reasons spelled out above, the SCC view takes the ultimate reality of spacetime as being field-theoretical in nature. That is, the spatio-temporal aspects of the world are constituted, characterized, and explained completely in terms of the gravitational field (connection). First, the manifold continuum is constituted and characterized by an infinite number of degrees of freedom of the field in a continuous way. Second, it is the relational aspect of the field, stemming from its universal coupling with all other physical entities,[17] represented by the metric that stipulates the spatio-temporal structure of the world. This field-theoretical framework has provided a firm foundation for the construction of a tenable theory of quantum gravity that is compatible with the general theory of relativity.

The understanding of the physical reality of quantum fields, as it is constructed in existing quantum field theories, in my view, can be summarized as follows.

1. A quantum field is a dynamical global substratum that is ever fluctuating,[18] locally excitable,[19] and quantum in nature.[20]
2. The substratum, however, is itself defined over (or ontologically supported by) a pre-existing background spacetime, namely, a four-dimensional Minkowskian manifold with a fixed, classical chrono-geometrical structure.[21]
3. This global but structured background spacetime underlies (i) a global vacuum state of the field; (ii) an infinite number of degrees of freedom of the field, indexed

[17] Generally, the relational aspect of a field as a holistic structure is manifested in its interactions with test bodies that are placed in different locations in the field configuration. In the case of gravitational field, its interactions are characterized by its universal coupling, including its couplings with rods and clocks, which specify the behaviour of rods and clocks and thus give metric meaning to the relational aspect of the gravitational field.

[18] The intrinsic and primitive quantum fluctuations of a field's physical properties over a spacetime region is the ontological basis for the coupling of physics at different scales, which in turn is a conceptual basis for renormalization group organization of physics.

[19] A field can be locally excited by its intrinsic fluctuations or by external disturbances.

[20] It means that the local excitations of a field obey quantum principles, such as canonical commutation or anti-commutation relations and uncertainty relations (which is defined in terms of the variations of measurable properties over a spacetime region.

[21] 'Classical' here means that the structure cannot be treated in a quantum mechanical way. Otherwise, the whole theoretical structure of quantum field theory would collapse. For more on this see Cao (1999).

by the spacetime points; (iii) the localizability of each and every degree of freedom of the field; (iv) the causal (light-cone) structure and the quantum structures (commutation or anti-commutation relations). Note that the uncertainty principle and its metaphysical presupposition (and/or implication), namely the fluctuations of physical properties definable in a spacetime region, are anchored in such a fixed background spacetime. Without such a background, these notions would not even be definable.

2.3 QUANTUM GRAVITY

The above discussions have direct bearings on the construction of consistent quantum theories of gravity, in terms of fundamental ontological commitment as well as overall theoretical structure. Ontologically, the first step in the construction, as in the construction of any other fundamental physical theory, is to determine or find out what should be taken as the fundamental degrees of freedom that should be investigated quantum mechanically. In terms of overall theoretical structure, two questions have to be properly addressed. First, the question concerning the reconstruction of the classical limit, namely, the classical theory of general relativity should be derivable from the quantum theory of gravity. Second, the complicated relationship among kinematical structures, dynamical structures, and causal structures has to be clarified.

If we take the above discussions about physical reality seriously, then we have to take some substantial dynamical physical entities or their components to be the fundamental degrees of freedom and investigate their quantum behaviour. Classically, the relevant physical entity here is the connection. Thus if we can actively quantize it, we will have a desirable classical limit. This observation has provided strong motivation for taking connections as a starting point for constructing a quantum theory of gravity (Ashtekar 1986).

But there are some subtle problems, and thus the works by Ashtekar and his colleagues (the loop quantum gravity school[22]) may have to be interpreted in a different way. The relationship between quantum entities and classical entities is very complicated, more complicated than the notion of active quantization can conceptualize. For example, fermion fields have no classical limit, and gravitational fields cannot be quantized in the conventional way because the resultant theory gives only meaningless results. A crucial point here is that the entities in the quantum sector and the entities in the classical sector may not be the same entities only behaving differently on different energy scales.

If we take this point seriously, we have to give up the attempt of actively quantizing some classical degrees of freedom when it is not appropriate, for example, in the case of gravity. In its stead, a proper position to take is the quantum realist position, which is complemented by an emergentist understanding of the classical limit. More specifically, this position suggests that we should first take some fundamental degrees

[22] See for example, Smolin (2003), Rovelli (2004), and Perez (2003) and numerous references therein.

of freedom, which need not have a direct link to any classical entities, and investigate them quantum mechanically. Only when the theoretical structure of the quantum sector has been constructed in a consistent way should efforts be made to re-establish its links with classical phenomena by reconstructing some classical entities. These reconstructed classical entities may have totally different characteristics from their quantum counterparts, and thus should be legitimately regarded as something emergent when we move from one energy regime to another, similar to the emergence of ice from water in the phase transition.

If we construct a quantum theory of gravity in such a quantum-realist way, the constraint of quantum theory on the construction (violent quantum fluctuations that undermine the manifold's smooth topology which is the foundation of classical general relativity), that had seriously worried Wheeler (1973), may not even be relevant here.

It is irrelevant because the manifold involved in the construction of a quantum theory is not spacetime itself (which, as a phenomenal structure constituted by classical gravitational field, can only be classical), but a parameter spacetime, whose only function is to pin down the most general features of a theoretical structure, the individuality of which is constituted by the quantum gravitational field and the classical limit of which would be spacetime. Most important among these general features is the dimensionality, which can endure quantum fluctuations without itself undergoing any change. Whatever the characteristic features, resulting from quantum fluctuations, of differential and other topological structures a manifold possesses, they are constituted by the structure, behaviour pattern, and other relational aspects of the quantum gravitational field, or as the expression of the latter, and have no direct bearings on the classical, observable spacetime structures.

The same reasoning can be applied to the so-called problem of time in canonical formulations of quantum gravity. Many physicists and philosophers working on quantum gravity have tried hard to understand the 'deep' and 'mysterious' implications of the time problem. However, if we interpret the canonical formulations in a quantum realistic way, if we at the same time take an SCC interpretation of the general relativity, namely, that spacetime is constituted by the classical gravitational field, then we would realize that in the quantum realm, there is no spacetime, and thus no time at all. According to the emergentist view mentioned above that complements the quantum realist view, we do not have to have time at the quantum level to get a classical time. Time may emerge from some quantum structures that are not based on any temporal structures.

In terms of ontological commitment and quantum realism discussed above, the evolution of the loop quantum gravity school is very instructive. In fact, the transitions from taking connections as the fundamental degrees of freedom, to taking spin networks (as functions of connections and as a convenient basis in the Hilbert space of square-integrable functions of connections)[23] and then diffeomorphism

[23] A spin network is a labelled graph, whose edges and vertices are labelled by some spin variables, that is embedded into a non-dynamic four-dimensional differentiable manifold. For details, see Perez (2003).

equivalent classes of spin networks as the fundamental units for analysis, have revealed a trajectory of gradually moving away from something directly suggested by a classical entity to something intrinsically quantum in nature. That is, it is an evolution from a pursuit of active quantization to a quantum realist pursuit. But this recognition should not blind us from seeing the importance of the original commitment to the connections: all the subsequent transformed fundamental degrees of freedom, because of their links to the original ones, are physically real degrees of freedom rather than purely relational ones, and thus the links have provided the successive theoretical constructions with a firm ontological foundation.

In terms of overall theoretical structure in the construction of consistent quantum theories of gravity, the direct bearings of the positions presented in this article can also be illuminated by the developments in loop quantum gravity. Let me start with a discussion on the relationship between dynamical structures, causal structures, and kinematical structures, which is closely related with the theoretical constraint posed by general relativity upon any quantum theory.

As we have noticed above (the end of §2.2), a quantum theory requires a background spacetime, a Minkowskian spacetime with a fixed classical chrono-geometrical structure. However, by a closer examination, we find that the necessity for having a fixed Minkowskian spacetime structure lies, mainly, in its functions of defining local fields and their causal structure, and thus is dispensable if we can find other ways to localize quantum fields and to define their causal structure. That is, if we can define causal structures among fundamental degrees of freedom in a dynamical way, and reinterpret localization in a relationalist way understood in the spirit of the SCC view, then the constraint from general relativity that undermines the kinematical structure of conventional quantum field theory can be satisfactorily met in the construction of a consistent theory of quantum gravity, such as the spin-foam model.

I am quite fascinated by the recent developments along the line that has led to the spin-foam model. Technical obstacles aside, conceptually, a quite satisfactory picture has already emerged. The whole loop quantum gravity programme, that has culminated in the spin-foam model started its construction within the canonical approach; the fundamental reason for many scholars to have serious reservations about the approach, and to be in favour of the covariant approach, is that the former artificially separates space and time, and is thus in direct contradiction with the basic teachings of special and general relativity. But the relevant literature about the spin-foam model over the last few years has shown that this concern has in fact been addressed quite properly.

The canonical people, motivated by their dynamical concerns, in addressing the transition between the diffeomorphism equivalent classes of spin networks, have adopted the sum over histories approach.[24] This has profound implications, at least conceptually.

First, it is inherently four-dimensional, and thus paves the way for a covariant formulation of the canonical approach.

[24] Here this means summing over foams (each foam is a possible transition or a history), or over the graphs and labels of the spin network involved in transition. More can be found in Perez (2003).

Second, it gives a concrete model for defining causal structures in terms of dynamical processes (transitions) without involving any pregiven, fixed background manifold, let alone any fixed kinematical structures. From the SCC standpoint, this is philosophically very satisfactory. In addition, the causal structures derived from dynamical processes in this model are by no means global, but have incorporated all the local complexities of the underlying dynamical processes.

Third, the kinematical structures derived from the causal structures are accordingly intrinsically local and dynamical. In particular, the very notions of spatial and temporal are defined solely in terms of causal and acausal, and equal time slices are defined by spin-network states that are involved in a causal sequence. The result is that everything in the kinematical structures is piece-wise, namely, local and variable (or dynamical). But still, the classical notions of spacetime and space and time seem to be reconstructible, at least in principle. For example, it can be done through a coarse grained approximation to discrete quantum structures (see e.g. Smolin 2001).

Of course, I know that this is an over-optimistic remark. Two serious problems remain to be addressed. First, other fields have to be incorporated. This may still be considered as a technical obstacle. But then the other problem is definitely a very serious conceptual problem. That is that the very notion of causality, which is the foundation for all the merits I have just mentioned, is not properly defined in the spin-foam model. The trouble here is that causality can only be defined in each history, which is dynamical, local and well defined. But what about the overall causality, causality as the result of summing over the histories? I find it difficult to have any clear ontological, metaphysical, or conceptual picture as to what the resulting notion of causality is. Different from causality in each foam, the resulting causality cannot be said to be local because there is no clear-cut local sense to be defined. However, to be fair to the sum over history people, I should mention that the same question can also be raised for its prototype, namely the Feynman path integral approach that is widely used in quantum physics. What is the ontological basis of these paths? Do they really exist? Although nobody can answer these questions, the Feynman approach works extremely well. Then why should we bother in the case of the spin-foam model? Here is the difference. The Feynman approach in conventional quantum physics does not touch upon the underlying causal and kinematical structures. Thus the question we can raise is only a disguised version of the same question that is common to all formulations of quantum physics, namely the superposition of quantum states. But this is not the case in the case of the spin-foam model. Here, in addition to this common question of superposition, the sum over histories has also blurred our clear sense of causality and that of temporality and spatiality. This is obviously a very serious conceptual problem, which has to be properly addressed before we can have a proper interpretation of the spin-foam model. Still, judging from a structural and constructive realist position, this model is the most promising, so far, among all attempts at constructing a consistent model of quantum gravity.

ACKNOWLEDGEMENTS

The first draft of this article, titled 'Philosophical Issues in Attempts at a Quantum Theory of Gravity', was presented at the *Boston Colloquium for Philosophy of Science on Perspectives on Quantum Gravity: A Tribute to John Stachel*, 6 March 2003. I am grateful to comments made by other participants and audience. The discussions and exchanges during and after the colloquium with John Stachel and Abhay Ashtekar are particularly helpful in clarifying my own positions.

REFERENCES

Ashtekar, A. (1986) "New Variables for Classical and Quantum Gravity". *Physical Review Letters*, 57: 2244–7.

—— (2003) Private communication.

Bergmann, P. G. (1957) "Topics in the Theory of General Relativity". In *Brandeis University Summer Institute of Theoretical Physics*. Waltham, Mass.: Brandeis University (pp. 1–44).

Brunetti, R., K. Fredenhagen, and M. Köhler (1996) "The Microlocal Spectrum Condition and Wick Polynomials of Free Fields on Curved Spacetimes". *Commun. Math. Phys.* 180 (633).

Cao, T. Y. (1985) 'An Intellectual History of 20th Century Field Theories'. Fellowship dissertation submitted to Trinity College, Cambridge, August 1985.

—— (1997) *Conceptual Development of 20th Century Field Theories*. Cambridge: Cambridge University Press.

—— (1999) "Introduction: Conceptual Issues in Quantum Field Theory". In T. Yu Cao (ed.), *Conceptual Foundations of Quantum Field Theory*. Cambridge: Cambridge University Press (pp. 1–27).

—— (2001) "Prerequisites for a Consistent Framework of Quantum Gravity". *Studies in the History and Philosophy of Modern Physics*, 32(2): 181–204.

—— (2003a) "Structural Realism and the Interpretation of Quantum Field Theory", *Synthese*, 136(1): 3–24.

—— (2003b) "Can We Dissolve Physical Entities into Mathematical Structures?" *Synthese*, 136(1): 57–71.

—— (2003c) "Philosophical Issues in Attempts at a Quantum Theory of Gravity". Presented at at the *Boston Colloquium for Philosophy of Science on Perspectives on Quantum Gravity: A Tribute to John Stachel*, 6 March 2003.

Chakravartty, A. (1998) "Semirealism". *Studies in the History and Philosophy of Science*, 29: 391–408.

Dorato, M. (2000) "Substantivalism, Relationalism, and Structural Spacetime Realism". *Foundations of Physics*, 30(10) 1605–28.

French, S., and J. Ladyman (2003) "Remodelling Structural Realism: Quantum Physics and the Metaphysics of Structure", *Synthese*, 136(1): 31–56.

Grunbaum, A. (1977) "Absolute and Relational Theories of Space and Space-Time". In J. Earman, C. Glymour, and J. Stachel (eds.), *Foundations of Space-Time Theories*. University of Minnesota Press (pp. 303–73).

Hawking, S., and W. Israel (eds.) (1979) *General Relativity: An Einstein Centenary Survey*. Cambridge: Cambridge University Press.

Isham, C. J., R. Penrose, and D. W. Sciama (eds.) (1975) *Quantum Gravity: An Oxford Symposium*. Oxford: Clarendon Press.

Kay, B. S. (1996) "Quantum Fields in Curved Spacetime: Non-Global Hyperbolicity and Locality". In S. Doplicher, R. Longo, J. Roberts, and L. Zsido (eds.), *Proceedings of the Conference on Operator Algebras and Quantum Field Theory*. Rome: Accademia Nationale dei Lincei.

Komar, A. (1958) "Construction of a Complete Set of Independent Observables in the General Theory of Relativity". *Physical Review*, 111: 1182–7.

Ladyman J. (1998) "What is Structural Realism?" *Studies in the History and Philosophy of Science*, 29: 409–24.

Perez, A. (2003) "Spin Foam Model for Quantum Gravity". ArXiv:gr-qc/0301113 v2.

Rovelli, C. (2004) *Quantum Gravity*. Cambridge: Cambridge University Press.

Saunders, S. (2003) "Indiscernibles, General Covariance, and Other Symmetries: The Case for Non-Reductive Relationalism". In Abhay Ashtekar et al. (eds.), *Revisiting the Foundations of Relativistic Physics*. Kluwer: Dordrecht (pp. 151–73).

Smolin, L. (2001) *Three Roads to Quantum Gravity*. London: Weidenfeld & Nicolson.

—— (2003) "How Far are we from the Quantum Theory of Gravity?" hep-th/0303185 v2.

Stachel, J. (1980) "Einstein's Search for General Covariance, 1912–1915". Presented at the *9th International Conference on General Relativity and Gravitation*, Jena, 1980. Reprinted in D. Howard and J. Stachel (eds.), *Einstein and the History of General Relativity*. Boston: Birkhäuser (pp. 63–100).

—— (1986) "What a Physicist Can Learn from the Discovery of General Relativity". In R. Ruffini (ed.), *Proceedings of the Fourth Marcel Grossmann Meeting on General Relativity*. Amsterdam: Elsevier (pp. 1857–62).

—— (1993) "The Meaning of General Covariance". In John Earman et al. (eds.), *Philosophical Problems of the Internal and External Worlds*. Pittsburgh: University of Pittsburgh Press (pp. 129–60).

—— (2002) "Structure, Individuality and Quantum Gravity". Presented at PSA 2002, Milwaukee, 8 Nov. 2002.

Stein, H. (1967) "Newtonian Space-Time". *Texas Quarterly*, 10: 174–200. Reprinted in R. Palter (ed.), *The Annus Mirabilis of Sir Isaac Newton*. Cambridge, MA: MIT Press, 1970 (pp. 258–84).

Strawson, P. F. (1959) *Individuals*. London: Methuen.

Wald, R. (1994) *Quantum Field Theory in Curved Spacetime and Black Hole Thermodynamics*. Chicago: University of Chicago Press.

Wheeler, J. (1962) *Geometrodynamics*. New York: Academic Press.

—— (1973) "From Relativity to Mutability". In J. Mehra (ed.), *The Physicist's Conception of Nature*. Dordrecht: Reidel (pp. 202–47).

3

Structure, Individuality, and Quantum Gravity

John Stachel

ABSTRACT

After reviewing various interpretations of *structural realism*, I adopt a definition that allows both relations between things that are already individuated (which I call 'relations between things') and relations that individuate previously un-individuated entities ('things between relations'). Since both spacetime points in general relativity and elementary particles in quantum theory fall into the latter category, I propose a principle of maximal permutability as a criterion for the fundamental entities of any future theory of 'quantum gravity'; i.e. a theory yielding both general relativity and quantum field theory in appropriate limits. Then I review a number of current candidates for such a theory. First I look at the effective field theory and asymptotic quantization approaches to general relativity, and then at string theory. Next, a discussion of some issues common to all approaches to quantum gravity based on the full general theory of relativity argues that *processes*, rather than *states* should be taken as fundamental in any such theory. A brief discussion of the canonical approach is followed by a survey of causal set theory. The chapter ends by suggesting a new approach to the question of which spacetime structures should be quantized.

3.1 WHAT IS STRUCTURAL REALISM?

The term 'structural realism' can be (and has been) interpreted in a number of different ways.[1] I assume that, in discussions of structuralism, the concept of 'structure' refers to some set of relations between the things or entities that they relate, called *the relata*. Here I interpret 'things' in the broadest possible sense: they may be material objects, physical fields, mathematical concepts, social relations, processes, etc.[2] People have used the term 'structural realism' to describe different approaches to the nature of

[1] For a recent survey, with references to earlier literature, see the symposium 'Structural Realism and Quantum Field Theory', Symons (2003), which includes papers by Tian Yu Cao, Steven French and James Ladyman, and Simon Saunders.

[2] In this section, 'thing' is used in a sense that includes processes. Note that from §3.5.1 on, it is used in a more restricted sense, in which 'thing' is contrasted with 'process'.

the relation between things and relations. These differences all seem to be variants of three basic possibilities.

I. There are only relations without relata.[3]

As applied to a particular relation, this assertion seems incoherent. It only makes sense if it is interpreted as the metaphysical claim that ultimately there are only relations; that is, in any given relation, all of its *relata* can in turn be interpreted as *relations*. Thus, the totality of structural relations reduces to relations between relations between relations ... As Simon Saunders might put it, it's relations all the way down (e.g. Saunders 2003).[4] It is certainly true that, in certain cases, the relata can themselves be interpreted as relations; but I would not want to be bound by the claim that this is always the case. I find rather more attractive the following two possibilities:

II. There are relations, in which the things are primary and their relation is secondary.
III. There are relations, in which the relation is primary while the things are secondary.

In order to make sense of either of these possibilities, and hence of the distinction between them, one must assume that there is always a distinction between the *essential* and *non-essential* properties of any thing,[5] For II to hold (i.e. things are primary and their relation is secondary), no essential property of the relata can depend on the particular relation under consideration; while for III to hold (i.e. the relation is primary and the relata are secondary), at least one essential property of each of the relata must depend on the relation. Terminology differs, but one widespread usage denotes relations of type II as *external*, those of type III as *internal*. One could convert either possibility into a metaphysical doctrine: 'All relations are external' or 'All relations are internal'; and some philosophers have done so. But, in contradistinction to I, there is no need to do so to make sense of II and III. If one does not, then the two are perfectly compatible.
Logically, there is a fourth possible case:

IV. There are things, such that any relation between them is only apparent.

This is certainly possible in particular situations. One could, for example, pre-programme two mechanical dolls (the things) so that each would move independently

[3] See Krause (2004: 1), for the phrase 'relations without the relata'.
[4] 'I believe that objects are structures; I see no reason to suppose that there are ultimate constituents of the world, which are not themselves to be understood in structural terms. So far as I am concerned, it is turtles all the way down' (Saunders 2003: 329).
[5] For example, in quantum mechanics, electrons are characterized by their essential properties of mass, spin, and charge. All other properties that they may exhibit in various processes—such as positions, momenta, or energies—are non-essential (see n. 19). For a discussion of their haecceity, see the next section. As this example suggests, the distinction between essential and non-essential properties—and indeed the distinction between elementary and composite entities—may be theory dependent (see Dosch et al. 2004). (Note that having the same essence is what characterizes a natural kind.)

of the other, but in such a way that they seemed to be dancing with each other (the apparent relation—I assume that 'dancing together' is a real relation between two people).

Again, one might convert this possibility into a universal claim: 'All relations are only apparent.' Leibniz's monadology, for example, might be interpreted as asserting that all relations between monads are only apparent. Since God set up a pre-established harmony among them, they are pre-programmed to behave as if they were related to each other. As a metaphysical doctrine, I find IV even less attractive than I. And if adopted, it could hardly qualify as a variant of structural realism, so I shall not mention IV any further.

While several eminent philosophers of science (e.g. French and Ladyman) have opted for version I of structural realism, to me versions II and III (interpreted non-metaphysically) are the most attractive. They do not require commitment to any metaphysical doctrine, but allow for a decision on the character of the relations constituting a particular structure on a case-by-case basis.[6] My approach leads to a picture of the world, in which there are entities of many different natural kinds, and it is inherent in the nature of each kind to be structured in various ways. These structures themselves are organized into various structural hierarchies, which do not all form a linear sequence (chain); rather, the result is something like a partially ordered set of structures. This picture is dynamic in two senses: there are changes in the world, and there are changes in our knowledge of the world.

As well as a *synchronic* aspect, the entities and structures making up our current picture of the world have a *diachronic* aspect: they arise, evolve, and ultimately disappear—in short, they constitute *processes*. And our current picture is itself subject to change. What particular entities and structures are posited, and whether a given entity is to be regarded as a thing or a relation, are not decisions that are forever fixed and unalterable; they may change with changes in our empirical knowledge and/or our theoretical understanding of the world. So I might best describe this viewpoint as a dynamic structural realism.[7]

3.2 STRUCTURE AND INDIVIDUALITY

A more detailed discussion of many points in this section is presented in Stachel (2002, 2005).

It seems that, as deeper and deeper levels of these structural hierarchies are probed, the property of inherent individuality that characterizes more complex, higher-level entities—such as a particular crystal in physics, or a particular cell in biology—is

[6] For further discussion of cases II and III, see Stachel (2002), which refers to case II as 'relations between things', and to case III as 'things between relations'.

[7] For further discussion of the structural hierarchy, see Stachel (2005). For many examples of such hierarchies in the physics, biology, and cosmology, see Ellis (2002). Although my concepts of entity and structure are meant to be ontological, the term 'ontic structural realism' has been given a different significance (see Ladyman 1998).

lost. Using some old philosophical terminology, I say that a level has been reached, at which the entities characterizing this level possess *quiddity* but not *haecceity*. 'Quiddity' refers to the essential nature of an entity, its natural kind; and—at least at the deepest level which we have reached so far—entities of different natural kinds exist, e.g. electrons, quarks, gluons, photons, etc.[8] What distinguishes entities of the same natural kind (quiddity) from each other, their unique individuality or 'primitive thisness', is called their 'haecceity'.[9] Traditionally, it was always assumed that every entity has such a unique individuality: a haecceity as well as a quiddity. However, modern physics has reached a point at which we are led to postulate entities that have quiddity but no haecceity that is inherent, i.e. independent of the relational structures in which they may occur. Insofar as they have any haecceity (and it appears that degrees of haecceity must be distinguished[10]), such entities inherit it from the structure of relations in which they are enmeshed. In this sense, they are indeed examples of the case III: 'things between relations' (Stachel 2002).

Since Kant, philosophers have often used position in space as a principle of individuation for otherwise indistinguishable entities; more recently, similar attempts have been made to individuate physical events or processes.[11]

A physical process occupies a (generally finite) region of spacetime; a physical event is supposed to occupy a point of spacetime. In theories, in which spacetime is represented by a continuum, an event can be thought of as the limit of a portion of some physical process as all the dimensions of the region of spacetime occupied by this portion are shrunk to zero. Classically, such a limit may be regarded as physically possible, or just as an ideal limit. 'An event may be thought of as the smallest part of a process ... But do not think of an event as a change happening to an otherwise static object. It is just a change, no more than that' (Smolin 2002a: 53). See §3.5.1 for further discussion of processes. It is probably better to avoid attributing physical significance to point events, and accordingly to mathematically reformulate general relativity in terms of sheaves.[12]

Individuation by means of position in spacetime works at the level of theories with a fixed spacetime structure, notably special-relativistic theories of matter and/or

[8] Believers in a unified 'Theory of Everything' will hope that ultimately only entities of one natural kind will be needed, and that all apparently different kinds will emerge from the relational properties of the one fundamental quiddity. String theory might be regarded as an example of such a theory; but, aside from other problems, its current framework is based on a fixed background spacetime, as will be discussed in §3.3.

[9] Runes (1962): 'Haecceity ... [is] ... A term employed by Duns Scotus to express that by which a quiddity, or general essence, becomes an individual, particular nature, or being'. Teller (1995, 1998), following Adams (1979), noted the utility of the term 'haecceity', and his suggestion has been followed by many philosophers of physics.

[10] For example, the electrons confined to a particular 'box' (i.e. infinite potential well) may be distinguished from all other electrons, if not from each other.

[11] See e.g. Auyang (1995), ch. 6.

[12] See Stachel and Iftime (2005) for further discussion of this point. For one such reformulation of differential geometry, see Mallios (1998), and for applications to general relativity, see Mallios (2006) and Mallios and Raptis (2003).

fields[13] but, according to general relativity, because of the dynamical nature of all spacetime structures,[14] the points of spacetime lack inherent haecceity; thus they cannot be used for individuation of other physical events in a general-relativistic theory of matter and/or non-gravitational fields. This is the purport of the 'hole argument' (see Stachel 1993, and references therein). The points of spacetime have quiddity as such, but only gain haecceity (to the extent that they do) from the properties they inherit from the metrical or other physical relations imposed on them.[15] In particular, the points can obtain haecceity from the inertio-gravitational field associated with the metric tensor: For example, the four non-vanishing invariants of the Riemann tensor in an empty spacetime can be used to individuate these points in the generic case (see ibid. 142–3).

Indeed, as a consequence of this circumstance, in general relativity the converse attempt has been made: to individuate the points of spacetime by means of the individuation of the physical (matter or field) events or processes occurring at them; i.e. by the relation between these points and some individuating properties of matter and/or non-gravitational fields. Such attempts can succeed at the macroscopic, classical level;[16] but, if the analysis of matter and fields is carried down far enough—say to the level of the sub-nuclear particles and field quanta[17]—then the particles and field quanta of differing quiddity all lack inherent haecceity.[18] Like the points of

[13] Actually, the story is more complicated than this. The points of Minkowski spacetime, for example, are themselves homogeneous, and some physical framework must be introduced in order to physically individuate them. Only after this has been done can these points be used to individuate other events or processes. The physical framework may be fixed non-dynamically (e.g. by using rods and clocks introduced a priori); or if fixed by dynamical process (e.g. light rays and massive particles obeying dynamical equations), the resulting individuation must be the same for all possible dynamical processes. This is better said in the language of fibre bundles, in which particular dynamical physical fields are represented by cross-sections of the appropriate bundle: if the metric of the base space is given a priori, the individuation of the points of the base space is either also so given, or is the same for all cross-sections of the bundle (see Stachel and Iftime 2005).

[14] Except for continuity, differentiability and local topology of the differentiable manifold.

[15] Again this is better said in the language of fibred manifolds, in which particular dynamical physical fields are represented by cross-sections of the manifold: one can now define the base space as the quotient of the total space by the fibration. Thus, even the points of the base space (let alone its metric) are not defined a priori, and their individuation depends on the choice of a cross-section of the fibred manifold, which will include specification of a particular inertio-gravitational field. For a detailed discussion, see Stachel and Iftime (2005).

[16] See e.g., Rovelli 1991.

[17] I reserve the term 'elementary particles' for fermions and 'field quanta' for bosons, although both are treated as field quanta in quantum field theory. I aim thereby to recall the important difference between the two in the classical limit: classical particles for fermions and classical fields for bosons. In this chapter, I am sidestepping the question of whether and when the field concept is more fundamental than the particle concept in quantum field theory, especially in non-flat background spacetimes (I discuss it in detail in Stachel (2006a), but see the next footnote and n. 45); but in the special-relativistic theories, a preparation or registration may involve either gauge-invariant field quantities or particle numbers.

[18] At the level of non-relativistic quantum mechanics for a system consisting of a fixed number of particles of the same type, this is seen in the need to take into account the bosonic or fermionic nature

spacetime, insofar as they have any individuality, it is inherited from the structure of relations in which these quanta are embedded. For example, in a process involving a beam of electrons, a particular electron may be individuated by the click of a particle counter.[19]

In all three of these cases—spacetime points or regions in general relativity, elementary particles in quantum mechanics, and field quanta in quantum field theory—insofar as the fundamental entities have haecceity, they inherit it from the structure of relations in which they are enmeshed. But there is an important distinction here between general relativity on the one hand and quantum mechanics and quantum field theory on the other: the former is background independent while the latter are not; but I postpone further discussion of this difference until §3.5.2.

What has all this to do with the search for a theory of quantum gravity? The theory that we are looking for must underlie both classical general relativity and quantum theory, in the sense that each of these two theories should emerge from 'quantum gravity' by some appropriate limiting process. Whatever the ultimate nature(s) (quiddity) of the fundamental entities of a quantum gravity theory turn out to be, it is hard to believe that they will possess an inherent individuality (haecceity) already absent at the levels of both general relativity and quantum theory (see Stachel 2005). So I am led to assume that, whatever the nature(s) of the fundamental entities of quantum gravity, they will lack inherent haecceity, and that such individuality as they manifest will be the result of the structure of dynamical relations in which they are enmeshed. Given some physical theory, how can one implement this requirement of no inherent haecceity? Generalizing from the previous examples, I maintain that the way to assure the inherent indistinguishability of the fundamental entities of the

of the particle in question by the appropriate symmetrization or antisymmetrization procedure on the product of the one-particle Hilbert spaces (see e.g. Haag 1996: 35–6, for more details). At the level of special-relativistic quantum field theory, in which interactions may change particle numbers, it is seen in the notion of field quanta, represented by occupation numbers (arbitrary for bosons, either zero or one for fermions) in the appropriately constructed Fock space; these quanta clearly lack individuality. (See e.g. Teller 1995; Haag 1996: 36–8.). At the level of quantum field theory in background curved spacetimes, 'a useful particle interpretation of states does not, in general, exist' (Wald 1994: 47).

[19] The macroscopic counter is assumed to be inherently individuated. It seems that, for such individuation of an object, a level of structural complexity must be reached, at which it can be uniquely and irreversibly 'marked' in a way that distinguishes it from other objects of the same nature (quiddity). My argument is based on an approach, according to which quantum mechanics does not deal with quantum systems in isolation, but only with processes that such a system can undergo. (For further discussion of this approach, see Stachel 1986, 1997). A process (Feynman uses 'process', but Bohr uses 'phenomenon' to describe the same thing) starts with the preparation of the system, which then undergoes some interaction(s), and ends with the registration of some result (a 'measurement'). In this approach, a quantum system is defined by certain essential properties (its quiddity); but manifests other, non-essential properties (its haecceity) only at the beginning (preparation) and end (registration) of some process. (Note that the initially prepared properties need not be the same as the finally registered ones.) The basic task of quantum mechanics is to calculate a probability amplitude for the process leading from the initially prepared values to the finally registered ones. (I assume a maximal preparation and registration; the complications of the non-maximal cases are easily handled.) (See §3.5.1 for a discussion of whether this interpretation of quantum mechanics is viable in the context of quantum gravity).

theory is to require the theory to be formulated in such a way that physical results are invariant under all possible permutations of the basic entities of the same kind (same quiddity).[20] I have named this requirement the *principle of maximal permutability*. (See Stachel and Iftime (2005) for a more mathematically detailed discussion.)

The exact content of the principle depends on the nature of the fundamental entities. For theories, such as non-relativistic quantum mechanics, that are based on a finite number of discrete fundamental entities, the permutations will also be finite in number, and maximal permutability becomes invariance under the full symmetric group. For theories, such as general relativity, that are based on fundamental entities that are continuously, and even differentiably related to each other, so that they form a differentiable manifold, permutations become diffeomorphisms. For a diffeomorphism of a manifold is nothing but a continuous and differentiable permutation of the points of that manifold.[21] So, maximal permutability becomes invariance under the full diffeomorphism group. Further extensions to an infinite number of discrete entities or mixed cases of discrete-continuous entities, if needed, are obviously possible.

In both the case of non-relativistic quantum mechanics and of general relativity, it is only through dynamical considerations that individuation is effected. In the first case, it is through specification of a possible quantum-mechanical process that the otherwise indistinguishable particles are individuated ('The electron that was emitted by this source at 11:00 a.m. and produced a click of that Geiger counter at 11:01 a.m.'). In the second case, it is through specification of a particular solution to the gravitational field equations that the points of the spacetime manifold are individuated ('The point at which the four non-vanishing invariants of the Riemann tensor had the following values: ... '). So one would expect the principle of maximal permutability of the fundamental entities of any theory of quantum gravity to be part of a theory in which these entities are only individuated dynamically.

Thomas Thiemann has pointed out that, in the passage from classical to quantum gravity, there is good reason to expect diffeomorphism invariance to be replaced by some discrete combinatorial principle:

> The concept of a smooth spacetime should not have any meaning in a quantum theory of the gravitational field where probing distances beyond the Planck length must result in black hole creation which then evaporate in Planck time, that is, spacetime should be fundamentally discrete. But clearly smooth diffeomorphisms have no room in such a discrete spacetime. The fundamental symmetry is probably something else, maybe a combinatorial one, that looks like a diffeomorphism group at large scales. (Thiemann 2001: 117)

[20] In this principle, the word 'possible' is to be understood in the following sense: if the theory is formulated in such a way that some dynamically independent permutations of the fundamental entities are possible, then the theory must be invariant under all permutations. Or the theory must be formulated in such a way that no such permutations are possible (see n. 18).

[21] Here, diffeomorphisms are to be understood in the active sense, as point transformations acting on the points of the manifold, as opposed to the passive sense, in which they act upon the coordinates of the points, leading to coordinate re-descriptions of the same point. See Stachel and Iftime (2005) for a more detailed discussion, based on the use of fibred manifolds and local diffeomorphisms.

In the next section, I shall look at the effective field theory approach to general relativity and asymptotic quantization; and then, in the following section, at string theory, both in the light of the principle of maximal permutability. §3.5 discusses some issues common to all general-relativity-based approaches to quantum gravity. I had hoped to treat loop quantum gravity in detail in this chapter, but the discussion outgrew my allotted spatial bounds; so just a few points about the canonical approach are discussed in §3.6, and the fuller discussion relegated to a separate paper (Stachel 2006a). §3.7 is devoted to causal set theory, and §3.8 sketches a possible new approach, suggested by causal set theory, to the question of what spacetime structures to quantize.

3.3 EFFECTIVE FIELD THEORY AND ASYMPTOTIC QUANTIZATION

The earliest attempts to quantize the field equations of general relativity were based on treating it using the methods of special-relativistic quantum field theory, perturbatively expanding the gravitational field around the fixed background Minkowski space metric and quantizing only the perturbations. By the 1970s, the first wave of such attempts petered out with the realization that the resulting quantum theory is perturbatively non-renormalizable. With the advent of the effective field theory approach to non-renormalizable quantum field theories, a second, smaller wave arose[22] with the more modest aim of developing an effective field theory of quantum gravity valid for sufficiently low energies (for reviews, see Donoghue 1995 and Burgess 2004). As is the case for all effective field theories, this approach is not meant to prejudge the nature of the ultimate resolution of 'the more fundamental issues of quantum gravity' (Burgess 2004: 6), but to establish low-energy results that will be reliable whatever the nature of the ultimate theory.[23]

The standard accounts of the effective field approach to general relativity take the metric tensor as the basic field, which somewhat obscures the analogy with Yang–Mills fields:

> Despite the similarity to the construction of the field strength tensor of Yang Mills field theory, there is the important difference that the [Riemannian] curvatures involve two derivatives of the basic field, $R \sim \partial\partial g$. (Donoghue 1995: 4)

But much of the recent progress in bringing general relativity closer to other gauge field theories, and in developing background-independent quantization techniques,

[22] Presumably (to continue the metaphor), the fashion for string theory pre-empted the waters, in which the second wave would otherwise have flourished. Of course, the effective field theory approach has been applied effectively to perturbatively renormalizable theories (see Dosch et al. 2004).

[23] 'Effective field theory is to gravitation as chiral perturbation theory is to quantum chromodynamics: appropriate at large distances, and impotent at short,' James Bjorken, 'Preface' to Rovelli 2004: p. xiii.

has come from giving equal importance (or even primacy) to the affine connection as compared to the metric (see §§3.6 and 3.8, and Stachel 2006a). Since the curvature tensor involves only one derivative of the connection, $R \sim \partial\Gamma$, this approach brings the formalism of general relativity much closer to the gauge approach used to treat all other interactions. From this point of view, one role of the metric tensor is to act as potentials for the connection $\Gamma \sim \partial g$. From this viewpoint, one can reformulate the starting point of general relativity as follows.

The equivalence principle demands that inertia and gravitation be treated as intrinsically united, the resulting inertio-gravitational field being represented mathematically by a non-flat affine connection Γ.[24]

If one assumes that this connection is metric, i.e. that the connection can be derived from a second-rank covariant metric field g, then according to general relativity such a non-flat metric field represents the chrono-geometry of spacetime.

But the effective field approach assumes that the true chrono-geometry of spacetime remains the Minkowski spacetime of special relativity, represented by the fixed background metric η.[25] There is a unique, flat affine connection $\{\}$ compatible with the Minkowski metric η,[26] and since the difference between any two connections is a tensor, $\Gamma - \{\}$—the difference between the non-flat and flat connections—is a tensor that serves to represents a purely gravitational field.

Contrary to the purport of the equivalence principle, inertia and gravitation have been separated with the help of the background metric and a kinematics based on the background fields has been introduced that is independent of the dynamics of this gravitational tensor. However, the background metric is unobservable: the effect of this gravitational field on all (ideal) rods and clocks is to distort their measurements in such a way that they map out the non-flat chrono-geometry associated with the g-field, which if we did not know better, we would be tempted to think of as the true chrono-geometry.[27] The points of the background metric (flat or non-flat—see n. 25) are then assumed to be individuated up to the symmetry group of this metric, which at most can be a finite-parameter Lie group (e.g. the ten parameter

[24] See Stachel (2006b) for a detailed discussion of this approach. Note that in this brief account I represent each geometric object by a single symbol, omitting all indices.

[25] Of course, one could start with any fixed background metric—flat or non-flat—and perform such a split between the inertial connection associated with this metric and the gravitational field tensor, the latter again being the difference between the inertio-gravitational connection associated with the g-field and the inertial connection associated with the background metric. But regardless of spacetime metric chosen, the kinematics of the theory is based on this fixed background metric; and this limits the spacetime diffeomorphisms, under which the theory is presumed invariant, to the symmetries (if any) of this metric.

[26] I use this symbol as a reminder that the Christoffel symbols define this connection.

[27] The reader may be reminded of the Lorentzian approach to special relativity: there really is a privileged inertial frame ('the ether frame'). But motion through the ether contracts all (ideal) measuring rods, slows down all (ideal) clocks, and increases all masses in such a way that this motion is undetectable. Clocks in a moving inertial frame synchronized by light signals, but forgetting about the effects of this motion on the propagation of light, will read the 'local time' and light will appear to propagate with the same speed c in all inertial frames. I often wonder why those who adopt the special-relativistic approach to general relativity don't abandon special relativity too in favour of the ether frame?

Poincaré group for the Minkowski background metric) acting on the points of spacetime.[28]

Since the full diffeomorphism group acting on the base manifold is not a symmetry group of the background metric,[29] this version of quantum gravity does not meet our criterion of maximal permutability. If we choose a background spacetime with no symmetry group, each and every point of the background spacetime manifold will be individuated by the non-vanishing invariants of the Riemann tensor. But if there is a symmetry group generated by one or more Killing vectors, then points on the orbits of the symmetry group will not be so individuated, but must be individuated by some additional non-dynamical method.[30]

Other diffeomorphisms can only be interpreted *passively*, as coordinate redescriptions of the background spacetime and inertial fields. They can be given an *active* interpretation only as *gauge transformations* on the gravitational potentials $h = g - \eta$.[31]

Since the effective field approach does not claim to be any more than a low-energy approximation to any ultimate theory of quantum gravity, rather than an obstacle to any theory making such a claim, this approach presents a challenge. Can such a theory demonstrate that, in an appropriate low-energy limit, its predictions match the predictions of the effective field theory for experimental results?[32] Since these experimental predictions will essentially concern low-energy scattering experiments involving gravitons, it will be a long time indeed before any of these predictions can be compared with actual experimental results; and the effective field theory approach has little to offer in the way of predictions for the kind of experimental results

[28] We merely mention the additional global complications: the topology chosen for the background spacetime (e.g. $\mathbb{R}^4$ for Minkowski spacetime) may not be the same as the topology associated with particular g-fields that solve the field equations (e.g. the Kruskal manifold for the Schwarzschild metric). As Trautman showed long ago, if one solves for the g-field of a point mass at the origin by successive approximations, at each order the solution field is regular except at the origin. But when the infinite series is summed, the Schwarschild solution is obtained with its quite different topology.

[29] This symmetry group includes only those diffeomorphisms generated by the Killing vectors of the background metric.

[30] In the extreme case of Minkowski spacetime, for which all the invariants of the Riemann tensor vanish, the individuation of all the points must be non-dynamical. However, the situation is not as bad as this comment might suggest. Singling out of one point as origin, of three orthogonal directions, of a unit of spatial distance and a unit of time serve to complete the individuation of all remaining points.

[31] That is, such a transformation can be looked upon either passively, as a transformation to a non-inertial frame of reference, in which inertial forces appear; or actively as a transformation of the gravitational field, producing additional gravitational forces. Of course, it can be divided into two parts, each of which is given a different one of the two interpretations. When the gravitational field is quantized and the background field is not, the need for a choice makes it evident that this interpretation of general relativity differs from the standard, diffeomorphism-invariant interpretation in more than matters of taste.

[32] Since the effective field theory approach is based on the assumption of a spacetime continuum, the challenge will take a radically different form depending on whether the more fundamental theory is itself based on the continuum concept, as is string theory; or whether it denies basic significance to the spacetime continuum, as do loop quantum gravity and causal set theory.

that work on phenomenological quantum gravity is actually likely to give us in the near future.

In a sense, one quantum gravity programme has already met this challenge: Ashtekar's (1987) asymptotic quantization, in which only the gravitational in- and out-fields at null infinity—i.e. at $\Im^+$ (scri-plus) and $\Im^-$ (scri-minus)—are quantized. Without the introduction of any background metric field, it is shown how non-linear gravitons may be rigorously defined in terms of these fields as irreducible representations of the symmetry group at null infinity. This group, however, is not the Poincaré group at null infinity, but the much larger Bondi–Metzner–Sachs group, which includes the super-translations depending on functions of two variables rather than the four parameters of the translation group. This group defines a unique kinematics at null infinity that is independent of the dynamical degrees of freedom, and it is this decoupling of kinematics and dynamics that enables the application of more-or-less standard quantization techniques. Just as the quotient of the Poincaré group by its translation subgroup defines the Lorentz group, so does the quotient of the BMS group by its super-translation subgroup. Since, in both the effective field and asymptotic quantization techniques, experiments in which the graviton concept could be usefully invoked involve the preparation of in-states and the registration of out-states, there must be a close relation between the two approaches; although, as far as I know, this relation has not yet been elucidated in detail.

In summary, both the effective field theory and asymptotic quantization approaches avoid the difficulties outlined in the previous section by separating out a kinematics that is independent of dynamics. In the former case, this separation is *imposed by fiat* everywhere on the spacetime manifold by singling out a background spacetime metric and corresponding inertial field, with the expectation that the results achieved will always be valid to good approximation in the low-energy limit of general relativity. In the latter case, the separation is achieved only for the class of solutions that are asymptotically flat at null infinity (or more explicitly, the Riemann tensors of which vanish sufficiently rapidly in all null directions to allow the definition of null infinity). It is then proved that at null infinity a kinematics can be decoupled from the dynamics at null infinity due to the symmetries of any gravitational field there, and that this can be done without violating diffeomorphism invariance in the interior region of spacetime. Again, this approach presents a challenge to any background-independent quantization programme: derive the results of the asymptotic quantization programme from the full quantum gravity theory in the appropriate limit.

3.4 STRING THEORY

String (or superstring) theory applies the methods of special-relativistic quantum theory to two-dimensional timelike world sheets, called 'strings'.[33] All known (and

[33] 'String theory is an ordinary QFT but not in the usual sense. It is an ordinary scalar QFT on a 2d Minkowski space, however, the scalar fields themselves are coordinates of the ambient target

some unknown) particles and their interactions, including the graviton and the gravitational interaction, are supposed to emerge as certain modes of excitation of and interactions between quantized strings. The fundamental entities of the original (perturbative) string theory are the strings—two-dimensional timelike world sheets—embedded in a given background spacetime, the metric of which is needed to formulate the action principle for the strings. For that reason, the theory is said to be 'background dependent'. Quantization of the theory requires the background spacetime to be of ten or more dimensions.[34]

The theory is seen immediately to fail the test of maximal permutability since the strings are assumed to move around and vibrate in this background, non-dynamical spacetime. So the background spacetime, one of the fundamental constituents of the theory, is invariant only under a finite-parameter Lie subgroup (the symmetry group of this spacetime, usually assumed to have a flat metric with Lorentzian signature) of the group of all possible diffeomorphisms of its elements. Many string theorists, with a background predominantly in special-relativistic quantum field theory (attitudes are also seen to be background dependent), initially found it difficult to accept such criticisms; so it is encouraging that this point now seems to be widely acknowledged in the string community.[35] String theorist Brian Greene recently presented an appealing vision of what a string theory without a background spacetime might look like, but emphasized how far string theorists still are from realizing this vision:

> Since we speak of the 'fabric' of spacetime, maybe spacetime is stitched out of strings much as a shirt is stitched out of thread. That is, much as joining numerous threads together in an appropriate pattern produces a shirt's fabric, maybe joining numerous strings together in an appropriate pattern produces what we commonly call spacetime's fabric. Matter, like you and me, would then amount to additional agglomerations of vibrating strings—like sonorous music played over a muted din, or an elaborate pattern embroidered on a plain piece of material—moving within the context stitched together by the strings of spacetime. ... [A]s yet no one has turned these words into a precise mathematical statement. As far as I can tell, the obstacles to doing so are far from trifling. ... We [currently] picture strings as vibrating in space and through time, but without the spacetime fabric that the strings are themselves imagined to yield through their orderly union, there is no space or time. In this proposal, the concepts of space and time fail to have meaning until innumerable strings weave together to produce them.
>
> Thus, to make sense of this proposal, we would need a framework for describing strings that does not assume from the get-go that they are vibrating in a preexisting spacetime. We would need a fully spaceless and timeless formulation of string theory, in which spacetime emerges from the collective behavior of strings. Although there has been progress toward

Minkowski space which in this case is 10 dimensional. Thus, it is similar to a first quantized theory of point particles' (Thiemann 2002: 12).

[34] However, Thiemann (2004) asserts that the so-called Pohlmayer string can be quantized in any number of dimensions, including four-dimensionsal Minkowski space.

[35] See Stachel (2003: 31–2) for quotations to this effect from review articles by Michael Green and Thomas Banks.

this goal, no one has yet come up with such a spaceless and timeless formulation of string theory—something that physicists call a background-independent formulation [of the theory] ... Instead, all current approaches envision strings as moving and vibrating through a spacetime that is inserted into the theory 'by hand' ... Many researchers consider the development of a background-independent formulation to be the single greatest unsolved problem facing string theory. (Greene 2004: 487–8)

One of the main goals of the currently sought-for M-theory (see Greene 2004: 376–412) is to overcome this defect, but so far this goal has not been reached.

3.5 QUANTUM GENERAL RELATIVITY: SOME PRELIMINARY PROBLEMS

String theory attempts to produce a theory of everything, including a quantum theory of gravity that will have general relativity (or a reasonable facsimile thereof) as part of its classical limit. Most other approaches to quantum gravity start from classical general relativity. In this section, I shall discuss two related issues that arise in the course of any such attempt.

3.5.1 States or Processes: Which is Primary?

There has been a long-standing debate between adherents of covariant and canonical approaches to quantum gravity. The former attempt to develop a four-dimensionally-invariant theory of quantum gravity from the outset; the latter start from a $(3 + 1)$-breakup of spacetime, emphasizing three-dimensional spatial invariance and developing quantum kinematics before quantum dynamics. Christian Wüthrich has related this debate to

the philosophical debate between proponents of the endurance view of time and those of the perdurance view [which] reflects a disagreement concerning whether, and to what degree, time is on a par with spatial dimensions. (Wüthrich 2003: 1)

According to the former view, 'an object is said to endure just in case it exists at more than one time'. According to the latter view,

objects perdure by having different temporal parts at different times with no part being present at more than one time. Perdurance implies that two [space-like] hypersurfaces ... do not share enduring objects but rather harbour different parts of the same four-dimensional object. (Wüthrich 2003: 1)

I shall use slightly different terminology to make this important distinction. One approach to the quantum gravity problem places primary emphasis on the three-dimensional state of some thing; from this point of view, a process is just a succession of different states of this thing. (The relation of this succession of states to some concept of 'time' is a contentious issue). The other approach places primary emphasis on four-dimensional processes; from this point of view, a 'state' is just a particular spatial cross-section of a process and of secondary importance: all such cross-sections are equal, and each sequence of states represents a different 'perspective' on the same process.

In pre-relativistic physics, the absolute time provided a natural foliation of spacetime into spatial cross-sections. So, even if one favoured the 'process' viewpoint for philosophical reasons, there was little harm to physics—if not to philosophy—in using the alternate 'state' viewpoint. While the split into spaces was not unique (one inertial frame is as good as another), each inertial frame corresponding to a different preferred fibration of spacetime, they all shared a unique time (absolute simultaneity). In short, there was a unique breakup of four-dimensions into $(3 + 1)$. In special-relativistic physics, this is no longer the case: there are an infinite number of such preferred cross-sections (one for each family of parallel spacelike hyperplanes in Minkowski space). Not only is the split into spaces not unique (one inertial frame is still as good as another), but now they do not even agree on a unique time slicing (the relativity of simultaneity): there is a different foliation for each preferred fibration. In short, there is a three-parameter family of 'natural' breakups of four-dimensions into $(3 + 1)$. So, in special-relativistic physics, and quite apart from philosophical considerations, the 'process' approach has much to recommend it over the 'state' approach.

General relativity is an inherently four-dimensional theory of spacetime—even more so than special relativity. There is no 'natural' breakup of spacetime into spaces and times, such as the inertial frames provide in special relativity. There are no preferred timelike fibrations or spacelike foliations.[36] Any breakup of this four-dimensional structure into a $(3 + 1)$ form requires the (explicit or implicit) introduction of an arbitrary 'frame of reference',[37] represented geometrically by the introduction of a fibration and foliation of spacetime. Then one may speak about the 'state' of a thing on a given hypersurface and its evolution from hypersurface to hypersurface of the foliation (over some 'global time'). But such a breakup is always relative to some chosen frame of reference. There are no longer any preferred breakups in generally relativity; there is always something arbitrary and artificial about the introduction of such a frame of reference. The process approach seems rooted in general-relativistic physics, just as it is in quantum theory (see n. 19). No one has put the case more strongly than Lee Smolin:

> [R]elativity theory and quantum theory each ... tell us—no, better, they scream at us—that our world is a history of processes. Motion and change are primary. Nothing is, except in a very approximate and temporary sense. How something is, or what its *state* is, is an illusion. It may be a useful illusion for some purposes, but if we want to think fundamentally we must not lose sight of the essential fact that it 'is' an illusion. So to speak the language of the new physics we must learn a vocabulary in which *process* is more important than, and prior to, stasis. (Smolin 2002a: 53; my emphases)

[36] The only 'natural' foliation would be a family of null hypersurfaces, and null hypersurface quantization has had many advocates, starting with Dirac. For a survey, see Robinson (2003).

[37] Einstein emphasized the amorphous nature of such a frame by calling it a 'reference-mollusc' (see Einstein 1917; cited from Einstein 1961: 99).

Now the canonical formalism is based on the introduction of a fibration and foliation of spacetime,[38] and by its nature tends to shift attention from processes in spacetime to states of things in space.[39] Bryce DeWitt, in his final book, has put the case in the context of quantum field theory:

When expounding the fundamentals of quantum field theory physicists almost universally fail to apply the lessons that relativity theory taught them early in the twentieth century. Although they carry out their calculations in a covariant way, in deriving their calculational rules they seem unable to wean themselves from canonical methods and Hamiltonians, which are holdovers from the nineteenth century, and are tied to the cumbersome $(3 + 1)$-dimensional baggage of conjugate momenta, bigger-than-physical Hilbert spaces and constraints. (DeWitt 2003: i, p. v)

This has immediate implications for a theory of quantum gravity. Whether one should be looking for quanta of space or of spacetime seems to be one essential point of difference between the canonical loop quantum gravity approach and the covariant causal set approach. In his exposition of the canonical approach, Carlo Rovelli asserts: 'Spacetime is a temporal sequence of spaces, or a history of space'. He asks:

What are the quanta of the gravitational field? Or, since the gravitational field is the same entity as spacetime, what are the quanta of space? (Rovelli 2004: 18)[40]

The unremarked shift from 'quanta ... of spacetime' to 'quanta of space' is striking, but almost seems forced on Rovelli by the canonical, 'history of space' approach. On the other hand, discussing causal set theory, a 'process' approach to quantum gravity,[41] Fay Dowker states:

Most physicists believe that in any final theory of quantum gravity, space-time itself will be quantized and grainy in nature. ... So the smallest possible volume in four-dimensional

[38] All treatments of the canonical formalism mention the need for a foliation, but most do not mention the fibration, which is needed in order to understand the geometrical significance of the lapse and shift functions (see Stachel 1962, 1969).

[39] This is not meant to imply that the full four-dimensional invariance cannot be recovered in some variant of the canonical approach. For a brief discussion of 'proposals to make the canonical formulation more covariant', with references, see Thiemann (2002: 15–16). For a summary of a different approach, see Salisbury (2003).

[40] As will be seen in Stachel (2006a), this is not the only instance of Rovelli's wavering between 'space' and 'spacetime'. There I also discussed the interpretation of the discrete spectra, in loop quantum gravity, of the operators for three-volume and two-area on the initial hypersurface. Attributing direct physical significance to such mathematical results obtained *before* solution of the Hamiltonian constraints seems to violate the general-relativistic golden rule: 'no kinematics before dynamics.' I thank Fotini Markopoulou for helpful discussion of this point.

[41] The causal set approach, to be discussed in §§3.6 and 3.7, does not attempt a quantization of the classical theory. Rather, its aim is to construct a quantum theory of causal sets based on two features of classical general relativity that it takes as fundamental:

(1) The causal structure, which is replaced by a discrete causal set; and
(2) The four-volume element, which is replaced by the quantum of process.

It must then be shown that the classical equations can be recovered from some sort of limit of causal sets, or of an ensemble of such sets.

spacetime, the Planck volume, is 10^{-42} cubic centimetre seconds. If we assume that each of these volumes counts a single space-time quantum, this provides a direct quantification of the bulk. (Dowker 2003: 38)

3.5.2 Formalism and Measurability

There has always been a dialectical interplay between formalism and measurability in the development of quantum theory, first seen in the discussion about the physical interpretation of the commutation relations in non-relativistic quantum mechanics, see e.g. Heisenberg (1930) and later in the similar discussion in quantum electrodynamics.[42] This interplay was well expressed by Bohr and Rosenfeld in their classic discussion of the measurability of the components of the electric and magnetic field:

[O]ur task will thus consist in investigating whether the complementary limitations on the measurement of field quantities, defined in this way, are in accord with the physical possibilities of measurement. (Bohr and Rosenfeld 1933, cited from Rosenfeld 1979: 358)

By 'in this way' they mean: 'the field quantities are no longer represented by actual point-functions but by functions of space-time regions, which formally correspond to the average value of the idealized field components over the region in question' (ibid.), and that delta-function commutation relations at points must be replaced by commutation relations smeared over such (finite) regions of spacetime.

By 1933, quantum electrodynamics had been developed to a point that enabled Bohr and Rosenfeld 'to demonstrate a complete accord' between the formal commutation relations of the field components and the physical possibilities of their measurement. In the case of quantum gravity, the theory is still in a state of active exploration and development, and one may hope that investigation of the possibilities of ideal measurements of the variables basic to each approach can help to clarify still unresolved issues in the physical interpretation of the formalism, and perhaps even help in choosing between various formalisms.[43] It has been objected that a quantum theory of gravity requires a different interpretation of quantum theory, in which no external 'observer' is introduced. First of all, a quantum process does not require an 'observer' in any anthropomorphic sense. Once the results of a preparation and registration are recorded (see n. 19), the quantum process is over, whether anyone ever looks at ('observes') the results or not. Carlo Rovelli has argued well against combining the problem of quantum gravity with the problem of the interpretation of quantum mechanics:

I see no reason why a quantum theory of gravity should not be sought within a standard interpretation of quantum mechanics (whatever one prefers). Several arguments have been

[42] See e.g. Bohr and Rosenfeld (1933). Bergmann and Smith provide a careful discussion of the analogies and differences that arise when a Bohr-Rosenfeld-style analysis of measurability is attempted for the linearized Riemann tensor (see Bergmann and Smith 1982, which summarizes and extends their earlier papers).

[43] See Stachel (2001) for further comments on this question.

proposed to connect these two problems [quantum gravity and the interpretation of quantum mechanics]. A common one is that in the Copenhagen interpretation the observer must be *external*, but it is not possible to be external from the gravitational field. I think that this argument is wrong; if it was correct it would apply to the Maxwell field as well. We can consistently use the Copenhagen interpretation to describe the interaction between a macroscopic classical apparatus and a quantum-gravitational phenomenon happening, say, in a small region of (macroscopic) spacetime. The fact that the notion of spacetime breaks down at short scale within this region does not prevent us from having the region interacting with an external Copenhagen observer. (Rovelli 2004: 370).[44]

Rovelli has discussed the physical interpretation of the canonical formalism for a field theory describing some generic field, symbolized by Φ:

The data from a local experiment (measurements, preparation, or just assumptions) must in fact refer to the state of the system on the entire boundary of a finite spacetime region. The field theoretical space ... is therefore the space of surfaces Σ ['where Σ is a 3d surface bounding a finite spacetime region'] and field configurations Ψ on Σ. Quantum dynamics can be expressed in terms of an [probability] amplitude $W[\Sigma, \Psi]$. ... Notice that the dependence of $W[\Sigma, \Psi]$ on the geometry of Σ codes the spacetime position of the measuring apparatus. In fact, the relative position of the components of the apparatus is determined by their physical distance and the physical time elapsed between measurements, and these data are contained in the metric of Σ. (Rovelli 2004: 23).

From the 'process' viewpoint (see §3.5.1), this is an encouraging approach: what occurs in the spacetime region bounded by Σ constitutes a process, and an amplitude is only defined for such processes. However, as Rovelli emphasizes, this definition is broad enough to include a background-dependent theory, i.e. a theory with a fixed background spacetime. But his real subject of course is theories, such as general relativity, which are background independent:

Consider now a background independent theory. Diffeomorphism invariance implies immediately that $W[\Sigma, \Psi]$ is independent of Σ. ... Therefore, in gravity W depends only on the boundary value of the fields. However, the fields include the gravitational field, and the gravitational field determines the spacetime geometry. Therefore the dependence of W on the fields is still sufficient to code the relative distance and time separation of the components of the measuring apparatus! (Rovelli 2004: 23)

He summarizes the contrast between the two cases:

[44] I disagree on one major point. There is at least one big difference between the Maxwell field and the gravitational field: the non-universality of the electromagnetic charge-current vector versus the universality of gravitational stress-energy tensor. Because charges occur with two signs that can neutralize each other, a charge-current distribution acting as a source of an electromagnetic field can be manipulated by matter that is electrically neutral and so not acting as a source of a further electromagnetic field; and one can shield against the effects of a charge-current distribution. Because mass comes with only one sign, all matter (including non-gravitational fields) has a stress-energy tensor, no shielding is possible, and any manipulation of matter acting as a source of gravitational field will introduce an additional stress-energy tensor as a source of gravitational field. A glance at Bohr and Rosenfeld (1933) shows how important the possibility of neutralizing the charges on test bodies is for measurement of the (averaged) components of the electric field with arbitrary accuracy, for example. This difference may well have important implications for the measurement of gravitational field quantities. For detailed discussion, see Bergmann and Smith (1982).

What is happening is that in background dependent QFT we have two kinds of measurements: those that determine the distances of the parts of the apparatus and the time elapsed between measurements, and the actual measurements of the fields' dynamical variables. In quantum gravity, instead, distances and time separations are on an equal footing with the dynamical fields. This is the core of the general relativistic revolution, and the key for background independent QFT. (Rovelli 2004: 23)

This is a brilliant exposition of the nature of the difference between background-dependent and background-independent quantum field theories (QFTs). But it immediately raises a number of questions, most of which I discuss in Stachel (2006a). Here I shall mention only one of them.

Rovelli's interpretation of the canonical formalism seems based on the assumption that, in both background-dependent and background-independent QFTs, one can idealize a field measurement by confining its effects to the (three-dimensional) boundary of a (four-dimensional) spacetime region; and in particular to a finite region of (three-dimensional) space, while neglecting its finite (one-dimensional) duration in time. Yet, since the pioneering work of Bohr and Rosenfeld on the measurability of the components of the electromagnetic field (Bohr and Rosenfeld 1933), it has been known that this is not the case for the electromagnetic field in Minkowski spacetime.

It is ... of essential importance that the customary description of an electric field in terms of its components at each space-time point, which characterizes classical field theory and according to which the field should be measurable by means of point charges in the sense of electron theory, is an idealization which has only restricted applicability in quantum theory. This circumstance finds its proper expression in the quantum-electromagnetic formalism, in which the field quantities are no longer represented by true point functions but by functions of space-time regions, which formally correspond to the average values of the idealized field components over the region in question. The formalism only allows the derivation of unambiguous predictions about the measurability of such region-functions ... [U]nambiguous meaning can be attached only to space-time integrals of the field components (Bohr and Rosenfeld 1933: 358–61).

The aim of the paper was to show that these theoretical limits on measurability coincide with the actual possibilities of (idealized) measurements of these field quantities. Bohr and Rosenfeld demonstrate in some detail how test bodies occupying a finite region of space over a finite period of time—that is, test processes over finite regions of spacetime—can be used to measure electromagnetic field averages over these regions.[45]

[45] One may see in this work the germ of the algebraic approach to quantum field theory, which also is based on attaching primary significance to operators defined over finite regions of spacetime.

The primary physical interpretation of the [quantum field] theory is given in terms of local operations, not in terms of particles. Specifically, we have used the basic fields to associate to each open region O in space-time an algebra $A(O)$. Of operators in Hilbert space, the algebra generated by all $F(f)$, the fields 'smeared out' with test functions f having their support in the region O. We have interpreted the elements of $A(O)$ as representing physical operations performable within O and we have seen that this interpretation tells us how to compute collision cross sections once the correspondence $O \rightarrow A(O)$ is known. This suggests that the net of algebras A, i.e, the correspondence [given above] constitutes the intrinsic mathematical description of the theory. The

Schweber summarizes the point nicely:

In fact, Bohr and Rosenfeld in their classic papers on the question of the measurability of electromagnetic fields have already shown that only averages over small volumes of space-time of the field operators are measurable ... This is because of the finite size of the classical measuring apparatus and the finite times necessary to determine forces through their effects on macroscopic test bodies. (Schweber 1961: 421–2)

Dosch et al. (2004) generalize the point to all quantum field theories:

The quantum fields, in terms of which the theory is constructed, are operators that depend on space-time. This dependence on space-time however, shows the behavior of a generalized function or distribution. Therefore, a well-defined operator cannot be related to a definite space-time point x, but only to a space-time domain of finite extent (Dosch et al. 2004: 4).

Sorkin (1993) contrasts non-relativistic quantum mechanics and quantum field theory:

Now in non-relativistic quantum mechanics, measurements are idealized as occurring at a single moment of time. Correspondingly the interpretive rules for quantum field theory are often stated in terms of ideal measurements which take place on Cauchy hypersurfaces. However, in the interests of dealing with well-defined operators, one usually thickens the hypersurface, and in fact the most general formulations of quantum field theory assume that there corresponds to any open region of space-time an algebra of observables which—presumably—can be measured by procedures occurring entirely within that region.[46]

The question of whether measurements should be associated with (three-dimensional) things or (four-dimensional) processes recurs in loop quantum gravity, and will be discussed in Stachel (2006a).

3.6 CANONICAL QUANTIZATION (LOOP QUANTUM GRAVITY)

From the point of view of the principle of maximal permutability, the basic problem of the canonical formalism is that, by introducing a fibration and a foliation, it violates the principle.

Indeed, when the Cauchy problem is formulated in terms of Lie derivatives with respect to the vector field defined by the foliation and fibration (see Stachel 1962, 1969), it is already trivially diffeomorphism invariant in the sense that, if the fibration and foliation are dragged along together with all the initial data fields imposed on the spacetime, nothing is changed.

mentioned physical interpretation establishes the tie between spacetime and events. The role of 'fields' is only to provide a coordinatization of this net of algebras. (Haag 1996: 105).

This approach of course extensively utilizes the existence and properties of the background Minkowski spacetime. In the section on causal sets, I shall cite Haag's speculations about how it might be generalized in the absence of such a background spacetime.

[46] See the previous footnote.

But the fibration and foliation already assign a certain amount of individuality to the points of spacetime *before* any initial data fields are imposed on it. Points on the same hypersurface of the fibration and/or the same curve of the foliation are distinguished by these characteristics.

What remains of permutation invariance is the three-dimensional diffeomorphism invariance of the physically significant parts of the Cauchy data on the initial hypersurface; and the freedom to change from evolution of the initial data along one fibration and foliation to evolution along another, which is done by the choice of the lapse and shift functions.

There certainly is a price to pay for this loss of manifest permutation invariance of the theory under four-dimensional diffeomorphisms. Since the canonical formalism manifestly preserves only three-dimensional diffeomorphism invariance within each hypersurface of the foliation, it is perhaps not too surprising that many of its problems have to do with time and dynamics.

Speaking about loop quantum gravity, Perez states,

> The dynamics is governed by the quantum Hamiltonian constraint. Even when this operator is rigorously defined it is technically difficult to characterize its solution space. This is partly because the $(3 + 1)$ decomposition of spacetime (necessary in the canonical formulation) breaks the manifest 4-diffeomorphism invariance of the theory making awkward the analysis of the dynamics. (Perez 2003: p. R 45)

Perez continues:

> The difficulty in dealing with the scalar constraint [i.e. the Hamiltonian] is not surprising. The vector constraint, generating space diffeomorphisms, and the scalar constraint, generating time reparametrizations, arise from the underlying 4-diffeomorphism invariance of gravity. In the canonical formulation the $3 + 1$ splitting breaks the manifest four-dimensional symmetry. The price paid is the complexity of the time reparametrization constraint S (i.e. the Hamiltonian). (Perez 2003: p. R 49)

It is also clear that, by its nature, the canonical formalism favours a 'state of things' over a 'process' approach to quantum gravity. Whether this is a drawback; and if so, how it can be overcome are questions discussed in Stachel (2006a), together with a number of other questions raised by the modern canonical formulation of general relativity ('loop quantum gravity').

Here I shall only comment on a question raised in the previous section: the interplay between formalism and measurement. Ashtekar and Lewandowski (2004) describe the strategy adopted in loop quantum gravity:

> To pass to the quantum theory [of a fully constrained theory in its phase space formulation],[47] one can use one of the two standard approaches: i) find the reduced phase space of the theory representing 'true degrees of freedom' thereby eliminating the constraints classically and then construct a quantum version of the resulting unconstrained theory;[48] or ii) first

[47] For detailed discussions of what is called 'refined algebraic quantization' of a system with first class constraints, see Thiemann (2002: 22–5), and Thiemann (2001: §II.7, pp. 280–4).

[48] In this chapter, I shall not discuss alternative (i). Yet the possibilities of this alternative approach to quantization should be kept in mind. In particular, the $(2 + 2)$ formalism (for a recent

construct quantum kinematics for the full phase space ignoring the constraints, then find quantum operators corresponding to the constraints and finally solve the quantum constraints to obtain the physical states.[49] Loop quantum gravity follows the second avenue ... (Ashtekar and Lewandowski 2004: 51).

Note that, in this approach, the commutation relations are simply postulated. For the physical interpretation and validation of the formalism, the configuration variables and their conjugate momenta should be given some interpretation in terms of procedures for their individual measurement; and—a crucial second step—the limitations on the joint definability of pairs of such variables implied by the postulated commutation relations should coincide with the limitations on their joint measurability by these procedures (for a detailed discussion, see Bergmann and Smith 1982).

The question of the link between measurability and formalism seems especially acute in view of the claim by Ashtekar and Lewandowski that there is essentially only one possible representation of the algebraic formalism, and the suggestion by Thiemann that this mathematical uniqueness may have been achieved at too high a physical price.[50] As explained in detail in Stachel (2006a), in loop quantum gravity the holonomies associated with a certain three-connection are taken as the 'configuration' variables, with a corresponding set of so-called electric-flux 'momentum' variables; taken together, they obey a certain algebra. The main problem is that of finding the physically appropriate representation of this holonomy-flux algebra (Ashtekar and Lewandowski 2004: 41). They succeed in constructing one, and state:

Remarkably enough, uniqueness theorems have been established: there is a precise sense in which *this is the only diffeomorphism invariant, cyclic representation of the kinematical quantum algebra* ... Thus, the quantum geometry framework is surprisingly tight. ... there is a unique, diffeomorphism invariant representation ... (ibid. 41, their emphasis)[51]

One might have expected that, as in ordinary quantum mechanics, there is a complementary representation, based on the 'momentum' variables conjugate to the configuration ones (these are the 'electric fluxes' mentioned above), which essentially constitute the three-metric. But these variables, suitably smeared, do not commute;[52] consequently, [t]his result brings out a

summary with references to earlier literature, see d'Inverno 2003) offers a possible classical starting point for such an approach, in which the two degrees of freedom of the inertio-gravitational field are given a direct geometrical interpretation.

[49] For quantum kinematics, see e.g. Thiemann (2001: §I.3).

[50] The main point of the following discussion should be clear even to readers unfamiliar with the canonical formalism: it is claimed that, given the techniques that make the new canonical quantization possible, there is just one possible operator representation of the algebra of the basic canonical variables. If correct, then it is crucial to make sure that the possibilities for actual measurement of these variables reaches, but does not exceed, the limits set by this algebra.

[51] Put more technically: 'As we saw, quantum connection-dynamics is very "tight"; once we choose the holonomies $A(e)$ and the "electric fluxes" $P(S,f)$ as the basic variables, there is essentially no freedom in the background-independent quantization' (ibid.).

[52] As a consequence, and even more significant for physical measurements, 'area operators $\hat{A}_S$ and $\hat{A}_{S'}$ fail to commute if the surfaces S and S' intersect. ... Thus, the assertion that "the spin network basis diagonalizes all geometrical operators" ... that one sometimes finds in the literature is incorrect' (ibid. 46).

fundamental tension between connection-dynamics and geometrodynamics. ... [T]he metric representation does not exist. (Ibid. 46)

This uniqueness claim is striking indeed. In Lorentz-invariant quantum field theories, there are unitarily inequivalent representations of the basic algebra, which led to the algebraic approach to quantum field theory that de-emphasizes the role of the representations.[53] But in loop quantum gravity,

[T]hese results seem to suggest that, for background independent theories, the full generality of the algebraic approach may be unnecessary. (Ibid: 41)[54]

Some caution is advisable here; this result is based on the requirement of spatial diffeomorphism invariance. But, as Thiemann notes,

[t]here are, however, doubts on physical grounds whether one should insist on spatial diffeomorphism invariant representation because smooth and even analytic structure of [the three-manifold] which is encoded in the spatial diffeomorphism group should not play a fundamental role at short scales if Planck scale physics is fundamentally discrete. In fact, as we shall see later, Q[uantum] G[eneral] R[elativity] predicts a discrete Planck scale structure and therefore the fact that we started with analytic data and ended up with discrete (discontinuous) spectra of operators looks awkward. Therefore ... we should keep in mind that other representations are possibly better suited in the final picture ... (Thiemann 2002: 40; see also the quotation from Thiemann (2001) at the end of §3.2).

Are there other representations once one drops the demand for diffeomorphism invariance? If there are, will they be inequivalent (as in quantum field theory) or equivalent (as in non-relativistic quantum mechanics) to the Ashtekar–Lewandowski representation? In view of the physical importance of the answers to these technical questions, one would like to be certain that the formal success of this possibly unique quantization is accompanied by an equally successful physical interpretation of it. Ashtekar and Lewandowski are far from claiming that this has been accomplished. In discussing quantum dynamics, they note that:

the requirement of diffeomorphism invariance picks out a unique representation of the algebra generated by holonomies and electric fluxes. Therefore we have a single arena for background independent theories of connections and a natural strategy for implementing dynamics provided, of course, this mathematically natural, kinematic algebra is also 'physically correct'. (Ashtekar and Lewandowski 2004: 51)

Once these 'kinematic' steps have been taken, there still remains the final 'dynamical' step:

[53] See Haag (1996) for a full account.

[54] While the full generality of the algebraic approach might not be needed, it might still be worthwhile to investigate the possibility of a four-dimensional algebra that reduces to the holonomy-flux algebra on a spacelike hypersurface. If this four-dimensional algebra could be related to physical measurements, this might provide another approach to the question, raised above, of physically justifying the holonomy-flux algebra instead of simply postulating it. See n. 57 for Haag's comments on the possibility of a net of algebras on a partially ordered set.

We come now to the 'Holy Grail' of Canonical Quantum General Relativity, the definition of the Hamiltonian constraint. [T]he implementation of the correct quantum dynamics is not yet completed and one of the most active research directions at the moment. (Thiemann 2001: 150)

Another question can be raised in this connection. Might it be possible to go beyond simply postulating the three-dimensional commutation relations in loop quantum gravity, which after all do not take us beyond one spatial hypersurface? Various spin-foam models are candidates for a four-dimensional, dynamical version of the canonical approach. So it might be possible to introduce four-dimensional commutation relations in such models (insofar as these models can be freed from their origins in, and close ties to, the canonical formalism—see Stachel 2006a) and show that the three-dimensional canonical commutation relations postulated on a spacelike hypersurface can be derived from them.[55] Since the presence of a causal structure seems necessary for the formulation of four-dimensional commutation relations, causal spin-foam models might be good candidates for such an approach (see Stachel 2006a).

3.7 THE CAUSAL SET (CAUSET) APPROACH

I should first like to draw attention to an early work on this subject, which I have not seen cited in the recent literature on causal sets (Kronheimer and Penrose 1967).

This approach would seem to have already adopted much of the viewpoint suggested here.[56] Spacetime points are replaced by quanta of process, elements of four-volume of order l_P^4, and these are the basic entities in this approach.[57]

They are connected to each other by a causal (partial) ordering relation. The causal ordering relation enables us to define a causal past and a causal future for each element of the causal set, forming the discrete analogues of the forward and backward light cones of a point, which define the classical causal structure of a spacetime. The number of quanta of process in any given four-dimensional process determines its spacetime 'volume'. Together, they provide the causal-set analogue of the four-dimensional metric tensor. After introducing the idea of a 'labeled causet', in which each element of the causet is labelled by the sequence in which it is introduced, Sorkin et al. comment:[58]

[55] '[I]t would be very interesting to reconstruct in detail the hamiltonian Hilbert space, as well as kinematical and dynamical operators of the loop theory, starting from the covariant spinfoam definition of the theory. At present [this problem is not] under complete control' (Rovelli 2004: 362).

[56] For a popular survey, see Dowker (2003). For a more technical survey, see Sorkin (2003).

[57] Rudolf Haag's comments on how the algebraic approach to field theory might be modified in the absence of a background spacetime bear a close resemblance to the causal set approach: 'In a minimal adaptation of the algebraic approach and the locality principle one could keep the idea of a net of algebras which, however, should be labeled now by the elements of a partially ordered set L (instead of regions in $\mathbb{R}^4$). L could be atomic, with minimal elements (atoms) replacing microcells in spacetime' (Haag 1996: 348).

[58] See Brightwell et al. 2002: 8.

After all, labels in this discrete setting are the analogs of coordinates in the continuum, and the first lesson of general relativity is precisely that such arbitrary identifiers must be regarded as physically meaningless: the elements of spacetime—or of the causet—have individuality only to the extent that they acquire it from the pattern of their relations to the other elements. It is therefore natural to introduce a principle of 'discrete general covariance' according to which 'the labels are physically meaningless'. But why have labels at all then? For causets, the reason is that we don't know otherwise how to formulate the idea of sequential growth, or the condition thereon of Bell causality, which plays a crucial role in deriving the dynamics. Ideally perhaps, one would formulate the theory so that labels never entered, but so far, no one knows how to do this—anymore than one knows how to formulate general relativity without introducing extra gauge degrees of freedom that then have to be canceled against the diffeomorphism invariance. Given the dynamics as we can formulate it, discrete general covariance plays a double role. On one hand it serves to limit the possible choices of the transition probabilities in such a way that the labels drop out of certain 'net probabilities'. This is meant to be the analog of requiring the gravitational action-integral S to be invariant under diffeomorphisms (whence, in virtue of the further assumption of locality, it must be the integral of a local scalar concomitant of the metric). On the other hand, general covariance limits the questions one can meaningfully ask about the causet (cf. Einstein's 'hole argument'). It is this second limitation that is related to the 'problem of time', and it is only this aspect of discrete general covariance that I am addressing in the present talk.

I think there is a certain confusion here between the causet analogues of active and passive diffeomorphisms. Sorkin rightly emphasizes that 'the elements of spacetime—or of the causet—have individuality only to the extent that they acquire it from the pattern of their relations to the other elements.' Yet he goes on to speak about invariance under a relabelling of the elements of the causet as 'discrete general covariance'. But such a relabelling corresponds to a coordinate transformation, i.e. a passive diffeomorphism. What is important physically is an active permutation of the elements of the causet. But clearly all the physically relevant information about the causet is contained in the network of order relations between its elements and, as long as this is not changed, a permutation of the elements changes nothing. In my terminology, the elements of the causet have quiddity—crudely put, they are quanta of four-volume—but lack haecceity: nothing distinguishes one from the other except its position in the causet. So, in spite of Sorkin's terminology, there is no problem here. The chief defect of the causal set approach is that so far it is not really a quantum theory; that is, it has not been able to take the step from transition probabilities to transition probability amplitudes, which would allow a Feynman formulation of the theory, leading to 'a "sum over histories" quantum theory of causets' (Dowker 2004), i.e. the addition of the amplitudes for indistinguishable processes that begin with the same initial preparation and end with the same final registration.

Perhaps associated with this problem is the circumstance that the theory is not very closely linked to classical general relativity. It simply postulates certain things—such as the discreteness of processes (i.e. four-volumes): the number of causet elements gives the volume of the corresponding region of the approximating spacetime in

Planck units[59]—that one might hope to derive from a quantum version of general relativity, for example by an appropriate quantization of the conformal factor in the metric (i.e. the determinant of the metric tensor). In the last section, I shall offer some suggestions about how this might be done.

3.8 WHAT STRUCTURES TO QUANTIZE?

There are a number of spacetime structures that play an important role in the general theory of relativity (see Stachel 2003). The chrono-geometry is represented mathematically by a pseudo-metric tensor field on a four-dimensional manifold. The inertio-gravitational field[60] is represented by a symmetric affine connection on this manifold. Then there are compatibility conditions between these two structures (the covariant derivative of the metric with respect to the connection must vanish). As noted above, major technical advances in the canonical quantization programme came when the inertio-gravitational connection was taken as primary rather than the chrono-geometrical metric. It is possible to start with only the metric field and derive from it the unique symmetric connection (the Christoffel symbols) that identically satisfies the compatibility conditions. A second order Lagrangian (the densitized curvature scalar) then leads to the field equations in terms of the metric. This is the route that was first followed historically by Hilbert, and is still followed in most textbooks.

It is possible to treat metric and connection as initially independent structures, and then allow the compatibility conditions between them to emerge, together with the field equations (written initially in terms of the connection), from a first order, Palatini-type variational principle. In this approach, metric and connection are in a sense 'dual' to each other. Either of these methods may be combined with a tetrad formalism for the metric, combined with one or another mathematical representation of the connection, e.g. connection one-forms, or tetrad components of the connection.

But one can go a step further and decompose the metric and affine connection themselves. If one abstracts from the four-volume-defining property of the metric, one gets the conformal structure on the manifold. Physically, this conformal structure is all that is needed to represent the causal structure of spacetime. Similarly, if one abstracts from the preferred parametrization (proper length along spacelike, proper time along timelike) of the geodesics associated with an affine connection, one gets a

[59] See Dowker (2004). See also the more complete discussion in Sorkin (2003), which offers several heuristic arguments.

[60] This is often referred to simply as the gravitational field. But it must be emphasized that the deeper meaning of the equivalence principle is that there is no frame-independent separation between inertia and gravitation. This is as true of Newton's gravitational theory, properly interpreted as a four-dimensional theory, as it is of general relativity.

projective structure on the manifold.[61] Physically the projective structure picks out the class of preferred paths in spacetime.[62] Compatibility conditions between the causal and projective structures can be defined, which also guarantee the existence of a corresponding metric and compatible affine connection.[63]

But the important point here is that, as shown in Weyl (1921), in a manifold with metric, its conformal and affine structures suffice to determine that metric (up to an overall constant factor). We can use this circumstance to our advantage in the first order, so-called Palatini variational principle for general relativity. As mentioned above, in this case metric and affine connection are independently varied, and the metrical nature of the connection follows from the variation of the connection. But one can go further: break up the metric into its four-volume structure-determining part (essentially, the determinant of the metric) and its conformal, causal structure-determining part (with unit determinant); and break up the affine connection into its projective, preferred path-determining part (the trace-free part of the connection) and its preferred parameter-determining part (the trace of the connection).[64] Each of these four parts may then be varied independently. Such a breakup is of particular interest because, as noted in the previous section, the causal set theory approach to quantum gravity is based on taking the conformal structure and the four-volume structure as the primary constituents of the classical theory,[65] and then replacing them with discretized versions (ibid.): the causal set idea combines the twin ideas of discreteness and order to produce a structure on which a theory of quantum gravity can be based.

However, in causal set theory no attention seems to have been paid to the affine connection, and the possibility of finding quantum analogues of the traceless projective and trace parts of the connection. The answer to this question might lead to a link between the causal set approach and some quantized version of the dynamics of general relativity. It is easy to set up a conformal-projective version of the Palatini principle, in which the trace of the affine connection is dual to the four-volume structure, suggesting that the projective connection is dual to the conformal structure.[66]

So a covariant quantization based on this breakup might lead to a representation in which four-volume and conformal structures are the configuration variables—a sort

[61] Alternately, and perhaps better put, if one abstracts from the concept of parallel transport its property of preservation of ratios of parallel vectors, one gets the concept of projective transport, which preserves only the direction of a vector.

[62] Remember the distinction between curves, which are associated with a parametrization, and paths, which are not. Since the holonomies of the affine connection are independent of the parametrization of the curves, it is possible that the holonomies of the projective connection might be more suited to a four-dimensional version of loop quantum gravity.

[63] See Ehlers et al. (1972). For some further results clarifying the interrelations between conformal, (semi-)Riemannian, volume and projective geometrical structures, see Sanchez-Rodriguez (2001).

[64] For discussions of projective and conformal geometry, see Schouten (1954: 287–334).

[65] We can say that a spacetime is its causal structure (order) plus volume information (Dowker 2004).

[66] Further study has shown that the relations between the four structures are more complicated. The results of this joint work with Mihaela Iftime will be reported elsewhere.

of four-dimensional analogue of quantum geometro-dynamics; with the possibility of another representation in which the projective connection and the trace factor are the configuration variables—a sort of four-dimensional analogue of the loop quantum gravity representation. Of course such an approach would seem to require a full, four-dimensional quantization procedure rather than a canonical one.

As suggested above, such an approach might provide a way to derive the fundamental quantum of process in general relativity. Quanta of (proper) three-volume and (proper) time might then be regarded as 'perspectival' effects of 'viewing' quanta of four-volume by observers in different states of relative motion through spacetime.

ACKNOWLEDGEMENTS

I thank Carlo Rovelli for written comments, and Abhay Ashtekar and Thomas Thiemann for oral comments, on an earlier version of this chapter. Of course, I am solely responsible for my comments on their work.

I thank Mihaela Iftime, whose critical reading of the text led to a number of important clarifications and improvements.

REFERENCES

Adams, R. M. (1979) "Primitive Thisness and Primitive Identity". *Journal of Philosophy*, 76: 5–26.

Ashtekar, A. (1987) *Asymptotic Quantization*. Naples: Bibliopolis.

—— (1999) " 'Discussions' following 'Quantum Field Theory of Geometry' ". In T. Yu Cao (ed.), *Conceptual Foundations of Quantum Field Theory*. Cambridge: Cambridge University Press, (pp. 203–6).

—— and J. Lewandowski (2004) "Background Independent Quantum Gravity: A Status Report". ArXiv:gr-qc/0404018 v1.

Auyang, S. Y. (1995) *How is Quantum Field Theory Possible?* New York: Oxford University Press.

Baez, J. C., J. D. Christensen, T. R. Halford, and D. C. Tsang (2002) "Spin Foam Models of Riemannian Quantum Gravity". *Classical and Quantum Gravity*, 19: 4627–48.

Bergmann, P. G., and G. J. Smith (1982) "Measurability Analysis of the Linearized Gravitational Field", *General Relativity and Gravitation* 14: 1131–66.

Bohr, N., and L. Rosenfeld (1933) "Zur Frage der Messbarkeit der Elektromagnetischen Feldgrössen." *Mat-fys. Medd. Dan. Vid. Selsk* 12(8) (cited from the English translation, "On the Question of the Measurability of the Electromagnetic Field Quantities". In R. S. Cohen and J. Stachel (eds.), *Selected Papers of Leon Rosenfeld*. London: D. Reidel, 1979 (pp. 357–400).

Brightwell, G., F. Dowker, H. Garcia, J. Henson, and R. D. Sorkin (2002) "General Covariance and the 'Problem of Time' in a Discrete Cosmology". ArXiv:gr-qc/0202097 (to appear in the *Proceedings of the Alternative Natural Philosophy Association Meeting, August 16–21*, Cambridge, England).

Burgess, C. P. (2004) "Quantum Gravity in Everyday Life: General Relativity as an Effective Field Theory". **www.livingreviews.org/lrr-2004-5**.

Cao, T. Y. (ed.) (1999) *Conceptual Foundations of Quantum Field Theory*. Cambridge: Cambridge University Press.

DeWitt, B. (2003) *The Global Approach to Quantum Field Theory*, 2 vols. Oxford: Clarendon Press.
d'Inverno, R. A. (2003) "DSS 2 + 2". In Abhay Ashtekar et al. (eds.), *Revisiting the Foundations of Relativistic Physics: Festschrift in Honor of John Stachel.* Dordrecht: Kluwer Academic (pp. 317–47).
Donoghue, J. F. (1995) "Introduction to the Effective Field Theory Description of Gravity". ArXiv:gr-qc/9512024.
Dosch, H. G., V. F. Müller, and N. Serioka (2004) "Quantum Field Theory, Its Concepts Viewed from a Semiotic Perspective". **http://philsci-archive.pitt.edu/archive/00001624/01/ms.**
Dowker, F. (2003) "Real Time". *New Scientist*, 180: 36–9.
—— (2004) "Causal Sets as the Deep Structure of Spacetime". Transparencies for a lecture.
Ehlers, J., F. Pirani, and A. Schild (1972) "The Geometry of Free Fall and Light Propagation". In L. O'Raifertaigh (ed.), *General Relativity: Papers in Honor of J. L. Synge.* Oxford: Clarendon Press (pp. 63–84).
Einstein, A. (1917) *Über die spezielle und die allgemeine Relativitätstheorie (Gemeinverständlich).* Braunschweig: Friedr. Vieweg & Sohn.
—— (1961) *Relativity: The Special and General Theory*, 15th edn. trans. R. W. Lawson. New York: Crown Publishers.
Ellis, G. F. R. (2002) *The Universe around Us: An Integrative View of Science and Cosmology.* Available on the website **www.mth.uct.ac.za/~ellis/cosa.html.**
Filk, T. (2001) "Proper Time and Minkowski Structure on Causal Graphs". *Classical and Quantum Gravity*, 18: 2785–95.
Greene, B. (2004) *The Fabric of the Cosmos: Space, Time and the Texture of Reality.* New York: Alfred A. Knopf.
Haag, R. (1996) *Local Quantum Physics: Fields, Particles, Algebras.* 2nd edn. Berlin: Springer-Verlag.
Heisenberg, W. (1930) *The Physical Principles of the Quantum Theory.* Chicago: University of Chicago Press. Cited from the Dover Press reprint.
Krause, D. (2004) "Structures and Structural Realism". Available on the website **http://philsci-archive.pitt.edu/archive/00001558/01/OnticReal2004.pdf.**
Kronheimer, E. H., and R. Penrose (1967) "On the Structure of Causal Spaces." *Proceedings of the Cambridge Philosophical Society*, 63: 481–501.
Ladyman, J. (1998) "What is Structural Realism?" *Studies in the History and Philosophy of Science*, 29: 49–424.
Mallios, A (1998) "On an Axiomatic Treatment of Differential Geometry via Vector Sheaves. Applications". *Mathematica Japonica*, 48: 93–180.
—— (2006) *Modern Differential Geometry in Gauge Theories*, 2 vols. Boston: Birkhäuser.
—— and I. Raptis (2003) "Finitary, Causal and Quantal Vacuum Einstein Gravity". *International Journal of Theoretical Physics*, 42: 1479–1619.
Markopoulou, F. (2002) "Planck-Scale Models of the Universe". ArXiv:gr-qc/0210086 v1.
—— (2004) "Particles in a Spin Foam". (Transparencies for a talk at the QI/QG Workshop, Perimeter Institute, February 2004.) Available on the website **www.perimeterinstitute.ca/activities/scientific/cws/feb2004workshop/fotini.pdf.**
—— and L. Smolin (1997) "Causal Evolution of Spin Networks". ArXiv:gr-qc/9702025 v1.
Perez, A. (2003) "Spin Foam Models for Quantum Gravity". *Classical and Quantum Gravity*, 20: R43–R104.
Rideout, D. P., and R. D. Sorkin (2000) "Classical Sequential Growth Dynamics for Causal Sets". *Physical Review D*, 61: 024002.

Robinson, D. C. (2003) "Geometry, Null Hypersurfaces and New Variables". In Abhay Ashtekar et al. (eds.), *Revisiting the Foundations of Relativistic Physics: Festschrift in Honor of John Stachel.* Dordrecht: Kluwer Academic (pp. 349–60).

Rosenfeld, L. (1979) *Selected papers of Leon Rosenfeld*, ed. R. S. Cohen and J. Stachel. Dordrecht: D. Reidel.

Rovelli C. (1991) 'What is observable in classical and quantum gravity?' *Classical and Quantum Gravity*, 8: 97–316.

____(1999) "'Localization' in Quantum Field Theory: How Much of QFT is Compatible with What we Know about Space-time?" In T. Yu Cao (ed.), *Conceptual Foundations of Quantum Field Theory.* Cambridge: Cambridge University Press (pp. 207–30).

____(2003) "Loop Quantum Gravity". *Physics World*, November 2003: 37–42.

____(2004) *Quantum Gravity.* Cambridge: Cambridge University Press.

Runes, D. D. (ed.) (1962) *Dictionary of Philosophy.* Totowa, NJ: Littlefield, Adams & Co.

Salisbury, D. C. (2003) "Gauge Fixing and Observables in General Relativity". ArXiv:gr-qc/0310095 v1.

Samuel, J. (2000) "Is Barbero's Hamiltonian Formulation a Gauge Theory of Lorentzian Gravity?" *Classical and Quantum Gravity*, 17: L141–L148.

Sanchez-Rodriguez, I. (2001) "Intersection of *G*-structures of first or second order". In *Differential Geometry and Its Applications; Proc. Conf. Opava (Czech Republic), August 27–31, 2001.* Opava: Silesian University (pp. 135–40).

Saunders, S. (2003) "Structural Realism, Again". In Symons 2003: 127–33.

Schouten, J. A. (1954) *Ricci-Calculus*, 2nd edn. Berlin: Springer-Verlag.

Schweber, S. (1961) An Introduction to Relativistic Quantum Field Theory. Evanston, IL: Row, Peterson.

Smolin, L. (2002a) *Three Roads to Quantum Gravity.* Lymington: Basic Books.

____(2000b) "Technical Summary of Loop Quantum Gravity". Available on the website **www.qgravity.org/loop/**.

Sorkin, R. D. (1993) "Impossible Measurements on Quantum Fields". In B. L. Hu and T. A. Jacobson (eds.), *Directions in General Relativity*, ii: *Papers in Honor of Dieter Brill.* Cambridge: Cambridge University Press (pp. 293–305).

____(1997) "Forks in the Road, on the Way to Quantum Gravity". *International Journal of Theoretical Physics*, 36: 2757–81.

____(2003) "Causal Sets: Discrete Geometry". ArXiv:gr-qc/0309009 v1.

Stachel, J. (1962) *Lie Derivatives and the Cauchy Problem in Generalized Electrodynamics and General Relativity.* Ph.D. dissertation, Stevens Institute of Technology.

____(1969) "Covariant Formulation of the Cauchy Problem in Generalized Electrodynamics and General Relativity". *Acta Physica Polonica*, 35: 689–709.

____(1986) "Do Quanta Need a New Logic?" In R. G. Colodny (ed.), *From Quarks to Quasars: Philosophical Problems of Modern Physics.* Pittsburgh: University of Pittsburgh Press (pp. 229–347).

____(1993) "The Meaning of General Covariance: The Hole Story". In J. Earman et al. (eds.), *Philosophical Problems of the Internal and External Worlds.* Pittsburgh: University of Pittsburgh Press (pp. 129–60).

____(1997) "Feynman Paths and Quantum Entanglement: Is There Any More to the Mystery?" In R. S. Cohen, M. Horne, and J. Stachel (eds.), *Potentiality, Entanglement and Passion-at-a-Distance: Quantum Mechanical Studies for Abner Shimony*, vol. ii. Dordrecht: Kluwer Academic Publishers (pp. 244–56).

Stachel, J. (2001) "Some Measurement Problems in Quantum Gravity". Unpublished text based on lectures given at the *Relativity Center*, Pennsylvania State University.

____ (2002) " 'The Relations between Things' versus 'The Things between Relations': The Deeper Meaning of the Hole Argument". In D. B. Malament (ed.), *Reading Natural Philosophy: Essays in the History and Philosophy of Science and Mathematics*. Chicago: Open Court (pp. 231–66).

____ (2003) "A Brief History of Space-Time". In I. Ciufolini, D. Dominic, and L. Lusanna (eds.), *2001: A Relativistic Spacetime Odyssey: Experiments and Theoretical Viewpoints on General Relativity and Quantum Gravity. Proceedings of the 25th Johns Hopkins Workshop on Current Problems in Particle Theory*. Singapore: World Scientific (pp. 15–34).

____ (2005) "Structural Realism and Contextual Individuality". in Y. Ben-Menahem (ed.), *Hilary Putnam*. Cambridge: Cambridge University Press.

____ (2006a) "Some Problems of Loop Quantum Gravity". Preprint.

____ (2006b) "The Story of Newstein, or Is Gravity Just Another Pretty Force?" In Jurgen Renn and Matthias Schimmel (eds.), *The Genesis of General Relativity*, vol. iv. *Gravitation in the Twilight of Classical Physics: The Promise of Mathematics*. Berlin: Springer 2006 (pp. 1041–78).

____ & M. Iftime (2005) "Fibered Manifolds, Natural Bundles, Structured Sets, G-Spaces and All That: The Hole Story from Space-Time to Elementary Particles". ArXiv:gr-qc/0505138.

Symons, J. (ed.) (2003) *Symposium on 'Structural Realism and Quantum Field Theory'. Synthese*, 136(1).

Teller, P. (1995) *An Interpretative Introduction to Quantum Field Theory*. Princeton: Princeton University Press.

____ (1998) "Quantum Mechanics and Haecceities". In E. Castellani (ed.), *Interpreting Bodies: Classical and Quantum Objects in Modern Physics*. Princeton: Princeton University Press (pp. 114–41).

Thiemann, T. (2001) "Introduction to Modern Canonical Quantum General Relativity". ArXiv:gr-qc/0110034 v1.

____ (2002) "Lectures on Loop Quantum Gravity". ArXiv:gr-qc/0210094 v1.

____ (2004) "The LQG-String: Loop Quantum Gravity Quantization of String Theory I. Flat Target Space". ArXiv:hep-th/0401172 v1.

Wald, R. M. (1994) *Quantum Field Theory in Curved Spacetime and Black Hole Thermodynamics*. Chicago: University of Chicago Press.

Weyl, H. (1921) "Zur Infinitesimalgeometrie: Einordnung der projektiven und konformen Auffassung". *Nachrichten der Königlichen Gesellschaft der Wissenschaften zu Göttingen, Mathematisch-physikalische Klasse*: 99–112.

____ (1922) "Zur Infinitesimalgeometrie: p-dimensional Fläche im n-dimensionalen Raum". *Mathematische Zeitschrift*, 12: 154–60.

Wüthrich, C. (2003) "Quantum Gravity and the 3D vs. 4D Controversy". Available on the website **http://alcor.concordia.ca/~scol/seminars/conference/Wuthrich.pdf**.

4

Points, Particles, and Structural Realism

Oliver Pooley

> Even if we are able to decide on a canonical formulation of our theory, there is the further problem of metaphysical underdetermination with respect to, for example, whether the entities postulated by a theory are individuals or not ...
>
> We need to recognise the failure of our best theories to determine even the most fundamental ontological characteristic of the purported entities they feature ... What is required is a shift to a different ontological basis altogether, one for which questions of individuality simply do not arise. Perhaps we should view the individuals and nonindividuals packages, like particle and field pictures, as different *representations* of the same structure. There is an analogy here with the debate about substantivalism in general relativity. (Ladyman 1998)

In his paper 'What is Structural Realism?' (1998) James Ladyman drew a distinction between *epistemological* structural realism (ESR) and *metaphysical* (or ontic) structural realism (OSR). In recent years this distinction has set much of the agenda for philosophers of science interested in scientific realism. It has also led to the emergence of a related discussion in the philosophy of physics that concerns the alleged difficulties of interpreting general relativity that revolve around the question of the ontological status of spacetime points. Ladyman drew a suggestive analogy between the perennial debate between substantivalist and relationalist interpretations of spacetime on the one hand, and the debate about whether quantum mechanics treats identical particles as individuals or as 'non-individuals' on the other. In both cases, Ladyman's suggestion is that a structural realist interpretation of the physics—in particular, an *ontic* structural realism—might be just what's needed to overcome the stalemate. The purported analogy between the physics of spacetime points and the physics of quantum particles has been further articulated and defended by Stachel (2002), by Saunders (2003), and by French and Rickles (2003).

The main thesis of this chapter is that, whatever the interpretative difficulties of generally covariant spacetime physics are, they do not support or suggest structural realism. In particular, I hope to show that there is in fact no analogy that supports a similar interpretation of the metaphysics of spacetime points and of quantum particles. But the story is not simple, and a certain amount of stage setting is required.

4.1 WHAT IS STRUCTURAL REALISM?

The genesis of contemporary structural realism is well known: John Worrall (1989) proposed a realist interpretation of science with the intention of doing justice to two opposing arguments. On the one hand there is the 'no miracles' argument: the empirical success of science is held to be miraculous unless it has succeeded in correctly describing the reality behind the phenomena it saves. On the other hand, there is the 'pessimistic meta-induction', most famously, and forcefully, put by Laudan (1981): theories that have been superseded in Kuhnian scientific revolutions are now judged to be radically mistaken in their claims about the reality behind the phenomena, and this *despite* their often having enjoyed exceptional empirical success. It is only rational, therefore, to expect our current theories to suffer a similar fate come the next scientific revolution. Worrall's structural realism is designed to be realist enough so as to do justice to the 'no miracles' intuition and yet agnostic enough to avoid falling prey to the pessimistic meta-induction. In particular, continuity through theory change at the level of form or structure is supposed to license the belief that science is succeeding in characterizing the *structure* of reality (and hence its empirical success is not miraculous) even if science is radically wrong in its description of the fundamental *nature* of reality.

Ladyman asked whether this is epistemology or metaphysics. If it is epistemology, he argued, then there are problems if the doctrine is cashed out in the most obvious way, in terms of a theory's Ramsey sentence. Epistemological structural realists, like John Worrall and Elie Zahar, have since explicitly endorsed the Ramsey sentence strategy, and have sought to defend it against Ladyman's objections.

Recall that the Ramsey sentence T^* of a theory T is formed from T by replacing all the theoretical predicates that occur in the theory with predicate variables and then existentially quantifying. The epistemological structural realist holds that the cognitive content of a theory is fully captured by its Ramsey sentence. Two, related, results appear to pose a problem for such a position.

The implications of the first for structural realism were raised against Russell's structuralist position by Newman (1928); they have been discussed in the context of the recent debate by Demopoulos and Friedman (1985). Essentially the same result lies at the heart of Putnam's paradox. The problem is that it is theorem of set theory and second-order logic that any consistent proposition to the effect that a certain set of properties and relations exist, no matter what structural constraints are placed upon this set, will be true of any domain, provided that the domain has the right cardinality (and the upward Lowenheim–Skolem theorem entails that the only cardinality constraint concerns whether or not the domain is finite). The second result is due to Jane English (1973): any two Ramsey sentences that are incompatible cannot have all of their observational consequences in common. The conclusion to be drawn appears to be that if a Ramsey sentence is empirically adequate, then it is true. It doesn't really make substantive claims about the structure of reality beyond the phenomena since such claims, because they are made only in terms of existential quantification, will always be true (providing that the realm of the unobservable has the right cardinality).

Worrall and Zahar's response (2001) is to stress that the Ramsey sentence is formed by replacing only the *theoretical* predicates with variables; the *observational* predicates remain in place. But it is far from clear that this solves the problem. It certainly means that a consistent Ramsey sentence doesn't simply make a statement about the cardinality of the set of individuals in the world; it does have empirical consequences. But, putting aside Newman's specific target in Russell, the charge was never that structural realism conceived in terms of a theory's Ramsey sentence made only trivial claims about the world. The charge is that it collapses into *empiricism*; that the realist claim that science is in some way latching onto reality beyond the phenomena so as to do justice to the 'no miracles' intuition has been lost.[1]

In fact, in further articulating their response, Worrall and Zahar face a dilemma. They rely on a distinction between observational and theoretical predicates, but let us ask how this might relate to a distinction between (directly) observable and unobservable, or inferred, *entities*. Grant such a partition of the domain of the world, and one has two possibilities: (1) an observational predicate is one that is satisfied only by observable entities or (2) observational predicates can be satisfied by unobservable entities.[2] If one opts for (1), then clearly the Newman problem applies in full force to the question of which *unobservable* entities fall in the extensions of the theoretical predicates. If one opts for (2), then one can ask with what right does the structural realist claim to be able to keep fixed the *extensions* of the observational predicates, given that such predicates apply to unobserved entities whose existence is conjectural.

And even if we grant that the extensions of the observational predicates are to be held fixed, and that there is no subdomain of the world to which none of them apply, the structural realist is still not entitled to assume that holding true the Ramsey sentence will give a fix on the extensions of the theoretical properties and relations over which it quantifies that is sufficient to satisfy realist intuitions. A complete fix will be achieved if the original theory is categorical—if all of its models are isomorphic—but (formalizations of) real scientific theories, even together with reports of all relevant empirical data, will not, of course, be categorical.

In fact, Worrall and Zahar seem to be prepared to acknowledge that there is no more to the content of a Ramsey sentence than all of its consequences that contain only observational predicates. The reason why they do not consider that this means that their version of structural realism collapses into empiricism is that they distinguish between a theory's observational content—which they claim must be empirically *decidable*, if it is to count as genuinely observational—and the more inclusive set that also includes all of the empirical generalizations that the theory entails. They concede that the theory's Ramsey sentence itself might be amongst the latter. But even if to endorse empirical generalizations really is to 'go against the

[1] For a careful defence of the claim that the truth of a theory's Ramsey sentence is equivalent to the combination of its empirical adequacy together with a cardinality constraint, see Ketland (2004).

[2] The phenomenalistic strain in Zahar's philosophy might suggest that he is committed to (1).

canons of even the most liberal version of empiricism' (2001: 241), by itself it hardly amounts to scientific realism, structural or otherwise.[3]

Let me conclude this discussion of the problems facing epistemological structural realism by mentioning one further worry. It is clear that the Ramsifying structural realist has to tread a very fine line. The Ramsey sentence must be held to give us enough of a fix on the extensions of theoretical predicates so as to avoid the Newman problem. But to achieve too strong a fix would undermine the original motivation for structural realism. The reason is that the Ramsey sentence refers to, and quantifies over, exactly the same entities as the original theory. As French and Ladyman press: 'If the meta-induction is a problem about lack of continuity of reference then Ramsifying a theory does not address the problem at all' (2003: 33).

4.2 UNDERDETERMINATION

The idea of an *ontic* structural realism can provoke a definite sense of unease, for it is hard not to worry that something akin to mystery mongering is taking place. It is one thing to claim that our knowledge of the unobservable realm is limited to structural knowledge, that all we can know about unobservable objects are their structural properties. It is quite another thing to claim that *all there is* beyond the phenomena is structure. What can this claim *mean*?

It would, however, be quite unfair to accuse the most prominent defenders of ontic structural realism of mystery mongering. In order to get beyond the slogans and metaphors and to arrive at a characterization of, if not ontic structural realism itself, then of one of its expected achievements, it is time to consider the problem of underdetermination. This problem is central to Ladyman's positive case for OSR: whatever else it is, OSR is intended to dissolve various metaphysical disputes centring on underdetermination. It is supposed to provide a new metaphysical perspective with respect to which certain troublesome, apparently irresolvable, choices simply do not arise.

Ladyman distinguishes between two quite distinct ways in which a *single* theory might be said to be empirically underdetermined.[4] The first type arises due to the

[3] Although it is perhaps close to the version of structural realism that I take Ladyman, in particular, to advocate. This has no use for a notion of reality 'behind' the phenomena and involves only the 'minimal metaphysical commitment' to 'mind independent modal relations between phenomena (both possible and actual)' (French and Ladyman 2003: 46). Where Ladyman would differ from Worrall and Zahar, presumably, is in claiming that these modal relations are not 'supervenient on the properties of unobservable objects and the external relations between them, rather this structure is ontologically basic' (ibid. 46; cf. also Ladyman 1998: 418).

[4] Traditional 'underdetermination of a theory by data' might be thought to involve two quite distinct theories which are nevertheless empirically equivalent, either with respect to all the observational evidence so far (weak underdetermination), or in principle (strong underdetermination). The types of underdetermination that Ladyman highlights, as will be seen, involve the interpretation of a *single* theory. In what follows I simply set aside the traditional problem, and use the term ***empirical underdetermination*** as a general term for the two *special* types of underdetermination that Ladyman discusses.

existence of different formulations of a single theory, formulations that can suggest radically different ontologies if interpreted realistically. This particular problem for the realist was stressed by Jones (1991). Call the type of underdetermination involved *Jones underdetermination.* Ladyman calls the second type of underdetermination *metaphysical underdetermination.* Roughly speaking, this type of underdetermination arises because of the existence of alternative (realist) interpretations of a single *formulation* of a theory, interpretations that again are supposed to involve radically incompatible ontologies. According to Ladyman the underdetermination between individual and non-individual interpretations of the quantum mechanics of identical particles, and between substantivalism and relationalism is of this type. Another nice example is the underdetermination between two rival conceptions of fields. Fields can be viewed either as substantival entities in their own right, with infinitely many degrees of freedom associated with their infinitely many point-like parts, or they can be viewed as consisting in the instantiation of a pattern of properties by spacetime points.

It is useful at this point to mention a parallel problem in the philosophy of mathematics. In 'What Numbers Could Not Be' (1965), Benacerraf highlighted the following difficulty facing anyone who would identify numbers with sets: there are many proposals, they are incompatible so at most one can be correct, but it seems that no cogent reason can be given for preferring one over another. If numbers are sets then which sets they are seems to be underdetermined. Defenders of structuralist views of mathematics view this difficulty that faces those who would take the reducibility of arithmetic to set theory as disclosing that numbers are really sets as one of the best supports for their philosophy. What all the set-theoretic constructions share is the same structure, and this structure is all that matters. Does this parallel reflect favourably on the ontic structural realist's attitude to empirical underdetermination?[5]

For cases of empirical underdetermination to support ontic structural realism, Ladyman had better be correct in saying that traditional realism goes beyond a commitment to structure precisely in terms of metaphysical commitments that are underdetermined by the evidence for them (Ladyman 1998: 418). On the face of it, and without being told what a commitment only to structure might involve, the claim does not look plausible. An obvious problem is that, on the most straightforward characterizations of structure (e.g. a set-theoretic one), most cases of different formulations of a theory will involve *different* structures.[6] Consider a model of a theory of Newtonian gravitation formulated using an action-at-a-distance force

Note that the division is not quite as clear cut as this summary suggests. Someone might argue, on the basis of an instance of strong traditional underdetermination that in fact we have two formulations of a single theory, not two theories. Conversely, someone might argue that Jones underdetermination (defined below) really shows us that what we were inclined to treat as two formulations of a single theory should in fact be regarded as two theories.

[5] I return to the parallel in §4.3. A crucial question will be whether the parallel supports the elimination of objects altogether.

[6] A fact which undermines, I think, the parallel between empirical underdetermination (at least of the Jones variety) and the underdetermination that motivates structuralist positions in the philosophy of mathematics.

and an empirically equivalent model of the Newton–Cartan formulation of theory. There is no (primitive) element of the second model which is structurally isomorphic to the flat inertial connection of the first model, and there are no (primitive) elements of the first model which are structurally isomorphic to the gravitational potential field, or the non-flat inertial structure of the second. Clearly a more sophisticated notion of structure is needed if it is to be something common to models of both formulations of the theory.

The claim that the structural realist might have the resources to be able to identify something beyond the empirical that is in common to different formulations of a single theory, and that he thus might be able to dissolve various interpretative problems by transcending the root underdetermination, is supposed to get some of its plausibility from the case of Schrödinger's and Heisenberg's original rival formulations of quantum mechanics. Through the work of Weyl and others, these formulations were soon recognized to be *different representations* of a single, mathematical *structure* in which states of a system correspond to rays in a Hilbert space, and observables correspond to operators (of the appropriate sort) that act on this space.[7] But there are at least three reasons to be sceptical that this example alone lends support to the idea that underdetermination in general, and Ladyman's metaphysical underdetermination in particular, motivate a radical ontic structural realism.

First, it is not clear that the ontic structural realist has a story to tell about this example. As is well known, whether anyone has come up with a truly successful *realist* interpretation—structural or otherwise—of the standard formulation of non-relativistic quantum mechanics that subsumes both Schrödinger's and Heisenberg's original formalisms is a controversial issue. Two of the potential candidates—GRW (Ghirardi–Rimini–Weber) collapse theory and de Broglie–Bohm pilot wave theory—break the unifying picture by preferring one basis with respect to which either genuine collapse occurs, or with respect to which the true beables are defined. And if more than one of the various interpretative options ultimately survives the many criticisms they all face, it would seem that quantum mechanics remains beset by underdetermination, albeit of a very different type to that involving wave versus matrix mechanics.

Second, this one example gives us little reason to suppose that whatever was achieved in this particular case will be, or even can be, repeated for other instances of Jones underdetermination. Consider the underdetermination that exists (relative to certain solutions) between Julian Barbour's Machian 3-space approach to general relativity (where the fundamental ontology consists of instantaneous 3-spaces and does not involve any primitive temporal notions), the traditional curved-spacetime formulation, and formulations involving spin-2 fields on a flat (or at least fixed) background spacetime. Here we have a clear case of different formulations of a theory that are associated with prima facie incompatible ontologies. A structural realist dissolution of this problem requires an explicit characterization of a mathematical

[7] In fact, Heisenberg's matrix mechanics and Schrödinger's wave mechanics were *not* strictly equivalent; see Muller (1997a, 1997b)!

framework that stands to each formalism as the abstract Hilbert space formalism of quantum mechanics stands to Schrödinger's wave mechanics and Heisenberg's matrix mechanics. Note that it is not enough that we have a good understanding (as we do) of the various mathematical relationships that exist between the formalisms. Ladyman's structural realist needs a single, unifying framework, which she can then interpret (in terms of an as-yet-to-be-articulated metaphysics of structure) as corresponding more faithfully to reality than do its various realist representations.

I am not optimistic that any such development is in the offing. It seems more likely that theoretical advances will favour one formulation over the others. String theory's triumph would, in many senses, vindicate the spin-2 picture. The success of loop quantum gravity (or of a variant, provided with the right sort of interpretation) could vindicate Barbour's advocacy of 3-space concepts over spacetime concepts.[8] The idea that underdetermination associated with different formulations is to be transcended by a more general framework with respect to which the different formulations are seen as different representations of a single, underlying reality might look suspiciously like an unwarranted generalization from a single, special case.

Third, the Heisenberg–Schrödinger example involves Jones underdetermination. However, the type of underdetermination that is supposed to be involved in the debates between the substantivalist and the relationalist, and between advocates of the 'individuals interpretation' of quantum particles and those who advocate a particles-as-non-individuals interpretation, is what Ladyman calls *metaphysical* underdetermination. Here it might seem more plausible that some interpretative stance according to which the rival viewpoints are merely different representations of the same reality will be possible. (But it is important to stress that those sympathetic to an ontic structural realism have yet to provide a positive characterization of any such position, so far they have only told us what the position is meant to achieve; see section 4.3 below.) But equally, one might wonder whether the underdetermination in question is one that should genuinely trouble the realist.

In fact, a genuine underdetermination between relationalism and substantivalism *is* one that should trouble the realist. This is because such an underdetermination *would* be an instance of Jones underdetermination. However, as things stand, there simply is no such underdetermination. The standard formulations of general relativity are straightforwardly substantivalist in that the metric field is (a) taken to represent a genuine and primitive element of reality and (b) most naturally interpreted as representing spacetime structure.[9] Now some believe that the hole argument calls this picture into question. But these same people also typically suggest that an alternative formulation, which would correspond to a genuinely relationalist world picture, should be sought (e.g. Earman 1989: ch. 9). I agree that relationalism

[8] Such discrimination between alternatives by theoretical advances gets some support from the history of physics, although the fact that underdetermination often reappears in a new guise means that the realist should not take too much comfort from this state of affairs.

[9] The second of these claims is contested by some (Earman and Norton 1987; Rovelli 1997). I say more about it below.

needs different physics (or at least a different formulation of the physics).[10] But I disagree that the formulation of general relativity (GR) that the realist naturally interprets along substantivalist lines is in trouble because of the hole argument, for there are different interpretative options available *within* the substantivalist camp.

And it turns out that this is the true location of Ladyman's metaphysical underdetermination. The opposition that he, and Stachel, characterize as between relationalism and substantivalism is really an opposition between *haecceitist* and *anti-haecceitist* substantivalism.[11] I will explain later what I mean by these terms, and why I believe that there is no real contest: anti-haecceitism is the clear winner. But even if one thought that there was a genuine choice to be made, and that interpreting the physics realistically failed, by itself, to make the choice, it is not clear why this should trouble the scientific realist. For, as we will see, there is a sense in which haecceitist substantivalism is simply an extension of anti-haecceitist substantivalism. Anti-haecceitist substantivalism represents a realist core position which it may or may not be correct to supplement. If this is the only choice to be made, it hardly constitutes an interesting threat to the scientific realist's belief in the existence of spacetime points.

Things are otherwise with the quantum mechanics (QM) of identical particles. For the benefit of those already *au fait* with the terminology, it turns out that, while the physics of identical particles strongly suggests anti-haecceitism, anti-haecceitism by itself does not suffice to explain all the peculiarities of the physics of quantum particles. The difference between the two cases is traceable to a difference in the physics. Perhaps unsurprisingly, the way in which *classical* GR is diffeomorphism invariant is rather different from the way in which the *quantum mechanics* of identical particles is permutation invariant.

The disanalogy highlights a sense in which it is misleading to present the quantum mechanical case as an instance of underdetermination between two realist interpretations, if the intended implication is that the two interpretations are equally viable. While French (1989; French and Redhead 1988) may have clearly demonstrated that the individuals interpretation of QM particles exists in logical space, it is not really a serious contender.[12] Equally, the non-individuals interpretation, if it

[10] Barbour's 3-space approach to GR constitutes a genuinely distinct interpretation, but not a relationalist one, since the fundamental ontology is substantival space (not spacetime).

[11] The positions have been labelled straightforward and sophisticated substantivalism by Belot and Earman (1991; 2001).

[12] Ladyman's claim that there is 'much dispute about whether or not quantum particles ... are individuals' (1998: 419) is unconvincing if to treat particles as individuals is to adopt the interpretative option delineated by French, and French and Redhead. I am not, of course, denying that exactly how one *should* conceive of quantum particles is a highly disputed question; rather I am only claiming that there is a fair degree of consensus that one should *not* conceive of quantum particles in certain ways. French (1998: 112, n. 62) himself states that it is not true that 'anything goes', but I understand him to take viewing the particle labels of the standard tensor product Hilbert space formalism as naming individuals as a genuine interpretative option. I don't believe it is.

truly accommodates the phenomena, is a more radical position than anti-haecceitist substantivalism.[13]

In the next section I briefly consider how far OSR's defenders have gone in attempting to characterize the position. Before doing so, I wish to raise two worries connected specifically to underdetermination. The first is that one might worry that dissolving the underdetermination is not desirable. Recall that I mentioned above the way in which the various alternative formulations of GR were linked to quite distinct attempts to advance beyond that theory. Having these alternatives in play might therefore serve a vital heuristic role in theoretical advance. Of course, from the perspective of OSR, the various formulations still exist; they are just now understood as different representations of the same structure. But perhaps, for advances to take place, it is important that the different formulations are considered to be genuinely distinct and exclusive alternatives. And perhaps, from the perspective of an advance, that the subsequently favoured alternative could be unified with the others in a single structure will seem like a happy accident; the overarching framework will appear to have a secondary status, rather than a fundamental one.

The second worry is that a radical structural metaphysics might make the underdetermination worse. Assuming such a metaphysics is possible (a big assumption), then *if* one adopts it, one will view the previous alternatives as different representations of the structure that one claims is fundamental. But is the fact that this is how it looks from the perspective of OSR enough to commend adopting that perspective? It seems likely that every side of the original underdetermination will be able to explain the other side's worldview. For example, if one believes in a dynamical spacetime connection, one can explain why things are as if spacetime were flat and gravity were a universal force. But why not also expect that the red-blooded realists will be able similarly to explain away the structural realist's perspective, just as GRW theory and Bohm theory can (or are supposed to be able to) explain the success of orthodox quantum theory? The defender of OSR can perhaps wield Ockham's razor, but the dialectical problem here, for the structural realist, is that this is exactly the type of consideration that might also favour one traditional realist interpretation over another. And since we do not yet have the structural realist metaphysics, or any guarantee that such a thing is conceivable, if we end up having to wield Ockham's razor in order to vindicate it, one might wonder why one should go looking for it in the first place.

4.3 WHAT *IS* ONTIC STRUCTURAL REALISM?

If it has been made clear what the relation of ontic structural realism to the problem of underdetermination is supposed to be, what has been said by way of a *positive*

[13] As Teller's discussion in his (2001) makes clear. As will become apparent, although I agree with Teller that the interpretative options that the substantivalist has in the face of the hole argument are not applicable to the case of identical QM particles, I differ with him both about how to characterize the substantivalist options, and about the difficulty of providing a successful realist interpretation of the quantum mechanics of identical particles.

characterization of a position that can do the job? The answer is, at this stage, not much, but in a recent paper French and Ladyman (2003) seek to further articulate their vision of OSR.

I have already mentioned Ladyman's claim that traditional realism goes beyond commitment to structure precisely in commitments that are underdetermined by the evidence. This suggests the tactic of attempting to identify exactly which elements of the realist's metaphysics are responsible for the underdetermination. French and Ladyman have a clear view:

> The locus of this metaphysical underdetermination is the notion of an object so one way of avoiding it would be to reconceptualise this notion entirely in structural terms. The metaphysical packages of individuality and non-individuality would then be viewed in a similar way to that of particle and field in QFT, namely as two different (metaphysical) representations of the same structure. (2003: 37)

The basic point in this quote, that OSR is to offer a perspective from which viewpoints previously taken to be alternatives are seen as representations of the same structure, has been well rehearsed above. What is new in this quote is the claim that an elimination (or, at least, a reconceptualization) of objects is the key. A little later, French and Ladyman are more specific:

> We regard the ontic form of SR as offering a reconceptualisation of ontology, at the most basic metaphysical level, which effects a shift from objects to structures. Now, in what terms does such a reconceptualisation proceed? This hinges on our prior understanding of the notion of an 'object' which has to do ... with the metaphysics of individuality. Given the above metaphysical underdetermination, a form of realism adequate to the physics needs to be constructed on the basis of an alternative ontology which replaces the notion of object-as-individual/non-individual with that of structure in some form. (Ibid.)

Before asking what we are to make of this talk of reconceptualizing objects in terms of structures, it is worth raising the following worry. In the previous section, a distinction was drawn between traditional and metaphysical underdetermination. The present proposal appears to be addressed only to the latter type (and then, rather specifically, to that involving the quantum physics of identical particles and, more controversially, the interpretation of spacetime points). Surely the notion of an object, and an object's individuality, is not the root cause of the underdetermination between, for example, spacetime formulations of GR and Machian geometrodynamics.

Putting aside this worry, let us ask what replacing the notion of objects with that of structure comes to. It is worth recalling the parallel with structuralist views in mathematics. Benacerraf's own view was that his argument to the effect that numbers could not be sets extended to support the conclusion that numbers could not be objects at all. But this point of view is not shared by many contemporary 'non-eliminative' mathematical structuralists, who hold that one can agree that a mathematical object is the very object that it is in virtue of its occupying its particular place in the relevant mathematical structure, without in any sense eliminating it as a genuine object.

One version of the non-eliminative view is Stewart Shapiro's *ante rem* structuralism, so-called after the analogous view concerning universals. This view takes 'structures, and their places, to exist independently of whether there are any systems of objects that exemplify them' (1997: 9). The indifference to whether there exist objects exemplifying the structures should not be taken to suggest an indifference to the existence of *mathematical* objects. Rather the mathematical objects are to be understood in terms of the 'places' in the structures; although they enjoy a somewhat secondary ontological status to the structures in which they are places, their existence is not being denied.

The idea of the independent existence of structures suggests an obvious comparison, viz. with the view that physical objects are nothing but bundles of collocated properties ('bundle theory'). Most variants of such a view are held to face the decisive objection that they entail an intolerably strong version of the principle of the identity of indiscernibles. But if one takes relations, and the structures that they form, seriously, one has the resources to frame a sophisticated bundle theory that entails only a relatively weak form of the identity of indiscernibles. According to such a view, it must always be possible to make out numerical diversity in relational terms that do not presuppose identity and difference, but this allows that two objects may nevertheless satisfy exactly the same open sentences with just one free individual variable.[14] Is this all that French and Ladyman have in mind when they talk of a structural reconceptualization of objects?

Although it is suggested by their quoting, apparently with approval, Cassirer's talk of electrons as the ' "points of intersection" of certain relations', and of entities being 'constituted' in terms of relations, there are two reasons why I doubt that it is what they intend. First, the new structural metaphysics was supposed to transcend questions of individuality/non-individuality. But the obvious interpretation of the bundle-theoretic proposals is that they *do* yield individuals: determinately numerically distinct particulars, albeit ones whose ontological status, and individuality, is secondary to, and dependent upon, that of properties and relations. The second reason, which is related to the first, is that there is no reason for a bundle theorist to have a problem with standard logic and set theory. Standard logic and set theory presuppose the existence of individuals that are determinately numerically distinct, but they do not presuppose that the individuality of these individuals is independent of the properties and relations that predicates can express, or of the sets that the individuals can form. And yet French and Ladyman do see standard logic as a barrier to articulating their view:

> How can you have structure without (non-structural) objects? Here the structuralist finds herself hamstrung by the descriptive inadequacies of modern logic and set theory which retains the classical framework of individual objects represented by variables and which are the subject of predication or membership respectively (cf. Zahar (1996)). In lieu of a more appropriate framework for structuralist metaphysics, one has to resort to a kind of 'spatchcock' approach, reading the logical variables and constants as *mere placeholders which allow us to define and describe the relevant relations which bear all the ontological weight.* (2003: 41)

Talk of 'mere placeholders' might suggest that their view is no more radical than the bundle-theoretic suggestion, but in a later footnote they are more explicit about

[14] The requisite formal treatment of identity goes back to Hilbert and Bernays (1934). It is advocated by e.g. Quine (1986: 63–4), and recently has been systematically applied to issues in the philosophy of physics by Simon Saunders (2000; 2003).

what they perceive to be the inadequacies of set theory, and the current unavailability of anything that serves the ontic structural realist's needs:

> [B]oth of these modes of representation—group theory and set theory—presuppose distinguishable elements, which is precisely what we take modern physics to urge us to do away with. If we are going to take our structuralism seriously, we should therefore be appropriately reflective and come up with thorough-going structural alternatives to group theory and set theory... [Krause's attempt to construct a 'quasi-set theory' (Krause 1992)] insofar as [it] is based on *objects* which do not have well defined identity conditions... represents a formalism of one side of our metaphysical underdetermination, rather than a structuralist attempt to avoid it altogether. What is needed is the construction of a fundamental formalisation that is entirely structural; we shall leave this to future works or future (cleverer) philosophers. (2003: 52; my emphasis)[15]

It is time to examine more closely whether the realist's commitment to objects really does lead to an objectionable form of underdetermination, as French and Ladyman maintain.

4.4 OBJECTS

Ladyman claims that 'traditional realism *does* involve acceptance of more than the structural properties of theoretical entities' (1998: 418). The realist's additional, metaphysical commitments are underdetermined by the empirical evidence but are supposed to be of such interpretative importance to the realist that our best theories fail to determine 'even the most fundamental ontological characteristics of the purported entities they feature' (ibid. 419–20). What are the realist's additional commitments? Are they as central to a traditional realism as Ladyman claims? Should we be troubled that facts concerning them are underdetermined by the physics itself?

The problem is supposed to concern whether the fundamental entities (spacetime points or quantum particles) that the realist posits are *individuals*, but what is it for an object to be an individual? This is, of course, a question that has been much discussed in this context (van Fraassen 1991; French 1998; Teller 1998; French and Rickles 2003) but I have to confess myself unhappy with the course that these discussions sometimes take, and with many of their presuppositions. For example, the problem is sometimes approached by suggesting that if the objects are individuals, then there is a limited range of options for understanding what their individuality consists in:

(i) objects might be individuated in virtue of their possessing some sort of haecceity,
(ii) the individuality of the substance or matter of an object might be held to account for the object's individuality,

[15] Note that talk of 'distinguishable elements' is ambiguous. I stated above that standard logic and set theory presuppose the *determinate distinctness* of its elements. I claim that logic and set theory presuppose that their elements are distinguishable in no stronger sense than this. The diffeomorphism invariance of GR in no way suggests that spacetime points fail to be distinguishable elements in this sense, but the quantum physics of identical particles does threaten the view that fundamental particles are numerically distinct.

(iii) objects might be individuated in terms of their spatio-temporal location, or
(iv) objects might be individuated in terms of the properties they possess (and perhaps, also, the relations they stand in).[16]

An object would then, presumably, be a 'non-individual' if none of these ways of understanding it as 'possessing individuality' were available.

Talk of individuality, and of individuation, in these contexts is highly obscure. A more promising approach to the question is to focus on questions of identity and non-identity. First, one can ask, within the context of a *single* situation, about the *numerical distinctness* of the objects featuring in the situation. Here there are two obvious questions: (ND1) are the objects determinately numerically distinct? And, if they are, (ND2) what (if anything) confers, or is the ground of, this numerical distinctness? Second, one can ask about the *trans-situation identity* and distinctness of objects in *different* situations (for example, situations that obtain at different times, or different counterfactual possibilities). Here some of the possible questions are:

(TT) if an object that exists at one time is the same object as a particular object that exists at another time, does anything account for, or ground, this identity, or is it a primitive fact?

(TW) If an object that exists in one possible situation is the same object as an object that exists in another possible situation does anything account for, or ground, this identity, or is it a primitive fact?

Related to these questions is a rather more specific question:

(P) are the objects such that there can be two, genuinely distinct, situations which differ solely in terms of a permutation of some of the objects involved?

When the situations in question are distinct possible worlds, (P) becomes a question concerning *haecceitism*. As I will use the term, haecceitism is the position that there are pairs of genuinely distinct possible worlds that differ solely in terms of a permutation of some of the objects that exist in[17] both possible worlds.[18] Anti-haecceitism (concerning a class of objects) is simply the denial that two possible worlds can differ solely in terms of a permutation of objects of that type. When the situations in question are understood as obtaining at different times within a single world, (P) is not a question about haecceitism.

Now clearly one's answers to the questions (TT) and (TW), which concern trans-temporal and trans-possibility identity respectively, will have a bearing on how one answers the corresponding versions of (P). If one thinks that there can

16 Cf. French and Rickles (2003: 223).

17 Read 'exist in' in such a way that it is compatible with counterpart-theoretic approaches to trans-world identity.

18 In this I comply with a usage that is standard in much recent philosophical literature. It is due to David Kaplan (1975), and is, for example explicitly followed by Lewis (1986: ch. 4). As will become clear, it should not be confused with a commitment to *haecceities*, however these are to be understood. Of course, belief in a certain robust type of haecceity might license haecceitism in the modal sense meant here.

be primitive facts concerning trans-temporal, or trans-world, identity, then it seems that one will be committed to the view that a permutation of objects is alone sufficient to yield genuinely different situations of the type in question.[19] The converse, however, does not hold, at least in the case of trans-temporal identity. One can hold that the state of a system at two different times differs solely by a permutation of the system's constituent objects without holding that the objects' trans-temporal identities are brute matters of fact, for one might think that the trans-temporal identities are determined by various trans-temporal relations that do not supervene on the intrinsic states of the system at the two times in question. The most obvious possibility, of course, is that the identities are underwritten by the continuity of the objects' trajectories. It seems plausible, however, that no such relations are available to ground haecceitistic differences in the absence of primitive trans-world identity.

Now questions of individuality appear originally to have entered discussions of the interpretation of many-particle quantum mechanics in terms of question (P), though it is perhaps not totally clear whether the trans-world or trans-temporal version was in question. Crudely put, the assumption of equiprobability together with the answer 'Yes' to (P)—states which differ solely over a permutation of the objects involved are genuinely distinct states—yields Maxwell–Boltzmann statistics (see page 114 below). Answering 'No', therefore, looks like a way of accounting for quantum statistics. It is quite possible that all that some of the founding fathers of quantum mechanics—such as Born, Heisenberg, and Pauli—meant by quantum particles lacking individuality was that (some version of) (P) should receive a negative answer. As we will see in §4.8, denying that a permutation yields a distinct situation is not by itself sufficient to explain the full peculiarities of the quantum mechanics of identical particles.

Let us return to the question of the metaphysical commitments of traditional, object-positing realism. We have listed four putative accounts of individuality, (i) to (iv), and we have reviewed a set of questions—(ND1), (ND2), (TT), (TW), and (P)—concerning object identity. I wish to urge the following point of view: it is sufficient for a certain class of objects to qualify as individuals that (ND1) gets answered 'Yes'—that in a given situation there are facts of the matter about the objects' numerical distinctness. In particular, I claim that answering 'No' to (P)—especially in its modal version, does not impugn the objects' status as genuine, substantial, individuals.

Here I disagree with Paul Teller. He is concerned to identify a minimalist sense of haecceity that is connected with the idea that a particular subject matter (e.g. that of a particular physical theory) concerns *things*. I take it that an entity should be counted as an individual just if it is a thing and has a haecceity in some properly minimalist sense. Teller proposes three 'tests' for whether a subject matter includes (minimalist) haecceities:

[19] In fact things are not quite so straightforward: primitive trans-world identity can perhaps be combined with a denial of purely haecceitistic differences if it is coupled with a strong enough essentialism; cf. Maudlin's response to the hole argument (1989, 1990).

1. Strict identity: ... there is a fact of the matter for two putatively distinct objects, either that they are distinct or, after all, that they are one and the same thing.
2. Labeling: ... the subject matter comprises things that can be referred to with names directly attaching to the referents; that is ... things can be named, or labeled, or referred to with constants where the names, labels, or constants each pick out a unique referent, always the same on different occurrences of use, and the names, labels, or constants do not function by relying on properties of their referents.
3. Counterfactual switching: ... the subject matter comprises things which can be counterfactually switched, that is just in case *a* being *A* and *b* being *B* is a distinct possible case from *b* being *A* and *a* being *B*, where *A* and *B* are complete rosters of, respectively *a*'s and *b*'s properties in the actual world. (1998: 121)

'Strict identity' corresponds to a positive answer to (ND1); 'counterfactual switching' corresponds to a positive answer to the possible worlds version of (P). Now although Teller does not claim that the three tests are necessarily 'different ways of getting at the same idea' (1998: 122), he clearly thinks that the connections are close enough for the tests to be usefully grouped together.[20] But if—as I claim—determinate intra-situation distinctness has no implications for trans-situation identity, then such grouping can only lead to confusion. It is only the denial of determinate intra-situation identity and distinctness that threatens the individuality—the genuine objecthood—of the putative entities in question. However, if questions of intra-situation distinctness and trans-situation identity are not distinguished, one might erroneously infer non-individuality from the denial of (primitive) trans-situation identity. Keeping the two notions distinct is crucial if one is to understand the difference between difficulties raised by the diffeomorphism invariance of classical GR, and by the (anti)symmetrization of the quantum states of identical particles. In the next section we will see that the former only has implications, via the hole argument, for trans-situation identity, whereas the latter, at least according to some, threatens determinate numerical distinctness.

I claim that the realist, in positing objects as individuals, is committed to the determinate intra-situation numerical distinctness of the entities posited. To posit 'non-individual' objects, if sense can be made of this, would be to posit a class of entities whose numerical distinctness was somehow not determinate.[21] These two positions—the positing of a class of determinately distinct objects and the positing of

[20] In fact, in his (2001: 377), Teller claims that (3) is a consequence of (1) and (2). It is true that (1) and (2) allow us to form descriptions which, if they both describe possibilities, describe possibilities that involve counterfactual switching (i.e. that differ merely haecceitistically). But it does not follow from this fact alone that such possibilities *exist*. Note that 'labeling' will only fail to be possible even though strict identity applies when *no* reference to the individuals is possible. But even in such situations 'reference' via variables is possible. Compare our ability to talk about abstract symmetrical structures, e.g. Black's sphere world. In such cases labels or names can be used in a generic sense, but do not refer to one, rather than the other, of the objects related by the symmetry (cf. Teller 2001: 367).

[21] And some think that sense *can* be made of this: see Dalla Chiara et al. (1998).

a class of non-individual objects respectively—represent *core* realist positions. They do not go beyond an acceptance of the 'purely structural' properties of the entities in question (what properties could be more structural than the determinateness or otherwise of numerical distinctness?). And to go so far, and no further, is hardly an '*ersatz* form of realism', but rather a realism worthy of the name.

If the phenomena covered by a particular theory really were indifferent between these two, radically different metaphysics, then one would have an interesting case of genuine metaphysical underdetermination. Perhaps this is what we face in the quantum mechanics of identical particles. It is inappropriate, however, to regard the spacetime points of diffeomorphism-invariant generally relativistic physics as non-individuals in this sense; the physics simply gives us no scope to do so. In any case, the alleged metaphysical underdetermination discussed in the previous sections was not solely an underdetermination between these two core realist positions. Rather it concerned *further* metaphysical commitments to which the realist may or may not sign up.

Of course, the realist *is* perfectly entitled to sign up to other metaphysical commitments, and he may well endorse particular answers to questions such as (TW) and (P). In particular, it is clear that the four ways of understanding 'individuality', (i)–(iv), mentioned above might well be held to underwrite particular answers to such questions. The haecceities of (i), for example, might be thought both to ground numerical distinctness and to underwrite haecceitistic differences; the properties and relations of (iv) might be thought to ground numerical distinctness in such a way as to rule out haecceitistic differences. A 'bare particular' view of objects is a specific example of a metaphysics in line with (i), or perhaps (ii); the sophisticated bundle theory of the last section is a specific example (though not the only example) of a metaphysics in line with (iv).

To sum up, there are two points that I wish to make at this juncture. First any 'underdetermination' at this level surely should not trouble the scientific realist. It looks as if a quite general, metaphysical debate is being played out in the context of the entities of physics. Why should the fact that these two options arise in the context of the interpretation of, for example, spacetime points trouble the scientific realist? In this context, realism is a commitment to the existence of spacetime points, as determinately distinct, substantial individuals. It is hardly a failure of our best theory of them that it fails to determine whether they are bundles of properties, or bare particulars. In fact, it is not clear that the theory *is* indifferent to the choice, for the hole argument shows precisely that conceiving of spacetime points as something akin to bare particulars has the unwelcome consequence of a thoroughgoing indeterminism.

The second point is that it has long been recognized that the choice between the bare particulars view (according to which relata, and their numerical diversity, are ontologically prior to their properties and relations) and a (sophisticated) bundle theory (according to which relations are ontologically 'prior' to, and 'constitute' their relata) presents us with a false dichotomy. A 'no priority view' seems to many to be far more plausible than either. One can endorse the structuralist claim that the numerical diversity of certain objects is grounded in their being situated in a relational structure without reducing these objects to the properties and relations themselves. Equally, and conversely, one can claim that facts of numerical diversity are not so grounded, without going on to claim that they nevertheless *are* grounded by mysterious haecceities or substrata. Instead, one can just take such facts as primitive and as in need of no further metaphysical 'explanation'. It is time to see how these issues play out in the context of the debate concerning substantivalism.

4.5 SOPHISTICATED SUBSTANTIVALISM

In debates concerning the nature of spacetime, *substantivalism* is simply the view that spacetime and its pointlike parts exist as fundamental, substantial entities. This realist view would appear to be what follows from a fairly literal-minded reading of the mathematical formalism of the standard formulations of relativistic physics. For example, the models of general relativity are typically taken to be n-tuples of the form $\langle M, g, \phi_1, \ldots, \phi_{n-2}\rangle$ that satisfy Einstein's field equations. M is a four-dimensional differential manifold and g is a pseudo-Riemannian metric tensor. M and g, taken together, are naturally understood as representing substantival spacetime: the elements of M represent spacetime points, and g encodes the spatio-temporal relations in which they stand. The fields ϕ_i represent the material content of spacetime.

This simple story is supposed to be threatened by Einstein's hole argument. In its modern guise, which it owes to Stachel and to Earman and Norton, it points out that if $\mathcal{M}_1 = \langle M, g, \phi_1, \ldots, \phi_{n-2}\rangle$ is a model of a generally relativistic theory, then the theory's diffeomorphism invariance entails that $\mathcal{M}_2 = \langle M, d^*g, d^*\phi_1, \ldots, d^*\phi_{n-2}\rangle$ is also a model, for any diffeomorphism d (d^*g etc. are the pull-backs and push-forwards of the original fields under the action of the d). According to the argument, the substantivalist is committed to the view that $\mathcal{M}_1$ and $\mathcal{M}_2$ represent distinct possible worlds. It is then pointed out that this commits the substantivalist to a radical form of indeterminism. In the light of the previous section, it will be clear that what interests us is the claim that substantivalism—viewing spacetime points as genuine individuals—entails that $\mathcal{M}_1$ and $\mathcal{M}_2$ represent two distinct physically possible worlds. Before addressing this issue, however, I should briefly consider another threat to the substantivalist understanding of GR, namely that g should be understood, not as representing spacetime structure, but as a 'gravitational' field, much like any other material field.

Carlo Rovelli is someone who advocates such a view:

> In the physical, as well as philosophical literature, it is customary to denote the differential manifold *as well as* the metric/gravitational field ... as spacetime, and to denote all the other fields (and particles, if any) as matter. But ... [i]n general relativity, the metric/gravitational field has acquired most, if not all, the attributes that have characterized matter (as opposed to spacetime) from Descartes to Feynman: it satisfies differential equations, it carries energy and momentum, and, in Leibnizian terms, it *can act and also be acted upon*, and so on. ...
>
> Einstein's identification between gravitational field and geometry can be read in two alternative ways:
>
> i. as the discovery that the gravitational field is nothing but a local distortion of spacetime geometry; or

ii. as the discovery that *spacetime geometry is nothing but a manifestation of a particular physical field*, the gravitational field.

The choice between these two points of view is a matter of taste, at least as long as we remain within the realm of nonquantistic and nonthermal general relativity. I believe, however, that the first view, which is perhaps more traditional, tends to obscure, rather than enlighten, the profound shift in the view of spacetime produced by general relativity. (1997: 193–4)

When seeking to decide between these two views it should also be borne in mind that the 'metric/gravitational field' has also retained *all* of the attributes that lead us to view the analogous structures in pre-GR theories as codifications of spacetime structure. This point cannot be emphasized enough: there is a sense in which the variable, dynamic metric field g of generally relativistic theories plays *precisely* the same role as the flat, non-dynamic metric η of special relativistic theories. The sole difference between the two types of theory is that in one case spacetime is dynamical, and is governed by Einstein's field equations; in the other it is not. So the *sole* attribute that g has lost is the flip-side of one of the attributes that Rovelli claims it has gained, namely it is no longer *immutable* but is affected by matter. And the substantivalist will, of course, see this as making his realism about spacetime all the more plausible: as Rovelli says, spacetime now obeys the action–reaction principle (Anandan and Brown 1995).

What of Rovelli's contention that the metric of GR satisfies differential equations? The metric and affine structures of pre-GR theories also satisfy differential equations, albeit equations (such as the vanishing of the Riemann tensor) that are not of a great deal of physical interest.

What of the claim that the metric has acquired 'most, if not all' of the attributes that might lead us to regard it as matter? Rovelli elaborates the point as follows:

Let me put it pictorially. A strong burst of gravitational waves could come from the sky and knock down the rock of Gibraltar, precisely as a strong burst of electromagnetic radiation could. Why is the [second] 'matter' and the [first] 'space'? Why should we regard the second burst as ontologically different from the [first]? (1997: 193)

The attributes in question all arise from the fact that the metric is dynamical. Now this certainly supports the view that the metric represents a genuine entity, which does not enjoy an inferior ontological status to matter. But why go further and seek to assimilate it to matter? The phenomenon of gravitational waves certainly pushes one to regard whatever is represented by the metric field as a concrete, substantival entity. But why can't we interpret the potentially devastating effect of gravitational radiation as due to ripples in the fabric of spacetime itself?

Clearly we *can* give very different accounts of the rock's destruction by electromagnetic radiation and by gravitational radiation. According to the substantivalist, the parts of the rock of Gibraltar, as part of an extended rigid body, are being continually and absolutely accelerated away from their natural free-fall motions towards their common centre. The accelerating forces are the electromagnetic forces that account for the rock's rigidity. When the rock is hit by a strong burst of electromagnetic radiation, the natural motions of the parts of the rock do not (significantly) change. Rather the parts of the rock are *differently* accelerated by forces that overcome the counteracting forces between the parts of the rock. When the rock is hit by gravitational radiation, however, no additional accelerative forces are applied. Rather the natural motions are no longer towards the rock's centre but are radically divergent. So divergent, in fact, that the electromagnetic binding forces of the rock are no longer sufficient to accelerate the parts of the rock away from their natural trajectories.

The extent to which the metric should be assimilated to other fields is connected to the controversial question of whether gravitational waves, and more generally the metric, carry energy and momentum. The status of gravitational stress-energy is an intricate topic, but the case for drawing a distinction between it and the stress-energy of matter seems compelling (see Hoefer 2000, for an extended discussion). Scenarios where gravitational stress-energy seems most well defined typically involve island matter distributions in asymptotically flat spacetime. Exactly those cases, in other words, where the metric field can be analysed into the flat metric of special relativity (SR) plus a perturbation (see Norton 2000: §3 and the references therein). It is the perturbation, if anything, that corresponds to the 'gravitational field' and carries energy momentum. But Rovelli appears not to wish to view just the perturbation as representing a material field but instead wishes to brand the *entire* metric as 'material'. Gravitational stress-energy and gravitational waves do not force such an interpretation.

Let us grant, then, that a compelling case for regarding the metric field as just another material field has not been made, and that a straightforward, spacetime realist reading of generally relativistic theories remains viable. Our question now is whether such a reading also supports the view that two models related by a non-trivial diffeomorphism represent distinct possibilities. It is clear that this question is a version of the previous section's question (P). If our two diffeomorphic models, $\mathcal{M}_1$ and $\mathcal{M}_2$, are taken to represent two distinct physically possible worlds, then they are worlds which differ solely over a permutation of the spacetime points.

There is close to a consensus in the philosophical literature on Earman and Norton's hole argument that there is nothing anti-substantival about denying that there can be such distinct possible worlds (Butterfield 1989; Brighouse 1994; Rynasiewicz 1994; Hoefer 1996; an exception is Maudlin 1990). Following Belot and Earman, call any substantivalist position that denies haecceitistic differences, and regards $\mathcal{M}_1$ and $\mathcal{M}_2$ as two representations of the same possible world, *sophisticated substantivalism*.

Belot and Earman refuse to be convinced that the substantivalist can have it so easy. They see sophisticated substantivalists' responses either as lacking a 'coherent and plausible motivation', or as indicative of the 'insularity of contemporary philosophy of space and time' (1999: 167).

I would argue that it is Belot and Earman's refusal to take seriously the responses that they criticize which lacks a coherent and plausible motivation. In particular, their concern is that philosophers of physics appear to have cut themselves off from what physicists working in the area view as genuine conceptual problems. But the philosophers' defence of substantivalism, and their rejection of the dilemma posed by the hole argument, is not incompatible with taking seriously the concerns of working physicists. On closer scrutiny, and despite the lip service some physicists pay to it, the hole argument, and the debate between substantivalism and relationalism, turns out to have rather little to do with the issues of concern to physicists. To insist on reading the issue of substantivalism into these interpretative questions can only lead to confusion.

Two prime examples of physicists' concerns are (i) the notion of a background independent theory and (ii) whether a theory's observables should be diffeomorphism invariant.

Concerning (i), note that background independence has to do with whether a theory posits non-dynamical, absolute (background) fields, not with whether it sanctions haecceitistic differences. This feature of a theory has nothing to do with the hole argument, as Earman and Norton's application of that argument to 'local spacetime' formulations of background-dependent pre-GR theories illustrates (1987: 517–8).[22] However, it may well be connected to whether the appropriate Hamiltonian formulation of the theory is a constrained Hamiltonian theory for which the diffeomorphism group is a gauge group in a technical sense.[23]

Concerning (ii), the question of whether a theory's observables are diffeomorphism invariant needs further explication.[24] If it is taken to entail that no physical magnitude can take different values at different times (a version of the so-called 'problem of time'), then it is a *stronger* claim than the anti-haecceitistic claim that all diffeomorphic models represent the same physical situation.

Sophisticated substantivalism may be compatible with taking seriously physicists' concerns, but does it have a coherent motivation? The obvious thing to be said for the position is that one thereby avoids the indeterminism of the hole argument. This motivation is, of course, rather ad hoc. A less ad hoc motivation would involve a metaphysics of individual substances that does not sanction haecceitistic differences, perhaps because the individuals are individuated by—their numerical distinctness is grounded by—their positions in a structure. In the next section we will see that Stachel has recently sought to embed his response to the hole argument in exactly this type of more general framework. I hope enough has been said in this section and the previous one to indicate the coherence of such a point of view; it is perhaps a modest structuralism about spacetime points, but it is a far cry from the objectless ontology of the ontic structural realist.

There is one final line of defence of sophisticated substantivalism that needs to be undertaken. One might concede that in principle anti-haecceitism is compatible with spacetime points being substances, but nevertheless believe that the theoretical treatment of spacetime, read literally, strongly supports haecceitism. Is it not the case that the *natural* reading of $\mathcal{M}_1 = \langle M, g, \phi_1, \ldots \phi_{n-2}\rangle$ and $\mathcal{M}_2 = \langle M, d^*g, d^*\phi_1, \ldots d^*\phi_{n-2}\rangle$ interprets each point of M as representing the

[22] Of course, it becomes a moot question whether the fields representing spacetime structure should count as 'non-dynamical' background fields in the context of local spacetime formulations of pre-GR theories. In one sense they are dynamical, since they are held to obey field equations such as $R^a_{bcd} = 0$. In another sense, they are non-dynamical, since they do not vary (except globally) from model to model.

[23] See Earman (2003: 151–3); I hope to return to this topic on a future occasion.

[24] Consider, for example, Smolin's distinction between 'causal observables' and 'Hamiltonian constraint observables' in (2000).

very same point in each model, and therefore interprets the two models as attributing different properties to each point? Moreover, doesn't the mathematics of GR presuppose that the numerical distinctness of the points of M is independent of the properties and relations assigned to them by the fields g, ϕ_i? So if we're being literalistic realists about our theories, shouldn't we take a similar stance towards the individuality of spacetime points?

Something like this line of thought might well be responsible for what resistance there remains to sophisticated substantivalism's combination of anti-haecceitism and realism about spacetime points. But it does not stand up to scrutiny. For a start, it is not obvious that the numerical distinctness of the points of the mathematical object M *is* independent of their properties and relations. We have already had reason to consider structuralist approaches to mathematical objects. According to the mathematical structuralist, the individuality of the points of M *does* depend on their positions in the mathematical structure of which they are part. The mathematical structuralist, of course, needs to be able to give an account of the difference between the two models $\mathcal{M}_1$ and $\mathcal{M}_2$. Here the most obvious strategy is to point out that such a difference can be made out if the two are considered as distinct substructures embedded in a larger structure (cf. Parsons 2004: 68–9). (And if this line is taken, there is no reason, of course, to think that the substantivalist should postulate a counterpart in concrete reality of this larger (unspecified!) structure.)

If one remains attracted to these particular lines of haecceitist argument, a useful question to ask oneself is the following. Suppose, for the sake of argument, that the sophisticated substantivalist is right: individual spacetime points exist as basic objects, but possible spacetimes correspond to equivalence classes of diffeomorphic models of GR. How should the formalism of GR be modified to take account of the anti-haecceitism? (Note that this is not the demand for a *relationalist* reformulation that *does away with* spacetime points.) It should be clear that no such thing is needed. As soon as one is committed to the existence of a set of points with various geometrical properties, even if one is avowedly anti-haecceitistic, the most obvious way of representing such a set will be open to a haecceitistic misinterpretation.

In fact, the haecceitist substantivalist's mistake is a specific instance of a common type. Back in 1967, Kaplan identified the occurrence of essentially the same error in a rather different context:

> the use of models as representatives of possible worlds has become so natural for logicians that they sometimes take seriously what are really only artifacts of the model. In particular, they are led almost unconsciously to adopt a *bare particular metaphysics*. Why? Because the model so nicely separates the bare particular from its clothing. The elements of the universe of discourse of a model have an existence which is quite independent of whatever properties the model happens to tack onto them. (1979: 97)

It seems that the use of mathematical models as representatives of possible worlds has become so natural for some philosophers of physics that they too have been led almost unconsciously to endorse haecceitistic distinctions that are really only artefacts of the model.

4.6 STACHEL'S GENERALIZED HOLE ARGUMENT FOR SETS

According to John Stachel, the moral of the hole argument is that diffeomorphically related mathematical solutions to the field equations of GR (hereafter: *diffeomorphs*) do not represent physically distinct solutions. Although, in the past, he has referred to his position as a relationalist one, it is really, as I use the terms, a version of sophisticated substantivalism. He does not believe that to count diffeomorphs as representing the same physical solution one has to eliminate spacetime points.[25] According to Stachel, one would be forced to view diffeomorphs as representing distinct physical solutions if one took spacetime points to be 'individuated independently of the metric field'. One can maintain the existence of points and count diffeomorphs as representing the same physical solution if one assumes that 'the points of the manifold are *not* individuated independently of the $g_{\mu\nu}$ field; i.e., that these points inherit *all* their chronogeometrical (and inertiogravitational) properties and relations from that field' (Stachel 2002: 233).[26]

So far I am in agreement with Stachel. The talk of the points being 'individuated' is a little obscure. At one point Stachel talks equivalently of an entity being 'distinguishable' from entities of the same kind (p. 236). 'Distinguishable' can be understood in an epistemological or an ontological sense. It must be the latter that is in question, and I suggest that, minimally, it is the entities' determinate numerical distinctness that is at stake (recall questions (ND1) and (ND2) from section 4.3). In claiming that points are not individuated independently of the metric field, Stachel can be understood as claiming that their determinate distinctness from one another is grounded in their standing in the spatio-temporal relations to one another that they do. This in turn is held to prevent our interpreting diffeomorphically related models as representing two situations involving the very

[25] He does refer to fibre bundle formulations of theories in which one 'eliminates' an independently specified base space, replacing it with the quotient space of the total space by the fibres. With the theory so formulated, one cannot permute the fibres of the total space without thereby permuting the points of the base space. But this does not mean, as Stachel claims, that one cannot even generate the models that provide the basis of the original hole dilemma. Two models the cross-sections of which are related by a fibre-preserving diffeomorphism of the total space still represent *mathematically* distinct objects. Further, since such cross-sections are 'differently placed' relative to the fibres, and since the base space is simply defined as the quotient space of the total space by the fibres, these two models represent two distributions of structurally identical fields 'differently placed' on the base space (i.e. spacetime). Of course, Stachel deems all cross-sections related by fibre-preserving diffeomorphisms of the total space to be physically equivalent. I agree. But we can, with as much right, make the analogous claim of the models of the traditional formulation of the theory. Consideration of formulations in terms of fibred manifolds without an independently specified base space gains us nothing.

[26] In my view there is something at least misleading about talk of the points 'inheriting' their properties from this 'field'. It suggests that the field is some entity in its own right (the 'gravitational field'). My preferred view, as explained in the previous section, is that the mathematical metric field is simply a *specification* of the points' (spatio-temporal) properties. It does not represent some*thing* that bestows or engenders the points' properties. It stands to the points as, for example, red stands to red things (however that is!).

same points as occupying different positions in the very same network of spatio-temporal relations.

Now Stachel wishes to situate the hole argument, and its moral concerning the individuation of spacetime points, in a more general framework.[27] Rather than considering only sets of spacetime points, he considers an arbitrary set **S** of n entities. And rather than considering the spatio-temporal properties encoded by $g_{\mu\nu}$, he considers an arbitrary ensemble **R** of n-place relations.[28] Stachel calls the relational structure $\langle \mathbf{S}, \mathbf{R} \rangle$ a 'world'. What can be said about the 'individuation' of the elements of **S**? Stachel suggests that there are two possibilities (within which he draws further divisions, which I will ignore for now):

1. the entities are individuated (that is, [are] distinguishable from other entities of the same kind) prior to and without reference to the relations **R**...
2. the entities are not individuated (that is are indistinguishable among themselves) without reference to the relations **R**. (2002: 236)

Later Stachel goes on to adapt terminology he borrows from Marx, and dubs entities of kind (2) *reflexively defined entities*.

Now diffeomorphisms are simply a special class of permutations of the set of manifold points, those that preserve the manifold's topological and differential structure. And just as one can start with a given model $\langle M, g, \phi_i \rangle$ and consider the mathematically distinct model $\langle M, d^*g, d^*\phi_i \rangle$ generated by the action on the fields induced by a particular diffeomorphism d, one can start with $\langle \mathbf{S}, \mathbf{R} \rangle$ and consider a new structure $\langle \mathbf{S}, \mathbf{PR} \rangle$ generated by the action on the M relations $\mathrm{R}_1, \ldots \mathrm{R}_M$ in **R** induced by an *arbitrary permutation* $\mathrm{P} : \mathbf{S} \to \mathbf{S}$. The definition of $\mathrm{PR}_i \in \mathbf{PR}$ is obvious: (taking an n-place relation to be defined extensively as a set of n-tuples of elements of **S**, i.e. as a subset of $\mathbf{S}^n$) for any $s = \langle \mathrm{s}_1, \ldots, \mathrm{s}_n \rangle \in \mathbf{S}^n, s \in \mathrm{PR}_i$ just if $\mathrm{P}^{-1}s = \langle \mathrm{P}^{-1}(\mathrm{s}_1), \ldots, \mathrm{P}^{-1}(\mathrm{s}_n) \rangle \in \mathrm{R}_i$.

By analogy with the interpretative questions that arise in connection with $\langle M, g, \phi_i \rangle$ and $\langle M, d^*g, d^*\phi_i \rangle$, one might consider whether the structure $\langle \mathbf{S}, \mathbf{R} \rangle$'s being a possible world entailed that $\langle \mathbf{S}, \mathbf{PR} \rangle$ is also a possible world. And, if both are held to be possible worlds, one might consider whether $\langle \mathbf{S}, \mathbf{R} \rangle$ and $\langle \mathbf{S}, \mathbf{PR} \rangle$ should be interpreted as the same, or as distinct possible worlds. Of course, $\langle \mathbf{S}, \mathbf{R} \rangle$ and $\langle \mathbf{S}, \mathbf{PR} \rangle$ will, in general,[29] not be identical, just as $\langle M, g, \phi_i \rangle$ and $\langle M, d^*g, d^*\phi_i \rangle$ are, in general, distinct mathematical entities. If we wish nonetheless to talk of their 'being' the same possible world, then a distinction needs to be drawn between the structures

[27] In what follows I adopt as far as possible Stachel's convention of using boldface type when referring to sets, italic type when referring to n-tuples (and to numbers), and roman type when referring to elements of sets or of n-tuples.

[28] One might consider relations of any adicity. For any adicity $N < n$, Stachel claims that the restriction to n-place relations is no real restriction because there is a natural way of associating an n-place relation with any N-place relation. Assuming I have understood his description correctly, the procedure that Stachel outlines will associate a single n-place relation with some pairs of distinct relations (of different adicity). I do not claim that this conflation causes problems for Stachel.

[29] The qualification concerns the case when *all* of the relations in **R** are symmetric with respect to all permutations of S (i.e. if for any $\mathrm{R}_i, s \in \mathrm{R}_i$ just if $\mathrm{P}s \in \mathrm{R}_i, \forall \mathrm{P}$), a case that will be important in the context of quantum particles.

$\langle \mathbf{S}, \mathbf{R} \rangle$ and $\langle \mathbf{S}, \mathbf{PR} \rangle$, considered as mathematical objects, and the possible worlds they represent. Our second question then becomes whether $\langle \mathbf{S}, \mathbf{R} \rangle$ and $\langle \mathbf{S}, \mathbf{PR} \rangle$ *represent* the same, or different, possible worlds.[30]

Stachel talks of *permutable* and *generally permutable* worlds, theories, and even entities. Although this is not how Stachel defines the terms, I propose the following definitions. The structure $\langle \mathbf{S}, \mathbf{R} \rangle$ is a *permutable world* just if, if it represents a possible world then $\langle \mathbf{S}, \mathbf{PR} \rangle$ also represents a possible world, for every permutation P. The structure $\langle \mathbf{S}, \mathbf{R} \rangle$ is a *generally permutable world* just if, if it represents a possible world then $\langle \mathbf{S}, \mathbf{PR} \rangle$ represents the *same* possible world, for every permutation P. With one minor qualification, I can follow Stachel in his definition of a *theory* and of a *permutable theory*:

> A *theory* is a 'rule that picks out a class of worlds: in other words, a class of ensembles of n-place relations: $\mathbf{R}, \mathbf{R}', \mathbf{R}''$, etc., whose places are filled by the members of the same set **S** of n entities **a**; further, let it be a permutable theory ... [just in case], if **R** is in the selected class of worlds, so is PR for all P. (2002: 244)[31]

Despite their status as the analogues of diffeomorphic models of GR, Stachel nowhere explicitly considers two distinct *structures* such as $\langle \mathbf{S}, \mathbf{R} \rangle$ and $\langle \mathbf{S}, \mathbf{PR} \rangle$. Instead he seeks to define things in terms of expressions of the form '$R_i(a)$ holds' and '$\mathbf{R}(a)$ holds'.[32]

Now the claim that $R_i(a)$ holds is simply the claim that $a \in R_i$ (recall that a is a particular n-tuple and that we are considering only n-place relations). But what does '$\mathbf{R}(a)$ holds' mean? The obvious interpretation is that $a \in R_i$ *for all* $R_i \in \mathbf{R}$.[33] But note that, for generic sets **S** and ensembles of relations **R**, there will be *no* sequence $a \in S^n$ such that $\mathbf{R}(a)$ holds![34]

This might lead one to suspect that Stachel's attempts to define the ensemble of relations **PR**, and his notion of a permutable world, in terms of the expression $\mathbf{R}(a)$ are doomed, and indeed they are. **PR** is said to hold for a if and only if $\mathbf{R}(P^{-1}a)$ holds. Stachel is not explicit about whether this is required for every a, but either way

[30] A way of talking that Stachel quickly slips into.

[31] My reservation concerns talk of a relation's 'places' being 'filled' by the members of some set. If we are conceiving of relations on a given domain in purely extensional terms (cf. Stachel 2002: 237), then strictly it makes no sense to talk of a fixed relation as having places that might be variously filled.

[32] Note that, as is standard practice, Stachel here makes the letter R_i do double duty as a predicate letter (in the expression '$R_i(a)$') and as the name for the relation that is this predicate's extension.

[33] For the record, Stachel's own elucidation is that '$\mathbf{R}(a)$' stands for 'the entire ensemble of relations filled by that sequence [i.e. by the particular n-tuple a]' (2002: 237).

[34] It will fail to hold for all a whenever **R** contains two disjoint relations. Rather unfortunately for Stachel, the pair of binary relations (promoted to continuously infinite relations as per Stachel's recipe) on a set of spacetime points expressed by the predicates 'x is timelike related to y' and 'x is spacelike related to y' is just such an example of disjoint relations. It will also fail even if no relations are disjoint, just so long as, for example the intersection of two relations does not intersect with a third etc. There is another oddity worth noting. Even for structures $\langle \mathbf{S}, \mathbf{R} \rangle$ such that $\mathbf{R}(a)$ (interpreted in this way) holds, $\mathbf{R}(a)$ constitutes an incomplete, and very arbitrary, specification of the structure. We are told that for this particular a, $a \in R_i$ for all $R_i \in \mathbf{R}$, but this tells us next to nothing about **R**. To be given the *complete* state of the world, we need to be told, for *every* n-tuple, and for every R_i individually, whether or not the n-tuple is in the relation.

the definition is not equivalent to the (more standard) definition of **PR** given above. For consider, as an example, $\mathbf{S} = \{s_1, s_2\}$, $\mathbf{R} = \{R_1 = \{\langle s_1, s_1\rangle\}, R_2 = \{\langle s_2, s_2\rangle\}\}$ and $\mathbf{R}' = \{R'_1 = \{\langle s_1, s_1\rangle\}, R'_2 = \{\langle s_1, s_2\rangle\}\}$. Now $\langle \mathbf{S}, \mathbf{R}\rangle$ and $\langle \mathbf{S}, \mathbf{R}'\rangle$ are non-isomorphic structures, but because neither $\mathbf{R}(a)$ nor $\mathbf{R}'(a)$ holds for any ordered pair a of elements of **S**, $\mathbf{R}(a)$ holds iff $\mathbf{R}(P^{-1}a)$ holds.[35]

Stachel's definition of a permutable world is equally problematic. $\langle \mathbf{S}, \mathbf{R}\rangle$ is said to be permutable if, whenever $\mathbf{R}(a)$ is a possible state of the world, then $\mathbf{PR}(a)$ is also a possible state of the world, for every $n!$ permutations P of **a**.[36] In the light of the troubles noted in the last paragraph, let us simply take $\langle \mathbf{S}, \mathbf{PR}\rangle$ to be the isomorphic structure generated by P, in the way outlined above on page 105, and consider again the example $\mathbf{S} = \{s_1, s_2\}$, $\mathbf{R} = \{R_1 = \{\langle s_1, s_1\rangle\}, R_2 = \{\langle s_2, s_2\rangle\}\}$. Intuitively, if $\langle \mathbf{S}, \mathbf{R}\rangle$ is to be permutable, $\langle \mathbf{S}, \mathbf{PR}\rangle$ should be (should represent) a possible world. But since $\mathbf{R}(a)$ and $\mathbf{PR}(a)$ are *not* states of the worlds $\langle \mathbf{S}, \mathbf{R}\rangle$ and $\langle \mathbf{S}, \mathbf{PR}\rangle$ for any a, whether $\langle \mathbf{S}, \mathbf{R}\rangle$ counts as permutable will be independent of whether or not $\langle \mathbf{S}, \mathbf{PR}\rangle$ also counts a possible world. One might hope that by taking the relations to be defined intensively, rather than extensively, sense can be made of these definitions. A little reflection shows that this will not work either.[37]

The ensemble **PR**, then, simply cannot be defined in terms of $\mathbf{R}(a)$ in the way Stachel suggests, however one understands $\mathbf{R}(a)$. In fact, the obvious definition of **PR** in terms of expressions such as $R_i(a)$ is the following:

$$\forall P, \forall R_i \in \mathbf{R} \text{ and } \forall a \in \mathbf{S}^n, \mathrm{PR}_i(a) \text{ iff } R_i(P^{-1}a)$$

That is, it must be given in terms of the individual expressions $R_i(a)$, not via the single expression $\mathbf{R}(a)$. Why does Stachel introduce the expression $\mathbf{R}(a)$ at all? It figures prominently in his set-theoretic version of the hole argument. We are asked to consider the class of ensembles of relations $\mathbf{R}, \mathbf{R}', \mathbf{R}''$ picked out by some permutable theory, and then the question is put 'Could such a permutable theory pick out a

35 This will be true for every pair of ensembles of relations on the same domain such that, for each ensemble, no n-tuple of elements of the domain is a member of every relation; the ensembles do not even have to be equinumerous! The condition '$\mathbf{PR}(a)$ holds iff $\mathbf{R}(P^{-1}a)$ holds' therefore clearly fails to define **PR** in terms of **R**.

36 If there is a possible world in which $\mathbf{R}(a)$ holds, Stachel calls $\mathbf{R}(a)$ a possible state of the world, and if $\mathbf{R}(a)$ holds he calls it a state of the world (2002: 237–8). As Stachel notes, there will only be as many as $n!$ distinct permutations if a is a *non-duplicating* n-tuple. With what right does Stachel consider only *non-duplicating* n-tuples? This is an indication that he is not really interpreting '**R**' in '$\mathbf{R}(a)$' as expressing an ensemble of relations, extensively defined or otherwise.

37 Stachel provides the following example to illustrate his definitions. Our set is {Cat, Cherry} and $R(x_1, x_2)$ holds iff x_1 is on a mat and x_2 is on a tree. With the relation specified in this way, one can perhaps make sense of the same relation 'having its places filled' by different n-tuples, even when the relation does not in fact hold of the n-tuples in question (cf. 2002: 237). But this does not enable us to make sense of $\mathbf{PR}(a)$. For consider extending the example with another relation: $R'(x_1, x_2)$ holds iff x_1 is red and x_2 has claws. And consider the world in which the cat is on a mat (and the cherry is not), the cherry is on a tree (and the cat is not), the cherry is red (and the cat is not) and the cat has claws (and the cherry does not). The obvious interpretation of '**R**(Cat, Cherry) holds' is that the cat is on a mat and the cherry is on a tree *and* that the cat is red and that the cherry has claws, i.e. the claim that **R**(Cat, Cherry) holds is *false* at the specified world, and, let us suppose, at every possible world. And, more generally, it seems that we again have an ensemble **R** of relations for which $\mathbf{R}(a)$ never holds for any ordered pair a of our domain.

unique state of the world by first specifying a unique world, i.e., one **R**; and then specifying how any number m of its places *less than* $(n - 1)$ are filled?' (2002: 244) Not if the entities in question are not reflexively defined, claims Stachel, for then, '$\mathbf{R}(a)$ and $\mathrm{P}\mathbf{R}(a)$ represent different states of the world' (ibid.). But this is just false. Whenever all the relations in **R** are symmetric with respect to all permutations, we get back the identical structure ($\mathbf{R}(a)$ iff $\mathrm{P}\mathbf{R}(a)$), so there can be no question of their representing different worlds. And even setting aside the special case of symmetric ensembles of relations, we again have the unwarranted restriction to non-duplicating n-tuples, for otherwise there is the possibility that $\mathbf{R}(a)$ and $\mathrm{P}\mathbf{R}(a)$ are both states of the same structure because, although $\mathbf{R} \neq \mathrm{P}\mathbf{R}$, $a = \mathrm{P}a$.

I think enough has been said to show that, taken literally, Stachel's description of a set-theoretic hole argument is in terminal trouble. But it is equally clear how to make precise what he must have had in mind. We need the notion of a *complete description* of the structure $\langle \mathbf{S}, \mathbf{R} \rangle$, and also of a *partial* description, for the latter will be the analogue of a *specification* of a metric field on all of a differentiable manifold save for a compact region (the 'hole' of the hole argument).[38] One way of giving such a complete description is the following. We suppose that for every relation $\mathrm{R}_i \in \mathbf{R}$ we have a predicate symbol R_i (I follow Stachel in using the same letter for the predicate and the relation it expresses). And similarly, we suppose that we have a name a_k for every element of **S**. A complete description will be a conjunction of the following three formulas: (i) an $(n^2 \times M)$-place conjunction that includes, for every predicate letter R_i and for every n-place sequence a of the names for elements of **S**, either the formula $\mathrm{R}_i(a)$ or the formula $\neg\mathrm{R}_i(a)$; (ii) an $n!$-place conjunction $(\bigwedge_{i \neq j} \mathrm{a}_i \neq \mathrm{a}_j)$ stating that different names name different elements of **S**; and (iii) an n-place disjunction $(\forall y (\bigvee_{i=1}^{n} y = \mathrm{a}_i))$ stating that there are no elements of **S** other than $a_1, \ldots, a_k$.

Now we may introduce $\mathcal{R}(a)$ as an abbreviation for this conjunction, where a is some particular *non-repeating* n-place sequence of the n names a_k. $\mathcal{R}(\mathrm{P}^{-1}a)$ will then be (an abbreviation of) a complete description of the (mathematically) distinct but isomorphic structure $\langle \mathbf{S}, \mathbf{PR} \rangle$, and $\exists x_1 \ldots \exists x_n \mathcal{R}(x_1, \ldots, x_n)$ will be a structural description true of any structure isomorphic to $\langle \mathbf{S}, \mathbf{R} \rangle$.[39]

I suggest that it is expressions such as $\mathcal{R}(a)$ that Stachel needs for his hole argument for sets. Recall that he considers first specifying a unique world **R**, and then specifying how many any number m of its places *less than* $(n - 1)$ are filled. Given the extensive conception of relation that he began with, to specify **R** is *already* to specify, for every

[38] In what follows I am influenced by the notational conventions of §3 of Belot (2001), which provides an admirably clear and uncluttered account of the symmetries and permutations of abstract structures and our means of describing them in a first-order language. Stachel is at pains to stress the coordinate independence of the hole argument, which might partly account for his reluctance to engage in the necessary, though limited, semantic ascent.

[39] As Belot notes, no first-order theory (i.e. no set of sentences involving only variables and no names) can determine a structure up to isomorphism if n is infinite. Correspondingly, the prescription we are considering, which *does* determine the structure up to isomorphism, yields an infinitary formula, i.e. something that is not well-formed according to standard first-order logic.

n-tuple, whether or not it is a member of each relation in **R**.[40] Instead, what he should have considered was first specifying an isomorphism class of structures via $\exists x_1 \ldots \exists x_n \mathcal{R}(x_1, \ldots, x_n)$, and then specifying how m 'places' in this structure are filled via a formula such as $\exists x_1 \ldots \exists x_{(n-m)} \mathcal{R}(b, x_1, \ldots, x_{(n-m)})$, where b is some particular non-repeating m-place sequence of the names for the elements of **S**.

Let $a = (b, c)$ (a is an n-place sequence of names, b is an m-place sequence of names and c is an $(n-m)$-place sequence of names). If the theory we are considering is a permutable theory, then if the structure that corresponds to the description $\mathcal{R}(a) = \mathcal{R}(b, c)$ is a possible world, so too is the structure corresponding to the description $\mathcal{R}(\mathrm{P}a) = \mathcal{R}(b, \mathrm{P}^{(n-m)}c)$, i.e. we consider a permutation P that acts non-trivially only on the last $(n-m)$ members of the sequence a. If the elements of **S** are not reflexively defined (i.e. if these entities are individuated independently of the ensemble of relations **R**, i.e. if $\langle \mathbf{S}, \mathbf{R} \rangle$ and $\langle \mathbf{S}, \mathbf{P}^{-1}\mathbf{R} \rangle$ represent distinct possible worlds), then specifying only $\exists x_1 \ldots \exists x_{(n-m)} \mathcal{R}(b, x_1, \ldots, x_{(n-m)})$ fails to pick out a unique world, for it is compatible with both $\mathcal{R}(a)$ and $\mathcal{R}(\mathrm{P}a)$, which both describe (distinct) structures allowed by the theory (namely, $\langle \mathbf{S}, \mathbf{R} \rangle$ and $\langle \mathbf{S}, \mathbf{P}^{-1}\mathbf{R} \rangle$ respectively), structures that represent distinct possible worlds. On the other hand, if the elements of **S** *are* reflexively defined (if they are 'generally permutable entities' that are *not* individuated independently of the ensemble of relations), then specifying only $\exists x_1 \ldots \exists x_{(n-m)} \mathcal{R}(b, x_1, \ldots, x_{(n-m)})$ *is* sufficient to pick out a unique world because, although compatible with both $\mathcal{R}(a)$ and $\mathcal{R}(\mathrm{P}a)$, these describe only formally distinct, isomorphic structures which represent the same world. It is surely in this latter case that Stachel intends to apply the label 'generally permutable' to the theory, and to the worlds it picks out. This, I submit, is the proper set-theoretic analogue of the hole argument.[41]

4.7 STACHEL ON IDENTICAL PARTICLES

It is finally time to consider Stachel's application of all of this to quantum particles. I quote at length:

> One can immediate apply this result to the current discussion about the individuality of elementary particles ... One group maintains that each elementary particle retains its individuality, and that quantum statistics are merely the result of the fact that certain states ... that are accessible to systems of elementary particles that are not of the same kind, are for some reason inaccessible to systems of particles that are all of the same kind. The other group maintains that quantum statistics has its origin in the lack of individuality of elementary

[40] And to give only an intensive definition of an ensemble of relations will fail to pick out a particular isomorphism class of structures at all.

[41] A specification of the metric on all of a manifold except a 'hole' (once one has solved the equations and determined the metric in the hole up to isomorphism) corresponds, in effect, to the formula $\exists x_1 \ldots \exists x_{(n-m)} \mathcal{R}(b, x_1, \ldots, x_{(n-m)})$; i.e., one has specified an equivalence class of diffeomorphic solutions to Einstein's field equations and has further specified, for all but the hole, which manifold points have which spatio-temporal properties.

particles. As far as I know, no one has ... mentioned the possibility of extending the hole argument from the discussion of the individuality of space-time points to the discussion of the individuality of elementary particles, as I shall now do.

If we take the points of our set to represent *n* elementary particles of the same kind, then quantum-mechanical statistics imposes the requirement that all physical relations between them be permutable. Our set theoretical hole argument shows that, if we ascribe an individuality to the particles that is independent of the ensemble of permutable relations, then no model can be uniquely specified by giving all the *n*-place relations **R** between them unless we further specify *which* particle occupies *each* place in these relations ... (2002: 245)

In a footnote he expands on what he means by 'the requirement that all physical relations between [the particles] be permutable':

The relations will represent values of physical properties of the system of identical particles, which must remain invariant under all permutations of the particle labels. Since these physical properties are represented by bilinear functions of the state vector of the system, they will remain invariant *whether the state vector remains invariant under a permutation (bosons) or changes sign (fermions)* ... (ibid. 261; my emphasis)

The obvious interpretation of the claim that the *relations* between the particles are themselves permutable is that all permutations of the set $\mathbf{S}$ are *symmetries* of these relations. That is, for every R_i, $a \in \mathbf{S}^n$ and permutation P, $a \in \mathrm{R}_i$ iff $\mathrm{P}a \in \mathrm{R}_i$. In this special case we have $\langle \mathbf{S}, \mathbf{R} \rangle = \langle \mathbf{S}, \mathbf{PR} \rangle$; there just can be no question of the two structures representing distinct worlds because we do not have two structures: there is only one. Such a situation does indeed correspond to what we find in the case of the quantum mechanics of identical bosons and fermions. The quantum states of such systems are required to be either symmetrized or antisymmetrized. That is, for an arbitrary permutation of the particle labels, one gets back the very same state (up to a phase factor of -1 in the case of fermions, if the permutation is odd). Stachel's footnote suggests that he is indeed considering such symmetrized and antisymmetrized states. His characterization of the group who maintain that elementary particles retain their individuality but that 'certain states ... that are accessible to systems of elementary particles that are not of the same kind, are for some reason inaccessible to systems of particles that are all of the same kind' also strongly suggests that he has (anti)symmetrized states in mind. Moreover, when one talks of the 'permutation invariance' of the quantum mechanics of identical particles, one is typically referring to the fact that the *physically allowed states themselves* are permutation invariant (up to a phase).[42]

But now consider how this situation plays out in the context of the set-theoretic hole argument. First consider Stachel's 'states of the world' $\mathbf{R}(a)$ and $\mathbf{R}(\mathrm{P}a)$. These

[42] Strictly, this corresponds to the quantum mechanics of identical particles satisfying the *symmetrization postulate*. Sometimes 'permutation invariance' is used to refer the strictly weaker requirement that the expectation values of all *physical* observables be permutation invariant (see French and Rickles 2003: §§2–3). Note that (i) the more general possibilities allowed by the weaker interpretation do not appear to be realized in nature and that (ii) to require the permutation invariance of *states* (up to a possible phase factor of -1) is to impose the symmetrization postulate.

are indeed distinct states of the world, but for any world $\langle \mathbf{S}, \mathbf{R} \rangle$ allowed by quantum mechanics, if $\mathbf{R}(a)$ is a state of this world, then $\mathbf{R}(\mathrm{P}a)$ is also a state of the very same world. If the relations themselves are permutable then $\mathbf{R}(a)$ and $\mathbf{R}(\mathrm{P}a)$ are either both true, or both false. Consider, instead, the complete description $\mathcal{R}(a)$ introduced above. If $\mathcal{R}(a)$ is a description of a world allowed by quantum mechanics, then $\mathcal{R}(\mathrm{P}a)$ will be a description of *exactly the same world* (it will be a logically equivalent formula). Thus stipulating that $\exists x_1 \ldots \exists x_{(n-m)} \mathcal{R}(b, x_1, \ldots, x_{(n-m)})$ *does* suffice to pick out a unique world, *even if one believes that the individuality of quantum particles transcends the relational structure in which they are embedded.*

In fact, Stachel does not always talk as if the ensemble of relations in question is itself permutable (i.e. that it corresponds to a symmetric or antisymmetric state). Before applying the hole argument to quantum particles, he writes:

Some suggest that, if elementary particles are not individuated, then any attempt to label them is misguided. On the contrary, it is just an example of the usual method of coordinatization, introduced when treating any set of entities that are numerous, yet indistinguishable ... What is important to realize is that, in all such cases, no one coordinatization (labelling in this case) is preferred over another; and that it is precisely invariance of all relations under all permutations of the labels that guarantees this. It is entirely indifferent which six electrons out of the universe make up a particular carbon atom[.] They are individuated, as K-shell or L-shell electrons of the atom, for example, entirely by the ensemble of their relations to the carbon nucleus of the atom and to each other. Indeed, the notation for the electronic structure of an atom is based on this type of individuation. (2002: 243)

Here, again, Stachel talks of the 'invariance of all relations under all permutations of the labels'. However, he also claims that one can think of, for example, a K-shell electron as being individuated by the ensemble of the relations that hold between the electrons of the atom, and between the electrons and their nucleus. In a footnote he elaborates: 'as a result of the Pauli exclusion principle for fermions, each electron in an atom can be fully individuated by the set of its quantum numbers' (2002: 260). And later he claims that his set-theoretic hole argument shows that:

if we ascribe an individuality to the particles that is independent of the ensemble of permutable relations, then no model can be uniquely specified by giving all the n-place relations $\mathbf{R}$ between them unless we further specify just *which* particles occupies *each* place in the relations. For example, the rules for filling atomic shells in the ground state of an atom with electrons would have to be regarded as radically incomplete, since they do not tell us *which* electron has the different quantum numbers that characterize that state. (2002: 245–6)

At this point, we should distinguish two, quite distinct, ways of describing, for example, the electrons in a particular atom. First we might give its state as an antisymmetrized vector in a tensor product Hilbert space. It is this description that corresponds to an ensemble of *permutable* relations. And, despite what Stachel appears to suggest, one *cannot* think of the labels that feature in this description as (arbitrary coordinate) labels for electrons with particular quantum numbers, individuated by the ensemble of relations. First, precisely because they are permutable, the relations fail to individuate in the way Stachel appears to imply: every place in the ensemble of relations is exactly like every other place.[43] Secondly, and relatedly, in this formalism, each electron, i.e. the entities associated with *each* label, enters equally into the state associated with a particular set of quantum numbers, for example, those corresponding to a K-shell electron with a particular spin.[44] It is for this reason that it is held by some that 'attempts to label the electrons are misguided'. It is just hard to believe that the labels in the tensor product formalism are genuine labels (cf. Teller 1998: §5); they are certainly not an example of the 'usual method of coordinatization'.

The alternative description involves Fock space and the occupation number formalism (see e.g. van Fraassen 1991: 438–48). Here no particle labels are employed at all. Rather the formalism uses *occupation numbers*, 'numbers describing how many times each maximal property is instantiated, with no regard to "which" particle has which of the properties' (Teller 1998: 128). Such descriptions correspond to formulae of the type $\exists x_1 \ldots \exists x_n \mathcal{R}(x_1, \ldots, x_n)$ involving no names and only variables. Note that the $\mathcal{R}$ occurring in *this* description will not abbreviate the same complex relation as the $\mathcal{R}$ that occurs in a description of the permutable structure corresponding to a symmetrized state vector. The $\mathcal{R}$s of the latter type of description have every possible symmetry: $\mathcal{R}(a)$ and $\mathcal{R}(\mathrm{P}a)$ are logically equivalent formulae for every P. In contrast, the $\mathcal{R}$ of an occupation number description will have *no* symmetries: each place must correspond to a *different* maximal property.

If we now try to run the hole argument for an occupation number description we run into trouble, for there simply is nothing in the formalism that corresponds to further specifying which entities occupy which occupied states, for example, the various atomic shells in an atom. As soon as we do introduce things that formally look like such names—the labels of the tensor product Hilbert space formalism—we symmetrize, so that every label is associated with every occupied atomic shell. Either way, there appears to be no analogue of the hole argument for quantum particles. Stachel appears to conflate the two descriptions: on the one hand he talks about labels, and permutable relations (suggesting the labelled tensor product Hilbert space formalism), on the other hand he talks about different electrons being individuated by different sets of quantum numbers, in conformity with the Pauli exclusion principle (suggesting an occupation number description).

I hope to have made it clear that the diffeomorphism invariance of GR, and the permutation invariance of quantum mechanics are very different. In the first case, the

[43] This is not say that such relations cannot be held to individuate; although the relations are permutable, they may represent, e.g., symmetric yet irreflexive relations, which is enough to force the numerosity of the domain if one adopts the Hilbert and Bernays's definition of identity (see Saunders 2003: esp., 294–5). We can even think of *bosons* as being individuated by such relations when their symmetrized entangled state does not involve more than one copy of any single-particle state in each element of the superposition.

[44] These points are related to the claim that all identical particles, fermions just as much as bosons, violate the identity of indiscernibles, in every physically possible state (French and Redhead 1988). The claim is only true if the labels of the tensor product Hilbert space formalism are interpreted as genuine labels.

diffeomorphism invariance is a symmetry of the *theory*.[45] If $\langle M, g, \phi_i \rangle$ is a solution to the theory, then so is (the mathematically distinct) $\langle M, d^*g, d^*\phi_i \rangle$. But whether or not we take the interpretative step of regarding these two models as representing the same world, arbitrary diffeomorphisms are not symmetries of the *worlds* they represent (except in special cases where the metric has Killing vectors, in which case a small *subgroup* of the diffeomorphism group will be a symmetry group of the solution). Contrast this with the case of quantum mechanics. Starting with the tensor product Hilbert space formalism, permutations of particle labels *are* symmetries of the theory. For example, if Ψ_{12} is a physically possible state of two identical particles, then so is Ψ_{21}. But this is *because* $\Psi_{12} = (-)\Psi_{21}$. Permutations are symmetries of every *solution* of the theory, and that is how and why they are also symmetries of the theory. And if we consider instead the Fock space formalism, there simply are no particle labels to be permuted.

I conclude that the diffeomorphism invariance of GR and the permutation invariance of quantum mechanics are not formally analogous, and do not generate the same interpretative problems concerning the individuation of the putative subject matter of the theories. This is not, of course, to say that there are no similarities. To treat the points of the manifold of a solution of Einstein's field equations as akin to variables rather than names (a stance Maudlin (1989) has dubbed 'Ramseyfying substantivalism'), would be to regard the models of GR as akin to a Fock space description; it involves regarding diffeomorphic models as akin to syntactically distinct yet logically equivalent formulae. Conversely, if one can really think of the states associated with the occupation numbers of the Fock space formalism as *genuinely occupied by objects*, objects that are individuated by the properties attributed to them by these states (cf. Stachel's talk of electrons in an atom being individuated by their different quantum numbers), then what is to stop us naming them? Such names would not correspond to the labels of any quantum formalism, but they would correspond to the informal talk of physicists, who are happy to talk of *the* K-shell electron, for example, (assuming there is only one) or of *the* particle in the left-hand wing of the EPR apparatus.

In the next section I consider briefly whether this talk is really permissible. Before turning to that final topic, let us briefly consider French and Rickles's assessment of Stachel's analogy between points and particles. They write:

> Stachel ... understands the non-individuality of particles as their being individuated 'entirely in terms of the relational structures in which they are embedded' ... But then it is not clear what metaphysical work the notion of 'non-individuality' is doing, when we still have 'objects' which are represented by standard set theory (and this is precisely the criticism that can be levelled against attempts to import non-individuality into the spacetime context) ...
>
> Again the alternative, 'middle way' is to drop objects out of the ontology entirely, regarding both spacetime and particles in structural terms. Indeed, this appears to be the more appropriate way of understanding both Stachel's talk of individuating objects 'entirely in terms of relational

[45] The distinction between symmetries of theories, and of worlds, is discussed by Belot (2003: §4.2) and Ismael and van Fraassen (2003: 378); I am grateful to Paul Mainwood for emphasizing it to me.

structures in which they are embedded' ... However, rather than thinking of objects being individuated, we suggest they should be thought of as being structurally constituted in the first place. In other words, it is relational structures which are regarded as metaphysically primary and the objects as secondary or 'emergent'. (2003: 235)

In light of this section and the previous ones, the response I advocate to these suggestions should be clear. Why should non-individuality do any more work than (be anything more than) the denial of *primitive* individuality and haecceitistic differences? We have yet to be given a reason to think that standard set theory should *not* apply, at least to spacetime points.

4.8 IDENTICAL PARTICLES AND IDENTITY OVER TIME

A simple story is often retold in elementary discussion of quantum statistics. Suppose that we have two identical particles, a and b, and just two possible single-particle states L and R. We are told that if one 'thinks classically', one should expect four distinct states for the joint system:

1. $L(a)L(b)$
2. $L(a)R(b)$
3. $R(a)L(b)$
4. $R(a)R(b)$

And if we further suppose that each of these states is equally probable, then we get an instance of Maxwell–Boltzmann statistics: the possibility according to which one particle is in state L while one is in state R is twice as likely as each of the two possibilities according to which the particles are in the same state.

But quantum particles obey either Fermi–Dirac or Bose–Einstein rather than Maxwell–Boltzmann statistics. For example, in the Bose–Einstein case, the possibilities:

1. $L(2)R(0)$: two particles are in state L
2. $L(1)R(1)$: one particle is in state L and one particle is in state R
3. $L(0)R(2)$: two particles are in state R

are all equally likely. It seems that it is the supposition that $L(a)R(b)$ and $R(a)L(b)$ are distinct possibilities that led us to the wrong, classical, statistics. But isn't equating the possibilities $L(a)R(b)$ and $R(a)L(b)$ simply anti-haecceitism? It seems that the non-existence of haecceitistic differences between states involving identical quantum particles suffices to explain quantum statistics. Perhaps quantum statistics recommends exactly the same interpretative move as the hole argument in general relativity after all.

Unfortunately, things are not so simple. First, anti-haecceitism is compatible with Maxwell–Boltzmann statistics, as is shown by Huggett (1999). What is required, if one is to obtain such statistics while denying haecceitistic differences, is, crudely put, a continuum of possible microstates relative to a countable number of particles (ibid., §IV). It might be held that this result is just as well, for some see the Gibbs Paradox as motivating the denial of haecceitistic differences even in classical statistical mechanics (see e.g. Saunders 2003: 302), although this remains a matter of controversy. The difference between quantum statistical systems, and classical statistical systems, is then to be seen as arising precisely because, in the quantum case, one does not have a continuum of microstates.

I agree with this as far as it goes, but it seems that a puzzle remains. Why is the symmetrization postulate imposed; i.e. why, rather than simply stipulating that $L(a)R(b)$ and $R(a)L(b)$ represent the same state, do we take the appropriate quantum state to be $(1/\sqrt{2})\,(L(a)R(b) + R(a)L(b))$?

The answer has to do with a distinction drawn in section 4.3. Statistics are manifest *over time* in frequencies. To regard the states $L(a)R(b)$ and $R(a)L(b)$ as distinct in the context of the *persisting* particles a and b is not necessarily to sign up to haecceitistic differences. A single solution might involve the instantaneous state $L(1)R(1)$ at two different times t_1 and t_2. If it makes sense to ask whether the particle that occupies state L at t_1 is the same as the particle that occupies the state L at t_2, then we have a legitimate, *non-haecceitistic* reason for distinguishing between the instantaneous states $L(a)R(b)$ and $R(a)L(b)$. According to a point of view that goes back at least to Reichenbach (1956), the difference between classical and quantum statistics bears on such questions concerning *identity over time*, rather than haecceitism.

As a very simple example, consider our two one-particle states L and R, which, let us suppose, at regular time intervals, $t_1, t_2, \ldots$, are instantiated by two 'particles'.[46] Consider the two cases where the frequencies exhibited over time correspond to (i) Maxwell–Boltzmann statistics and (ii) Bose–Einstein statistics. The frequencies of scenario (i) can be explained in terms of a very simple dynamical model. It involves two persisting particles such that (a) at each time the probability of each particle occupying each state is $\frac{1}{2}$ and (b) the likelihood of their occupying each state at each time is independent of which state the other particle occupies at that time. Scenario (ii), on the other hand, is most simply explained by postulating that at each time the three possible instantaneous states $L(2)R(0)$, $L(0)R(2)$, and $L(1)R(1)$ are equally likely; there is no additional fact of the matter concerning whether the 'particle stage' that instantiates L at one time constitutes a stage of the same persisting particle as the particle stage that instantiates L at some other time. Of course, one *can* combine such additional facts about persistence with the Bose–Einstein statistics of scenario (ii). But if one does, the dynamics of the two particles can no longer be independent of each other but must involve 'causal anomalies' if the correct frequencies are to be recovered (Reichenbach 1956: 69–71 (in Castellani edn.)).[47]

[46] I am indebted, at this point, to a conversation with Nick Huggett.

[47] There is a rather interesting application of this type of example to the debate between perdurantists and endurantists. In response to the scenarios involving rotating homogeneous matter that endurantists press against perdurantists, Sider (2001: 224–36) has suggested that one might exploit the Mill–Ramsey–Lewis account of laws to pick out a preferred genidentity relation in favourable cases. (For a comprehensive discussion of arguments concerning rotating homogeneous matter in the context of the perdurantist–endurantist debate, see Butterfield 2004.) Now we can envisage two spatio-temporal Humean mosaics for which the Mill–Ramsey–Lewis prescription

The previous paragraph suggests the following possibility. Perhaps one can view the instantaneous *temporal stages* of quantum particles as genuine individuals, individuated by their sets of quantum numbers. The only problem with introducing particle labels as the names of such objects is that they illicitly introduce primitive *trans-temporal* identities between the particles that exist at one moment and those that exist at another. We are only forced to (anti)symmetrize the state in order to 'rub out' these illegitimate trans-temporal identities. The entities that exist at any given instant are not really to be thought of as each in an identical mixed state (it is not the case that every electron is currently equally a part of me, part of you, and part of this page of the paper you are reading). We would then, again, have a strong analogy between spacetime points and the fundamental ontology of identical particle quantum mechanics; both would be instantaneous entities fully individuated by their properties and relations.

The problem with this suggestion is that the non-commutative algebra of observables prevents our interpreting even the instantaneous ontology of quantum mechanics as a determinate set of reflexively defined individuals. The difficulty is that the choice of particular maximal properties to characterize the quantum particles is to a large extent arbitrary. Stachel states that 'each electron in an atom can be fully individuated by the set of its quantum numbers' (2002: 260). If these quantum numbers are to fully individuate, then the electron's component of spin in some direction must be included, conventionally the 'z'-direction. If such a set of properties really did individuate, we should be able to talk about, for example, *the* S-shell electron whose spin in the z-direction is $+\hbar/2$. But of course, we cannot, for if we could, symmetry would require that we could also talk about the S-shell electron whose spin in the x-direction is $+\hbar/2$. And we could then ask whether the S-shell electron whose spin in the z-direction is $+\hbar/2$ was the same electron as the S-shell electron whose spin in the x-direction is $+\hbar/2$. But this last question is illegitimate. There being a fact of the matter about its answer would contravene quantum mechanics' violation of Bell's inequalities. Unlike spacetime points in classical general relativity, quantum particles cannot be thought of as individuated by the relational structures in which they are imbricated. They are not even reflexively defined entities.

4.9 TWO MORALS FOR QUANTUM GRAVITY

In the previous two sections two conclusions were reached concerning identical particles in quantum mechanics. The first was that, since the *states* of identical particle quantum mechanics were permutation invariant, there could be no analogue of the hole argument that involved them. To run an analogue of the hole argument

yields probabilistic laws involving Maxwell–Boltzman and Bose–Einstein statistics respectively. As described in the paragraph above, the former favours a law formulated in terms of persisting particulars. However, there will be many ways of drawing the lines of persistence consistent with the statistics. Here is a case, then, where the simplest law favours introducing a genidentity relation, but fails to determine which particle stages should be regarded as genidentical.

one needs solutions of a permutation-invariant *theory* that are not themselves permutation invariant and which are thus interpretable, at least in principle, as representing physically distinct (although only haecceitistically distinct) states of affairs. The second, more tentative, conclusion, was that if a theory involves a non-commutative algebra of observables, then there is at least a prima facie problem facing those who would interpret the ontology of the theory as involving a single, determinate set of reflexively defined entities.

Both of these conclusions would appear to be applicable to loop quantum gravity (LQG), the first straightforwardly so. The *states* of loop quantum gravity satisfy the so-called diffeomorphism constraint. This means that they are (3-)diffeomorphism invariant: the states do not distinguish the points of the 3-manifold in terms of which they are, notionally, defined. In LQG the points of the spatial 3-manifold have a status exactly analogous to particle labels in identical particle quantum mechanics.[48]

The space of states that satisfy both the Gauss constraint and the diffeomorphism constraint is spanned by a basis of states that are labelled by *abstract* spin networks, or knots, where a knot is an equivalence class of graphs embedded in a manifold under diffeomorphisms. It is to these states (often, and somewhat confusingly, also referred to simply as spin-network states) that popular accounts of LQG typically refer (see Rovelli 2001: 110–11). The nodes of the graph can be thought of as quanta of volume—as elementary chunks of space—and the links as quanta of area separating these volumes. This picture suggests the following thought: might we regard the links and nodes of abstract spin networks as representing genuine entities (i.e. elementary volumes and surfaces of space), entities that are reflexively defined by the network of relations in which they stand?

The obvious worry with this proposal concerns the second conclusion mentioned above. The spin-network basis is just one basis for the space of states that satisfy the Gauss constraint. Other possible bases will provide us with a set of states that are not interpretable as networks of volumes and areas (the volume and area operators will not be diagonalized by these other bases). If non-commuting observables do not allow quantum particles to be straightforwardly interpreted as reflexively defined objects, the same will be true of the elementary quanta of loop quantum gravity.

ACKNOWLEDGEMENTS

Previous incarnations of this chapter were presented at the University of Leeds, the Universidad Autonoma de Barcelona, the LSE, the University of Oxford, and the University of Illinois at Chicago. I am grateful to members of all those

[48] Dean Rickles (2005) suggests that an analogue of the hole argument can be constructed in the context of LQG. The states involved in his construction are spin-network states that solve the Gauss constraint but which do not solve the diffeomorphism constraint (or the Hamiltonian constraint). They are therefore *not* solutions of the (quantum) Einstein equations. To claim, therefore, that 'Einstein's equation cannot determine where spin-networks are in the manifold' is misleading. Einstein's equation (or rather its quantum version) determines exactly where on the manifold a spin network is: it is smeared all over the manifold in a diffeomorphic-invariant fashion. See Pooley (2006).

audiences, in particular to Joe Melia, Carl Hoefer, Mauricio Suarez, James Ladyman, Michael Redhead, Nick Huggett, and Steve Savitt, for comments and criticism. I am also grateful to Brian Leftow, Justin Pniower, Harvey Brown, Simon Saunders, Jan Westerhoff, and Paul Mainwood for discussions about related matters, to an anonymous referee, and to the editors of this volume, not least for their patience. I am especially grateful to Jeremy Butterfield for extensive and careful comments on a draft of this chapter, which led to numerous improvements and which should have led to more. This chapter was composed during the tenure of a British Academy Postdoctoral Fellowship; I gratefully acknowledge the support of the British Academy.

REFERENCES

Anandan, J., and H. R. Brown (1995) "On the Reality of Space-time Geometry and the Wavefunction". *Foundations of Physics*, 25: 349–60.

Belot, G. (2001) "The Principle of Sufficient Reason". *Journal of Philosophy*, 118: 55–74.

—— (2003) "Notes on Symmetries". In K. Brading and E. Castellani (eds.), *Symmetries in Physics: Philosophical Reflections*. Cambridge: Cambridge University Press (pp. 393–412).

—— and J. Earman (1999) "From Metaphysics to Physics". In J. Butterfield and C. Pagonis (eds.), *From Physics to Philosophy*. Cambridge: Cambridge University Press (pp. 166–86).

—— —— (2001) "Pre-Socratic Quantum Gravity". In C. Callender and N. Huggett (eds.), *Physics Meets Philosophy at the Planck Scale*. Cambridge: Cambridge University Press (pp. 213–55).

Benacerraf, P. (1965) "What Numbers Could Not Be". *Philosophical Review*, 74: 47–73.

Brighouse, C. (1994) "Spacetime and Holes". In D. Hull, M. Forbes, and R. Burian (eds.), *Proceedings of the 1994 Biennial Meeting of the Philosophy of Science Association*, vol. i. East Lansing, MI: Philosophy of Science Association (pp. 117–25).

Butterfield, J. N. (1989) "The Hole Truth". *British Journal for the Philosophy of Science*, 40: 1–28.

—— (2004) "On the Persistence of Homogeneous Matter". Available at **http://philsci-archive.pitt.edu/archive/00002381/**.

Dalla Chiara, M. L., R. Giuntini, and D. Krause (1998) "Quasiset Theories for Microobjects: A Comparison". In E. Castellani (ed.), *Interpreting Bodies*. Princeton: Princeton University Press (pp. 142–52).

Demopoulos, W., and M. Friedman (1985) "Bertrand Russell's *The Analysis of Matter*: Its Historical Context and Contemporary Interest". *Philosophy of Science*, 52: 621–39.

Earman, J. (1989) *World Enough and Space-Time: Absolute versus Relational Theories of Space and Time* Cambridge, MA: MIT Press.

—— (2003) "Tracking down Gauge: An Ode to the Constrained Hamiltonian Formalism". In K. Brading and E. Castellani (eds.), *Symmetries in Physics: Philosophical Reflections*. Cambridge: Cambridge University Press (pp. 140–62).

—— and J. Norton (1987) "What Price Substantivalism? The Hole Story". *British Journal for the Philosophy of Science*, 38: 515–25.

English, J. (1973) "Underdetermination: Craig and Ramsey". *Journal of Philosophy*, 70: 453–62.

French, S. (1989) "Identity and Individuality in Classical and Quantum Physics". *Australasian Journal of Philosophy*, 67: 432–46.

—— (1998) "On the Whithering away of Physical Objects". In E. Castellani (ed.), *Interpreting Bodies*. Princeton: Princeton University Press (pp. 93–113).

____ and J. Ladyman (2003) "Remodelling Structural Realism: Quantum Physics and the Metaphysics of Structure". *Synthese*, 136: 31–56.

____ and M. L. G. Redhead (1988) "Quantum Physics and the Identity of Indiscernibles". *British Journal for the Philosophy of Science*, 39: 233–46.

____ and D. Rickles (2003) "Understanding Permutation Symmetry". In K. Brading and E. Castellani (eds.), *Symmetries in Physics: Philosophical Reflections*. Cambridge: Cambridge University Press (pp. 212–38).

Hilbert, D., and P. Bernays (1934) *Grundlagen der Mathematik*, vol. i. Berlin: Springer-Verlag.

Hoefer, C. (1996) "The Metaphysics of Space-time Substantivalism". *Journal of Philosophy*, 93: 5–27.

____ (2000) "Energy conservation in GTR". *Studies in History and Philosophy of Modern Physics*, 31: 187–99.

Huggett, N. (1999) "Atomic Metaphysics". *Journal of Philosophy*, 96: 5–24.

Ismael, J., and B. C. van Fraassen (2003) "Symmetry as a Guide to Superfluous Theoretical Structure". In K. Brading and E. Castellani (eds.), *Symmetries in Physics: Philosophical Reflections*. Cambridge: Cambridge University Press (pp. 371–92).

Jones, R. (1991) "Realism about What?" *Philosophy of Science*, 58: 185–202.

Kaplan, D. (1975) "How to Russell a Frege–Church". *Journal of Philosophy*, 72: 716–29.

____ (1979) "Transworld Heir Lines". In M. J. Loux (ed.), *The Possible and the Actual*. Ithaca, NY: Cornell University Press (pp. 88–109).

Ketland, J. (2004) "Empirical Adequacy and Ramsification". *British Journal for the Philosophy of Science*, 55: 287–300.

Krause, D. (1992) "On a Quasi-set Theory'. *Notre Dame Journal of Formal Logic*, 33: 402–11.

Ladyman, J. (1998) "What is Structural Realism?" *Studies in the History and Philosophy of Science*, 29: 409–24.

Laudan, L. (1981) "A Confutation of Convergent Realism". *Philosophy of Science*, 48: 19–48.

Lewis, D. K. (1986) *On the Plurality of Worlds*. Oxford: Basil Blackwell.

Maudlin, T. (1989) "The Essence of Space-time". In A. Fine and J. Leplin (eds.), *Proceedings of the 1988 Biennial Meeting of the Philosophy of Science Association*, vol. ii. East Lansing, MI: Philosophy of Science Association (pp. 82–91).

____ (1990) "Substances and Space-time: What Aristotle would have said to Einstein". *Studies in History and Philosophy of Science*, 21: 531–61.

Muller, F. A. (1997a) "The Equivalence Myth of Quantum Mechanics—Part I". *Studies in History and Philosophy of Modern Physics*, 28: 35–61.

____ (1997b) "The Equivalence Myth of Quantum Mechanics—Part II". *Studies in History and Philosophy of Modern Physics*, 28: 219–47.

Newman, M. H. A. (1928) "Mr Russell's Causal Theory of Perception". *Mind*, 37: 137–48.

Norton, J. D. (2000) "What can we Learn about the Ontology of Space and Time from the Theory of Relativity?" **http://philsci-archive.pitt.edu/archive/00000138/**. Paper presented at the International Ontology Congress, San Sebastian, Spain, October 2000.

Parsons, C. (2004) "Structuralism and Metaphysics". *Philosophical Quarterly*, 54: 56–77.

Pooley, O. (2006) "A Hole Revolution, or Are We Back Where We Started?" in *Studies in History and Philosophy of Modern Physics* 37: 372–80.

____ (in preparation) "What's so Special about General Relativity? Some Skeptical Remarks about the Status of General Covariance". Paper presented at the ESF Conference on Philosophical and Foundational Issues in Spacetime Theories, Oxford, March 2004.

Quine, W. V. (1986) *Philosophy of Logic* (2nd edition). Cambridge, MA: Harvard University Press.

Reichenbach, H. (1956) *The Direction of Time*. Berkeley and Los Angeles: University of California Press.

Rickles, D. P. (2005) "A New Spin on the Hole Argument". *Studies in the History and Philosophy of Modern Physics*, 36: 415–34.

Rovelli, C. (1997) "Halfway through the Woods: Contemporary Research on Space and Time". In J. Earman and J. Norton (eds.), *The Cosmos of Science*. Pittsburgh: University of Pittsburgh Press. (pp. 180–223).

——(2001). "Quantum Spacetime: What do we Know?" In C. Callender and N. Huggett (eds.), *Physics Meets Philosophy at the Planck Scale*. Cambridge: Cambridge University Press (pp. 101–22).

Rynasiewicz, R. A. (1994) "The Lessons of the Hole Argument". *British Journal for the Philosophy of Science*, 45: 407–36.

Saunders, S. W. (2002) "Indiscernibles, General Covariance, and Other Symmetries". In A. Ashtekar, D. Howard, J. Renn, S. Sarkar, and A. Shimony (eds.), *Revisiting the Foundations of Relativistic Physics: Festschrift in Honour of John Stachel*. Dordrecht: Kluwer. Preprint available: **http://philsci-archive.pitt.edu/archive/00000459**.

——(2003) "Physics and Leibniz's Principles". In K. Brading and E. Castellani (eds.), *Symmetries in Physics: Philosophical Reflections*. Cambridge: Cambridge University Press. (pp. 289–307).

Shapiro, S. (1997) *Philosophy of Mathematics: Structure and Ontology*. Oxford: Oxford University Press.

Sider, T. (2001) *Four Dimensionalism: An Ontology of Persistence and Time*. Oxford: Oxford University Press.

Smolin, L. (2000) "The Present Moment in Quantum Cosmology: Challenges to the Arguments for the Elimination of Time". In R. Durie (ed.), *Time and the Instant*. Manchester: Clinamen Press (pp. 112–43).

Stachel, J. (2002) '"The Relations between Things" Versus "The Things between Relations": The Deeper Meaning of the Hole Argument'. In D. B. Malament (ed.), *Reading Natural Philosophy: Essays in the History and Philosophy of Science and Mathematics*. Chicago: Open Court (pp. 231–66).

Teller, P. (1998) "Quantum Mechanics and Haecceities". In E. Castellani (ed.), *Interpreting Bodies*. Princeton: Princeton University Press (pp. 114–141).

——(2001) "The Ins and Outs of Counterfactual Switching". *Noûs*, 35: 365–93.

van Fraassen, B. C. (1991) *Quantum Mechanics: An Empiricist View*. Oxford: Oxford University Press.

Worrall, J. (1989) "Structural Realism: The Best of Both Worlds?" *Dialectica*, 43: 99–124.

——and E. Zahar (2001) "Ramseyfication and Structural Realism". In E. Zahar, *Poincaré's Philosophy: From Conventionalism to Phenomenology*. Chicago: Open Court. (Chapter Appendix IV: pp. 236–51).

Zahar, E. (1996) "Poincaré's Structural Realism and his Logic of Discovery". In G. Heinzmann et al. (eds.), *Henri Poincaré: Akten Des Internationale Kongresses in Nancy*. Berlin: Academic Verlag, and Paris: Albert Blanchard.

5

Holism and Structuralism in Classical and Quantum General Relativity

Mauro Dorato and Massimo Pauri

ABSTRACT

The main aim of our chapter is to show that interpretative issues belonging to classical General Relativity (GR) might be preliminary to a deeper understanding of conceptual problems stemming from ongoing attempts at constructing a quantum theory of gravity. Among such interpretative issues, we focus on the meaning of general covariance and the related question of the identity of points, by basing our investigation on the Hamiltonian formulation of GR as applied to a particular class of spacetimes. In particular, we argue that the adoption of a specific gauge-fixing within the canonical reduction of Arnowitt–Deser–Misner metric gravity provides a new solution to the debate between substantivalists and relationists, by suggesting a *tertium quid* between these two age-old positions. Such a third position enables us to evaluate the controversial relationship between entity realism and structural realism in a well-defined case study. After having indicated the possible developments of this approach in Quantum Gravity, we discuss the structuralist and holistic features of the class of spacetime models that are used in the above-mentioned canonical reduction.

5.1 INTRODUCTION: TWO STRANDS OF PHILOSOPHY OF PHYSICS THAT OUGHT TO BE BROUGHT TOGETHER

In recent philosophy of science, there have been *two* interesting areas of research that, independently of each other, have tried to overcome what was beginning to be perceived as a sterile opposition between two contrasting philosophical stances.

The first area of research involves the age-old opposition between the so-called spacetime *substantivalism*, according to which spacetime exists over and above the physical processes occurring in it, and *relationism*, according to which spatio-temporal relations are derivative and supervenient on physical relations obtaining among events and physical objects. The plausible claim that substantivalism and relationism, as they were understood before the advent of relativity or even before the electromagnetic view of nature, simply do not fit in well within the main features of the general theory of relativity, is reinforcing the need of advancing a *tertium quid* between these two positions, which tries in some sense to overcome the debate by incorporating some claims of both sides (Dorato 2000).

On the second front, discussions fuelled by the historical work of Thomas Kuhn have generated a contrast between those who believe that assuming the approximate truth of scientific theories is the best explanation for the predictive and explanatory success of science (viz., the scientific realists) and those who insist that the history of science is so replete with the corpses of abandoned entities (the phlogiston, the caloric, the ether, etc.) that one should believe only in the humanly observable consequences of our best scientific theories (viz., the instrumentalists). As an attempt to overcome this opposition and save the history of science from a complete incommensurability between successive theories, John Worrall (1989) has recently recuperated some forgotten lessons left to us by Poincaré, by pointing out that *structural realism* (i.e. belief in the relational content denoted by our mathematically expressed laws of nature) is 'the best of all possible worlds'. While giving some content to the view that there is (structural) continuity in the history of science, and therefore justifying a claim typically endorsed by the realists, structural realism *à la* Worrall was also meant as a warning against believing in non-directly observable physical entities. Discussing the example already put forth by Poincaré, Worrall remarked that while Fresnel's equations were later incorporated by Maxwell's synthesis, the ether-based models used by him to mathematically describe light have since been abandoned.

In this chapter, we aim to bring together these two strands of philosophical research by claiming that a certain form of *structural spacetime realism* (a view that we refer to as 'point structuralism') may offer the desired *tertium quid* between substantivalism and relationism. As we will see, such a solution emerges naturally from the Hamiltonian formulation of the general theory of relativity (GR), as applied to a definite class of solutions[1], which is important not just to shed light on the above debate but also to clarify—with the help of a well-defined case study—some philosophical problems that are currently affecting the literature on structural realism in general. Most importantly, taking a stance on the meaning of general covariance within classical GR seems to us a precondition also to develop a satisfactory quantum theory of gravity.[2]

Given these aims, our chapter is organized as follows. In §5.2, we try to clarify the relationships among the various forms of scientific realism that are currently discussed in the philosophical literature. Together with the question of clarifying the nature of a *physical* (versus a purely *mathematical*) structure, we believe that these issues are a precondition to understand the impact of *structural* realism on the issue of the *identity of point-events* in classical GR. After a brief review of the Hole Argument in §5.3.1, in §5.3.2 we show how a specific gauge fixing in the canonical reduction of ADM metric gravity, based upon a new use of the so-called Bergmann–Komar 'intrinsic pseudo-coordinates', can help us to formulate a new structural view of GR. In §5.4, we indicate some possible developments of this view on the status of 'points'

[1] The Christodoulou–Klainermann continuous family of spacetimes (Christodoulou and Klainermann 1999).

[2] See also Belot and Earman (1999).

in Quantum Gravity. In §5.5 and §5.5.1 we draw some philosophical conclusions from the preceding discussion by showing how our *point structuralism* represents an overcoming of both traditional substantivalist and relationist views of the spacetime of GR. Such a *point structuralism*, however, does *not* dissolve physical entities into mathematical structures, as it entails a robust realism toward the metric field and a weaker form of entity realism toward its 'point-events', as well as a theory-realist attitude toward Einstein's field equations.

5.2 MANY 'REALISMS' OR ONE?

With the progressive sophistication of our philosophical understanding of science, the issue of scientific realism seems to have undergone a process of complication that is not unlike the growth of a living cell or the development of an embryo. As evidence for this claim, note that nowadays there are at least *four* different ways of characterizing scientific realism, namely *theory realism*, *entity realism*, and, more recently, *structural scientific realism*, where the latter characterization, in its turn, has originated a division between the so-called *epistemic* structural realists and the *ontic* structural realists.

A theory realist defends the claim that the theories of a mature science and its laws are true in the limits of the approximation of a physical model or, in short, *approximately true* (whatever 'approximately' may mean in this context, a difficult problem that here we will not address). Entity realists claim that entities that are not *directly* observable with the naked eye (quarks, electrons, atoms, molecules, etc.) but are postulated by well-confirmed scientific theories exist in a mind-independent fashion. Epistemic structural realists claim, with Poincaré, that while real objects will always be hidden from our eyes, 'the true relations between these objects are the only reality we can attain': 'les rapports véritables entre ces objets sont la seule réalité que nous puissions atteindre' (Poincaré 1905: 162). Ontic structural realists, on the contrary, hold that we can only know structures or relations because they are all there is (Ladyman 1998).

One may wonder whether these various forms of realism are logically independent of each other, as many philosophers have claimed. We believe that they are not. For example, it is not at all clear whether it is really possible, *pace* Hacking (1983), to defend any form of entity realism without also endorsing some form of theory realism.[3]

Analogously, it is highly controversial whether one may have structural realism without also embarking on *theory realism* or *entity realism* of some form. For a necessarily brief defence of the implication from structural realism to theory realism, consider the following remarks. If: (i) the only reality we can know (as the epistemic structural realist has it) is the relations instantiated by existing but unknowable

[3] For a forceful defence of the view that unless we trust theoretical laws, we cannot choose between alternative explanations of the data in terms of rival models of theoretical entities, see Massimi (2004).

entities described by mathematically expressed laws (Worrall 1989, see also Morganti 2004) and (ii) the relations expressed by the equations of mathematical physics represent the *only* element of continuity across scientific revolutions, then clearly one has some evidence to assume that at least such equations and the theories they constitute are *approximately true* (even Poincaré admitted that 'les équations différentielles sont toujours vraies': ibid.). To the extent that (i) stability of equations through theory change is evidence for their approximate truth and (ii) realism about laws entails theory realism, Worrall's position seems to presuppose some form of realism about theories, an implication that in the philosophical literature has gone strangely unnoticed.

For a defence of the implication from structural realism to entity realism, suppose with the *ontic* structural realists that the relations referred to by mathematically formulated laws are knowable just because they exhaust what exists, so that entity realism is false. Alternatively, suppose with the *epistemic* structural realism that entity realism is epistemically unwarranted. In both cases, how can we endorse the existence of relations without also admitting the existence of something that such relations relate (their *relata*), namely something carrying intrinsic, non-relational properties?[4] Chakravartty (1998) and Cao (2003a), for instance, agree with the epistemic realist that our *knowledge* about unobservable entities is essentially structural, but refuse to dissolve physical entities into mathematical structure, thereby classifying themselves as entity realists, and endorsing the view that *structural realism entails entity realism.*[5] This is also the view we want to defend by considering the case study of the ontological status of point-events in classical GR: point-events are structurally individuated by some parts of the metric field, but (i) the metric field exists as an extended entity together with its point-events (entity realism), even though their identity depends on the choice of a global laboratory, and (ii) the law governing its behaviour must be regarded as approximately true.

5.2.1 What, Exactly, is a Physical Structure?

It should be clear at this point that a decisive progress on the issues concerning structural realism presupposes a clarification of the following, crucial question: 'what, exactly, is a physical structure?' As we will see, such a question is also crucial to address the problem of the nature of point-events in classical GR. Much seems to depend on how we want to understand a structure in *physical* terms, since for our purpose the definition of a *mathematical* structure can be taken as sufficiently clear, at least if the latter is regarded as *a system of differential equations* plus abstract objects purporting to describe a physical system.

[4] This worry has been expressed also by Redhead in private conversation. See French and Ladyman (2003: 41).

[5] Conversely, it is much less controversial to agree on the fact that a theory realist must be committed to structural realism about scientific laws, as well as to the existence of unobservable entities, since *holding a theory as approximately true implies believing in the referential power of its assumptions about unobservable entities.*

The problem is that it seems very difficult even to define a *physical* structure without bringing in its constituents, and thereby granting them existence. We will take this difficulty as a preliminary argument in favour of the implication that we want to defend, namely that structural realism implies entity realism, let alone theory realism.

For instance, if we preliminarily regard a structure as 'a stable system of relations among a set of constituents', i.e. a class of entities (see Cao 2003a: 6–7; Cao 2003b: 111)—where 'entities' is deliberately left sufficiently vague in order to cover cases in which the members of the above class lack distinct individuality as is the case for quantum particles—we immediately take an important stance in the above debate. By adopting this definition, in fact, we are already presupposing the independent existence of entities [the constituents], thereby ruling out of the game a priori the ontic structural realism defended by French and Ladyman (2003). Analogously, the so-called 'partial structure approach' (Bueno et al. 2002), according to which a structure is a set of individuals *together* with a family of partial relations defined over the set, seems to run an analogous risk, because the definition of a partial structure includes a set of individuals.

In particular, what is unclear to us is whether it makes sense to consider a physical, 'holistic structure as [*ontologically*] prior to its constituents', as Cao has it (2003c: 111), by simply arguing that its constituents, 'as placeholders, derive their meaning or even their existence from their function and place in the structure'. While the thesis of *meaning* holism may be uncontroversial but clearly irrelevant in the present ontological discussion, one must ask how a place-holder can have any ontological function in an evolving network of relationships without possessing at least some intrinsic *non-relational properties*. While we can imagine that place-holders with different intrinsic properties can contribute the *same* function in the holistic network, so that the structural, relational properties empirically underdetermine the intrinsic properties of the place-holders, it seems that no place-holder can even *have* a function without possessing *some* intrinsic properties.[6] An ordinary key and a magnetic strip on a card do certainly have *different* intrinsic properties, and knowing only about their common effect-type (opening doors),[7] it may be difficult to find out about their difference. However, if they have to yield the same function, the key and the magnetic card must each have their different intrinsic properties.

In a word, we believe that in order to clarify the meaning of 'structure' in the philosophy of physics in general and in the philosophy of space and time in particular, it is essential to revert to the original meaning that 'structuralism' had in linguistics or anthropology. In such contexts, structuralism referred to a sort of *holistic thesis about the identity of the members of a set of stable relations*, and was not conceived, as today it sometimes is, as an attempt to eliminate the constituents. In other words, in what follows we will assume that the interesting and clearer question to be posed in our

[6] A property is intrinsic or non-relational if and only if its attribution does not presupposes the existence of any other entity. For instance, 'being a father' is clearly extrinsic or relational, while 'being square' is intrinsic.

[7] Notice that the doors in the two cases are differently constituted.

context involves a conflict between *two ways of understanding the identity of individual entities constituting a physical system.* The first views the identity of physical entities (in our case, spacetime points) as being constituted by the whole system in which they are physically embedded. The second accords each such individual constituent a distinct identity and individuality, in such a way that the constituents carry different intrinsic or monadic properties which enable us to distinguish them.

So construed, the dispute between the structural realist and the entity realist becomes one between an ontology of quasi-entities that lack any intrinsic individuating property and only possess relational (extrinsic) properties on the one hand (ontic, relational holism) and an ontology of individuals carrying intrinsic and monadic property on the other (entity realism). The crucial question to ask then becomes: given a certain physical theory, to what extent do the relational properties of a set of constituents contribute to fix their identity?

In the next sections we will try to answer this question by showing how the structural and holistic identity of spacetime points in GR does not force us to abandon the typical entity realist's attitude toward the metric field. However, since the identity of its point-events will be shown to depend relationally on the choice of a global laboratory, one's entity realist attitude toward them will have to be of a weaker form. In any case, such an attitude is compatible with the fact that the points of *a bare manifold*, by lacking intrinsic identity, are deprived, to put it with Einstein, of 'the last remnant of physical objectivity'.

5.3 A CASE STUDY: THE HOLISTIC AND STRUCTURAL NATURE OF GENERAL-RELATIVISTIC SPACETIME IN A CLASS OF MODELS OF GR

5.3.1 The Hole Argument and its Consequences

In the recent years, the debate on spacetime substantivalism in GR has been revived by a seminal paper by John Stachel (1980), followed by Earman and Norton's philosophical argument against manifold spacetime substantivalism (1987). Both papers addressed Einstein's famous hole argument (*Lochbetrachtung*) of 1913–15 (Einstein 1914, 1916), which was soon to be regarded by virtually all participants to the debate[8] as being intimately tied to the nature of space and time, at least as they are represented by the mathematical models of GR.

In a nutshell, a mathematical model of GR is specified by a four-dimensional mathematical manifold $\mathcal{M}_4$ and by a metrical tensor field g, where the latter dually represents *both* the chrono-geometrical structure of spacetime *and* the potential for the inertial-gravitational field. Non-gravitational physical fields, when they are present, are also described by dynamical tensor fields, which appear as sources of the Einstein equations.

[8] For example: Butterfield (1989), Earman (1989), Maudlin (1990), Norton (1987, 1992, 1993), Stachel (1993).

The above-emphasized dual role of the metric field has recently generated a conceptual debate, that can be summarized by the following question: *which is the best candidate to interpret the role of space and time in GR, the manifold or the (manifold plus) the metric?* Those opting for the bare manifold $\mathcal{M}_4$ (like Earman and Norton) correctly point out that g cannot be understood as interpreting the role of the 'empty spacetime' of the traditional debate: by embodying the potential of the gravitational field, g is to be regarded as a (special) type of 'physical field'. Those opting—much more reasonably, in our opinion—for 'the manifold plus the metric field' (Maudlin 1990; Stachel 1993) also correctly point out that the metric provides the chrono-geometrical structure as well as, most significantly, the causal structure of spacetime. To the extent that one can see good arguments for *both* options—or even if, as we believe, the second option is the only plausible one—such an 'ambiguous' role of the metric seems to provide one of the main arguments to claim that the early-modern debate between substantivalists and relationists is now 'outmoded', because in GR it does not admit of a clear formulation (Rynasiewicz 1996).

Before agreeing on this sceptical remark, however, it is appropriate to go over the hole argument one more time, in order to show how it should really be tackled and what implications our proposed solution has for the above debate. Let us assume that $\mathcal{M}_4$ contains a *hole* $\mathcal{H}$: that is, an open region where all the non-gravitational fields vanish. On $\mathcal{M}_4$ we can define an *active* diffeomorphism $D_A : x' = D_A x$ (see, for example, Wald 1984) that re-maps the points inside $\mathcal{H}$, but blends smoothly into the identity map outside $\mathcal{H}$ and on the boundary. If we consider the transformed tensor field $g' \equiv D_A^* g$ (where D_A^* denotes the action of a diffeomorphism on tensor fields) then, by construction, for any point $x \in \mathcal{H}$ we have (in the abstract tensor notation) $g'(D_A x) = g(x)$, but of course $g'(x) \neq g(x)$ (in the same notation). The crucial fact to keep in mind at this point is that the Einstein equations are generally covariant: this means that if g is one of their solutions, so is the *drag-along* field g'.

What is the correct interpretation of the new field g'? Clearly, the transformation entails an *active redistribution of the metric over the points of the manifold in* $\mathcal{H}$, so the crucial question is whether and how the points of the manifold are primarily *individuated.* Now, *if* we think of the points of $\mathcal{H}$ as *intrinsically individuated physical events*, where 'intrinsic' means that their identity is independent of the metric—a claim that is associated with *manifold substantivalism*—then g and g' must be regarded as *physically distinct* solutions of the Einstein equations (after all, $g'(x) \neq g(x)$ at the *same* point x). This is a devastating conclusion for the *causality* (or, in other words, the *determinism*) of the theory, because it implies that, even after we completely specify a physical solution for the gravitational and non-gravitational fields outside the hole—in particular, *on a Cauchy surface for the initial value problem*—we are still unable to predict uniquely the physical solution within the hole. Clearly, if general relativity has to make any sense as a *physical* theory, there must be a way out of this foundational quandary, *independently of any philosophical consideration.*

According to Earman and Norton (1987), the way out of the *hole argument* lies in abandoning manifold substantivalism, even if they do not endorse a specific relationist view: they claim that if diffeomorphically related metric fields were to represent different physically possible worlds, then GR would turn into an *indeterministic* theory. And since the issue of whether determinism holds or not at the *physical, empirical* level cannot be decided by opting for a *metaphysical* doctrine like manifold substantivalism, they conclude that one should go for spacetime relationism.

Now, *if* relationism in GR were entailed by the claim that diffeomorphically related mathematical models don't represent physically distinct solutions, most physicists would count themselves as relationists. After all, the assumption that *an entire equivalence class of diffeomorphically related mathematical solutions represents only one physical solution* is regarded as the most common technical way out of the strictures of the hole argument (in the philosophical literature such an assumption is known, after Earman and Norton (1987), as *Leibniz equivalence*). However, we believe that it is not at all clear whether Leibniz equivalence really grinds corn for the relationist's mill, since the spacetime substantivalist can always ask: (1) why on earth should we identify *physical* spacetime with the bare manifold *deprived of the metric field*? (2) Why should we assume that the points of the mathematical manifold have an intrinsic *physical* identity independently of the metric field?[9]

In order to lay our cards on the table with respect to these (rhetorical) questions, we start from the latter in order to note an unfortunate ambiguity in the use of the term 'spacetime points': sometimes it refers to elements of the mathematical structure that is the first conceptual 'layer' of the spacetime model (the manifold), sometimes it refers to the points interpreted as *physical* events. To remedy this situation, we stipulate to use the term *point-events* to refer to physical events and simply *points* to refer to elements of the mathematical manifold. In this respect we just want to add that in the mathematical literature about topological spaces, it is implicitly assumed that their elements are already distinguished. Otherwise, one could not even state the Hausdorff condition, let alone define mappings, homeomorphisms, or active diffeomorphisms. It is well known, however, that the points of a *homogeneous* space (as the manifold would be prior to the introduction of the metric) cannot have any intrinsic *individuality*. As Hermann Weyl (1946) put it:

> There is no distinguishing objective property by which one could tell apart one point from all others in a homogeneous space: at this level, fixation of a point is possible only by a *demonstrative act* as indicated by terms like 'this' and 'there.'

Quite aside from the phenomenological stance implicit in Weyl's words, there is only one way to individuate points *at the mathematical level* that we are considering, namely by (arbitrary) *coordinatization*. By using coordinates, we transfer the individuality of *n*-tuples of real numbers to the elements of the topological set.

[9] It could be observed that such rhetorical questions lack bearing to a substantivalist about Minkowski spacetime, and that substantivalism should not be construed differently in different spacetime theories. However, it is not clear how pointlike events deprived of *every* physical quality can be regarded as possessing an intrinsic quality, or an *haecceity*. Also in the special theory of relativity, in order to individuate points of Minkowski spacetime, one has to rely on physical entities like *rods* and *clocks* or physical fields.

As to the first question above, we will have to limit ourselves to the following remarks: although the metric tensor field, *qua* physical field, cannot be regarded as the traditional *empty container* of other physical fields, we believe that it has *ontological priority* over all other fields. This pre-eminence has various reasons (Pauri 1996), but the most important is that the metric field tells all other fields how to move causally. In agreement also with the general-relativistic practice of not counting the gravitational energy induced by the metric as a component of the total energy, we believe that physical spacetime should be identified with the manifold endowed with its metric, thereby leaving the task of representing matter to the stress-energy tensor.

In consonance with this choice, Stachel[10] has provided a very enlightening analysis of the conceptual consequences of *modern* Leibniz equivalence. Stachel stresses that asserting that g and $D^*_A g$ represent *one and the same gravitational field* implies that *the mathematical individuation of the points of the differentiable manifold by their coordinates has no physical content until a metric tensor is specified.* Furthermore, if g and $D^*_A g$ must represent the same gravitational field, they cannot be physically distinguished in any way. Consequently, when we act on g with D^*_A to create the *drag-along* field $D^*_A g$, no element of physical significance can be left behind: in particular, *nothing* that could identify a point x of the manifold as the *same* point of spacetime for both g and $D^*_A g$. Instead, when x is mapped onto $x' = D_A x$, it *brings over its identity*, as specified by $g'(x') = g(x)$.

These remarks led Stachel to the important conclusion that vis-à-vis the physical point-events, the metric plays in fact the role of *individuating field.* More than that, even the topology of the underlying manifold cannot be introduced independently of the specific form of the metric tensor, a circumstance that makes Earman and Norton's choice of interpreting the mere topological and differentiable manifold as *spacetime* (let alone *substantival spacetime*) even more implausible. More precisely, Stachel suggested that this individuating role should be implemented by four invariant functionals of the metric, already considered by Komar (1955). However, he did not follow up on such a suggestion concretely, something that we will do in the next section, with the aim of further clarifying the nature of the physical point-events. We believe in fact that their status as the intrinsic elements of physical spacetime needs further analysis,[11] especially in view of the questions of structural realism and spacetime substantivalism that we raised before. In addition, our procedure will show why Stachel's original approach cannot be effective in its fully covariant form.

[10] See Stachel (1980, 1986, 1993).

[11] Let us recall that in 1984 Michael Friedman was lucidly aware of the unsatisfactory status of the understanding of the relation between diffeomorphic models in terms of Leibniz equivalence: 'Further, if the above models are indeed equivalent representations of the same situation (as it would seem they must do) then *how do we describe this physical situation intrinsically?* Finding such an *intrinsic* characterization (avoiding quantification over *bare* points) appears to be a non-trivial, *and so far unsolved mathematical problem.* (Note that it will not do simply to replace points with equivalence classes of points: for, in many cases, the equivalence class in question will contain *all* points of the manifold).' Friedman (1984: 663; our emphasis.)

5.3.2 The Dynamical Individuation of Point-Events

5.3.2.1 Pure Gravitational Field without Matter

It is well known that only some of the ten components of the metric are physically essential: it seems then plausible to suppose that only this subset can act as individuating field, and that the remaining components play a different role.

Bergmann and Komar (1960) and Bergmann (1960, 1962, 1977) introduced the notion of *intrinsic invariant pseudo-coordinates* already in 1960. These authors noted that for a vacuum solution of the Einstein equations, there are exactly four functionally independent *scalars* that can be written using the lowest possible derivatives of the metric. These are the four Weyl scalars (the eigenvalues of the Weyl tensor), here written in Petrov's compressed notation,

$$\begin{aligned} w_1 &= \mathrm{Tr}\,(gWgW), \\ w_2 &= \mathrm{Tr}\,(gW\epsilon W), \\ w_3 &= \mathrm{Tr}\,(gWgWgW), \\ w_4 &= \mathrm{Tr}\,(gWgW\epsilon W), \end{aligned} \tag{5.1}$$

where g is the *four*-metric, W is the Weyl tensor, and ϵ is the Levi–Civita totally antisymmetric tensor.

Bergmann and Komar then propose to build a set of *intrinsic pseudo-coordinates* for the point-events of spacetime as four suitable functions $I^{[A]}$ of the w_T

$$I^{[A]} = I^{[A]}\big[w_T[g(x), \partial g(x)]\big], \quad A = 0, 1, 2, 3. \tag{5.2}$$

Indeed, under the *non-restrictive* hypothesis that *no* spacetime symmetries are present—in an analysis of the physical individuation of points, we must consider *generic* solutions of the Einstein equations rather than the null-measure set of solutions with symmetries—the $I^{[A]}$ can be used to *label* the point-events of spacetime, at least locally. Since they are scalars, the $I^{[A]}$ are invariant under passive diffeomorphisms (they are rather like the so-called 'radar' coordinates and therefore do not define a coordinate chart in the usual sense), and by construction they are also constant under the *drag-along* of tensor fields induced by active diffeomorphisms.

At this stage, however, it is far from clear how to explicitly use these intrinsic coordinates to solve the puzzles raised by the hole argument, *especially in view of its connection with the Cauchy problem.* For it is essential to realize that the *hole argument is inextricably entangled with the initial-value problem of general relativity*, although, strangely enough, it has never been explicitly and systematically discussed in this context[12]. The main reason for this neglect is plausibly given by the fact that most authors have implicitly adopted the Lagrangian approach (or the *manifold way*), in which the initial-value problem turns out to be intractable because of the

[12] It is interesting to note, however, that David Hilbert had stressed this point already in 1917 (Hilbert 1917).

non-hyperbolic nature of Einstein's equations. This is also the main reason why we are obliged to turn to the Hamiltonian methods[13].

Three circumstances make the recourse to the Hamiltonian formalism especially propitious.

1. It is only within the Hamiltonian approach that can we separate the *gauge variables*—which carry the descriptive arbitrariness of the theory—from the *Dirac observables*, which are gauge-invariant quantities and are subject to hyperbolic (and therefore *causal* or *deterministic*) evolution equations.
2. In the context of the Hamiltonian formalism, we can resort to Bergmann and Komar's theory of 'general coordinate group symmetries' (1972) to clarify the significance of the *passive view* of active diffeomorphisms as *on-shell*[14] *dynamical symmetries* of the Einstein equations.
3. With respect to our main purpose of trying to understand the nature of point-events in classical GR, it is only within the ADM Hamiltonian formulation of GR that we can introduce a *specific gauge-fixing* (see §5.3.2.3) that can be invoked for their *physical (dynamical) individuation*[15].

5.3.2.2 Pure gravitational field: the ADM slicing of spacetime and the canonical reduction

The ADM (Arnowitt et al. 1962) Hamiltonian approach starts with a slicing of the four-dimensional manifold $\mathcal{M}_4$ into constant-time hypersurfaces Σ_τ, indexed by the *parameter time* τ, each equipped with coordinates σ^a (a = 1,2,3) and a three-metric ^3g (in components $^3g_{ab}$). In order to obtain the 4-geometry, we start at a point on Σ_τ, and displace it infinitesimally in a direction that is normal to Σ_τ. The resulting change in τ can be written as $d\tau = Nd\tau$, where N is the so-called *lapse function*. In a generic coordinate system, such a displacement will also shift the spatial coordinates: $\sigma^a(\tau + d\tau) = \sigma^a(\tau) + N^a d\tau$, where N^a is the *shift vector*. Then the interval between (τ, σ^a) and $(\tau + d\tau, \sigma^a + d\sigma^a)$ results: $ds^2 = N^2 d\tau^2 - {}^3g_{ab}(d\sigma^a + N^a d\tau)(d\sigma^b + N^b d\tau)$. The *configurational* variables N, N^a, $^3g_{ab}$ together with their ten conjugate momenta, index a twenty-dimensional phase space[16]. Expressed (modulo surface terms) in terms of the ADM variables, the Einstein-Hilbert action is a function of N, N^a, $^3g_{ab}$ and its first time derivative, or equivalently of N, N^a, $^3g_{ab}$ and the extrinsic curvature $^3K_{ab}$ of the hypersurface Σ_τ, considered as an embedded manifold.

[13] It is not by chance that the modern treatment of the initial value problem within the Lagrangian configurational approach (Friedrich and Rendall 2000) must in fact mimic the Hamiltonian methods.

[14] We distinguish *off-shell* considerations, made within the variational framework *before* restricting to the dynamical solutions, from *on-shell* considerations, made *after* such a restriction.

[15] The individuation procedure outlined here is based on the technical results obtained by Lusanna and Pauri (2004a, 2004b), see also Pauri and Vallisneri (2002): hereafter quoted as LP1, LP2 and PV, respectively.

[16] Of course, all these *variables* are in fact *fields*.

Since the original Einstein equations are not hyperbolic, it turns out that the canonical momenta are not all functionally independent, but satisfy four conditions known as *primary* constraints. Four other *secondary* constraints arise when we require that the primary constraints be preserved through evolution (the secondary constraints are called the *Super-Hamiltonian* $\mathcal{H}_0 \approx 0$, and the *super-momentum* $\mathcal{H}_a \approx 0$, $(a = 1, 2, 3)$ constraints, respectively). The eight constraints are given as functions of the canonical variables that vanish on the constraint surface[17]. The existence of such constraints implies that that not all the points of the twenty-dimensional phase space represent physically meaningful states: rather, we are restricted to the *constraint surface* where all the constraints are satisfied, i.e. to a twelve-dimensional $(20 - 8)$ *surface* which, on the other hand, does not possess the geometrical structure of a true phase space. When used as generators of canonical transformations, the eight constraints map points on the constraint surface to points on the same surface; these transformations are known as *gauge transformations* and form an eight-dimensional group.

To obtain the correct dynamics for the constrained system, we need to modify the Hamiltonian variational principle to enforce the constraints; we do this by adding the *primary* constraint functions to the Hamiltonian, after multiplying them by arbitrary functions (the *Lagrange–Dirac multipliers*). If, following Dirac, we make the reasonable demand that the evolution of all *physical variables* be unique—otherwise we would have *real* physical variables that are indeterminate and therefore neither *observable* nor *measurable*—then the points of the constraint surface lying on the same *gauge orbit*, i.e. linked by gauge transformations, *must describe the same physical state*.[18] Conversely, only the functions in phase space that are invariant with respect to gauge transformations can describe physical quantities.

To eliminate this ambiguity and create a one-to-one mapping between points in the phase space and physical states, we must impose further constraints, known as *gauge conditions* or *gauge fixings*. The gauge fixings can be implemented by arbitrary functions of the canonical variables, except that they must define a *reduced phase space* that intersects each gauge orbit exactly once (*orbit conditions*). The number of independent gauge fixings must be equal to the number of independent constraints (i.e. 8 in our case). The canonical reduction proceeds by a cascade procedure: the gauge fixings to the *super-Hamiltonian* and *super-momentum* come first (call it Γ_4); then the requirement of their time constancy fixes the gauges with respect to the primary constraints. Finally the requirement of time constancy for these latter gauge fixings determines the Lagrange multipliers. Therefore, the first level, Γ_4, of gauge fixing gives rise to a *complete* gauge fixing, call it Γ_8, which is sufficient to remove all the gauge arbitrariness.

[17] Technically, these functions are said to be *weakly* zero. Conversely, any *weakly* vanishing function is a linear combination of the *weakly* vanishing functions that define the constraint surface.

[18] Actually in GR, there are further and subtler complications concerning the geometric significance of the whole set of such transformations and the existence of geometrically inequivalent states (see LP1, LP2, and PV).

The Γ_8 procedure reduces the original twenty-dimensional phase space to a *copy* Ω_4 of the abstract *reduced phase space* $\tilde{\Omega}_4$ having four degrees of freedom per point ($12 - 8$ gauge fixings). Abstractly, this *reduced phase space* with its symplectic structure is defined by the quotient of the constraint surface with respect to the eight-dimensional group of gauge transformations and represents *the space of the abstract gauge-invariant observables of GR: two configurational and two momentum variables*. These observables carry the physical content of the theory in that they represent the *intrinsic degrees of freedom of the gravitational field* (remember that at this stage we are dealing with a pure gravitational field without matter).

The Γ_8-dependent *copy* Ω_4 of the abstract $\tilde{\Omega}_4$ is realized in terms of the symplectic structure (*Dirac brackets*) defined by the given gauge fixings and coordinatized by four *Dirac observables* [call such field observables $q^r(\tau, \vec{\sigma})$, $p_s(\tau, \vec{\sigma})$ (r,s = 1,2)]. The functional form of these Dirac observables provides a *concrete* realization of the *gauge-invariant abstract observables* in the given gauge Γ_8. Their expression, in terms of the original canonical variables, depends upon the chosen gauge, so that such observables, a priori, are neither tensors nor invariant under $_PDiff$. Yet, *off shell*, barring sophisticated mathematical complications, *any two copies of* Ω_4 *are diffeomorphic images of each other*. Note that the canonical reduction, which creates the distinction between gauge-dependent quantities and Dirac observables, is made *off shell*. After the canonical reduction is performed, the theory is completely determined: each physical state corresponds to one and only one set of canonical variables that satisfies the constraints and the gauge conditions.

It is important to understand qualitatively the geometric meaning of the eight infinitesimal *off-shell* Hamiltonian gauge transformations and thereby the geometric significance of the related gauge fixings. (i) The transformations generated by the four *primary* constraints modify the *lapse* and *shift* functions which, in turn, determine how densely the spacelike hypersurfaces Σ_τ are distributed in spacetime and also the *gravito-magnetism* conventions; (ii) the transformations generated by the three *super-momentum* constraints induce a transition on Σ_τ from a given three-coordinate system to another; (iii) the transformation generated by the *super-Hamiltonian* constraint induces a transition from a given a priori 'form' of the $3+1$ splitting of M^4 to another one, by operating deformations of the spacelike hypersurfaces in the normal direction.

The manifest effect of the related gauge fixings emerges only at the end of the canonical reduction and *after the solution of the Einstein–Hamilton equations has been worked out* (i.e. *on shell*), since the role of a complete gauge fixing Γ_8 is essentially that of choosing the *functional form* in which all the gauge variables depend upon the *Dirac observables*. In particular, the metric and the extrinsic curvature (and thereby also the complete definition of the Σ_τ embedding) are not completely defined until the Einstein–Hamilton equations are solved and the contribution of the Dirac observables calculated.

Given the geometrical meaning of the gauge fixings, a complete Γ_8 gauge fixing includes a choice of the conventions about *global simultaneity* and *gravito-magnetism*, together with the implicit definition of two *global congruences of timelike observers* and *an atlas of coordinate charts on the spacetime manifold*, in particular, *within the Hole* (see LP1 and LP2). In other words, a complete Γ_8 defines for every τ a *global, non-inertial, extended, spacetime laboratory with its coordinates* (hereafter denoted by the acronym *GLAB*). Concerning the physical interpretation of all the variables implied in the Hamiltonian approach to GR, it can be shown (see LP1) that, while the Dirac observables essentially describe *generalized tidal* effects of the gravitational field, the gauge variables, considered *off shell*, embody *generalized inertial effects* connected to the definition of the *GLAB* in which measurements take place, i.e. in which the gravitational phenomena manifestly *appear*[19]. Note that, unlike the special relativistic case, all the conventions are determined *dynamically*. In particular, different conventions within the same spacetime (the same 'universe') turn out to be simply *gauge-related options*.

In conclusion, it is only after the initial conditions for the *Dirac observables* have been arbitrarily selected on a Cauchy surface that one can determine dynamically the whole four-dimensional chrono-geometry. Of course, once Einstein's equations have been solved, the metric tensor and all of its derived quantities, in particular the light-cone structure, can be re-expressed in terms of *Dirac observables* in a gauge-fixed functional form[20].

Two important points must be stressed.

First, before the gauge fixings are implemented, in order to carry out the canonical reduction *explicitly*, we have to perform (*off shell*) a basic canonical transformation, the so-called Shanmugadhasan transformation, bringing from the original canonical variables to a new basis including the Dirac observables in a canonical subset in such a way that they have zero Poisson brackets with all the other variables. Now, the Shanmugadhasan transformation is highly *non-local* in the metric and curvature variables: even though, at the end, for any τ, the *Dirac observables* $q^r(\tau,\vec{\sigma}), p_s(\tau,\vec{\sigma})$, are *fields* indexed by the coordinate point σ^A, they are in fact *highly non-local functionals of the metric and the curvature over the whole surface* Σ_τ. We can write, *symbolically*:

$$
\begin{aligned}
q^r(\tau,\vec{\sigma}) &= \mathcal{F}_{[\Sigma_\tau]}{}^r\big[(\tau,\vec{\sigma})|\,{}^3g_{ab}(\tau,\vec{\sigma}),{}^3\pi^{cd}(\tau,\vec{\sigma})\big]\\
p_s(\tau,\vec{\sigma}) &= \mathcal{G}_{[\Sigma_\tau]_s}\big[(\tau,\vec{\sigma})|\,{}^3g_{ab}(\tau,\vec{\sigma}),{}^3\pi^{cd}(\tau,\vec{\sigma})\big], \qquad r,s=1,2;
\end{aligned}
\tag{5.3}
$$

where ${}^3g_{ab}$ and ${}^3\pi^{cd}$ are the 3-metric and the conjugated 3-momentum components, respectively.

[19] Such a *GLAB* is a non-rigid, non-inertial frame (the only one existing in GR) centred on the (in general) accelerated observer whose world-line is the origin of the 3-coordinates (Lusanna and Pauri 2004a). The gauge-fixing procedure determines the *appearance of phenomena* by determining uniquely the form of the inertial forces (Coriolis, Jacobi, centrifugal, ...) in each point of a GLAB. A crucial difference of this mechanism in GR with respect to the Newtonian case is the fact that the inertial potentials depend upon *tidal effects* (i.e. on the Dirac observables), besides the coordinates of the non-inertial frame.

[20] Conversely, as shown in PV, LP1, and LP2, in the absence of a dynamical theory of measurement, the epistemic circuit of GR can be approximately closed via an *experimental* three-step procedure that, starting from concrete radar measurements and using test-objects, ends up in a complete and *empirically coherent intrinsic individuating gauge fixing*.

Second: since the original canonical Hamiltonian in terms of the ADM variables is zero, it happens to be written solely in terms of the eight constraints and Lagrangian multipliers. This means, however, that this Hamiltonian generates purely harmless gauge transformations connecting different admissible spacetime 3+1 splittings, so that *it cannot engender any real temporal change* (in this connection see Earman (2002); Belot and Earman (1999, 2001)). The crucial point, however, is that, *in the case of the globally hyperbolic non-compact spacetimes*, defined by suitable boundary conditions and being asymptotically flat at spatial infinity, just like those of the class we are dealing with in this work[21], *internal mathematical consistency* entails that the generator of temporal evolution is the so-called *weak ADM energy*, which is obtained by adding the so-called DeWitt boundary surface term to the canonical Hamiltonian[22]. Indeed, this quantity *does generate real temporal modifications of the canonical variables*. Thus, the final Einstein–Hamilton–Dirac equations for the Dirac observables are

$$\dot{q}^r = \{q^r, H_{\text{ADM}}\}^*, \quad \dot{p}_s = \{p_s, H_{\text{ADM}}\}^*, \quad r, s = 1, 2, \tag{5.4}$$

where H_{ADM} is intended as the restriction of the *weak ADM energy* to Ω_4 and where the $\{\cdot, \cdot\}^*$ are the Dirac brackets.

Then, the initial value problem runs as follows: (1) selection of a complete Γ_8 (a *GLAB*); (2) assignment of the initial values of the Dirac observables on a Cauchy surface Σ_{τ_0}, in that *GLAB*; (3) solution of the Einstein–Hamilton–Dirac equations.

At this point, it is important to realize that the space of Cauchy data is partitioned into classes of *gauge-equivalent* data: all Cauchy data in a given class identify a single spacetime (a '4-geometry', or 'universe'). This entails that the *dynamical symmetries* of Einstein's equations fall in two classes only: (i) those acting within a single Einstein 'universe'; (ii) those mapping different 'universes' among themselves.

5.3.2.3 Pure gravitational field: the 'intrinsic individuating gauge', and the metrical fingerprint

We can now turn to briefly illustrate the process of the dynamical individuation of point-events. First of all we exploit a technical result by Bergmann and Komar (1960), namely the fact that the four Weyl scalar invariants (5.1), *once re-expressed in terms of the ADM variables*, turn out to be independent of the *lapse function* N and the *shift vector* N^a. This means that the *intrinsic pseudo-coordinates* are in fact functionals of the variables ${}^3g_{ab}$ and ${}^3K_{ab}$ only. Then we write

$$I^{[A]}[w_T(g, \partial g)] \equiv Z^{[A]}[w_T({}^3g, {}^3\pi)], \quad A = 0, 1, 2, 3, \tag{5.5}$$

(where the $Z^{[A]}$ represent the functions $I^{[A]}$ as re-expressed in terms of the *ADM variables*) and select a *completely arbitrary* coordinate system $\sigma^A \equiv [\tau, \sigma^a]$ *adapted* to

[21] As already said, it is the Christodoulou–Klainermann continuous family of spacetimes.

[22] The ADM energy is a Noether constant of motion representing the *total mass* of the 'universe', just one among the ten asymptotic Poincaré 'charges'. The mathematical background of this result can be found in Lusanna (2001) and references therein.

the Σ_τ surfaces[23]. Finally we apply the specific gauge fixing Γ_4 defined by

$$\chi^A \equiv \sigma^A - Z^{[A]}\big[w_T[(^3g(\sigma^B), {}^3\pi(\sigma^D))]\big] \approx 0, \quad A = 0, 1, 2, 3, \tag{5.6}$$

to the *super-Hamiltonian* (A = 0) and the *super-momentum* (A = 1,2,3) constraints. This is indeed a good gauge fixing provided that the functions $Z^{[A]}$ are chosen to satisfy the fundamental *orbit conditions* $\{Z^{[A]}, \mathcal{H}_B\} \neq 0, \quad (A, B = 0, 1, 2, 3)$, which ensure the independence of the χ^A and carry information about the Lorentz signature. At the end of the gauge fixing procedure Γ_8, the effect is that the values (i.e. the *evolution* throughout the mathematical spacetime $\mathcal{M}_4$) of the *Dirac observables*, whose dependence on space (and on parameter time) is indexed by the chosen coordinates σ^A, *reproduce* precisely σ^A as the Bergmann–Komar *intrinsic pseudo-coordinates, in the chosen gauge* Γ_8:

$$\sigma^A = Z^{[A]}[w_T(q^r(\sigma^B), \ p_s(\sigma^C)|\Gamma_8)], \quad A = 0, 1, 2, 3; \tag{5.7}$$

where the notation $w_T(q, p|\Gamma_8)$ represents the functional form that the Weyl scalars w_T assume in the chosen gauge.

In the language of constraint theory, after the canonical reduction is performed—and *only for the solutions of the equations of motion*—(5.7) becomes a *strong* relation[24]. *Such a strong relation is in fact an identity with respect to* σ^A, *and amounts to a 'definition' of the 'radar' coordinates* σ^A *as four scalars providing a physical individuation of any point–event, in the gauge fixed coordinate system, in terms of the intrinsic gravitational degrees of freedom.*

In this way each of the point–events of spacetime is endowed with its own *metrical fingerprint* extracted from the tensor field, i.e. the value of the four scalar functionals of the *Dirac observables* (exactly four!)[25]. The price that we have paid for this achievement is that we have broken general covariance! This, however, is not a drawback because every choice of 4-coordinates for a point (every gauge fixing, in the Hamiltonian language), in any procedure whatsoever for solving Einstein's equations, amounts to breaking general covariance by definition. On the other hand the whole extent of general covariance can be recovered by exploiting the gauge freedom.

At first, this result may sound surprising: *qua* diffeomorphism-invariant quantities, the intrinsic pseudo-coordinates can be forced within a 'radar' coordinate system corresponding to any experimental arrangement. From the Hamiltonian viewpoint, however, they are necessarily gauge-dependent functionals. It is not known, as yet, whether the sixteen canonical variables of the Shanmugadhasan basis—which include the *Dirac observables*—could be replaced by sixteen diffeomorphism-invariant quantities (scalars) which, in particular, would include *tensorial Dirac observables.*

[23] Note that the σ^A are in fact 'radar' (scalar) coordinates.

[24] This means that the relation is expressed by functions which not only vanish on the constraint surface but also have all vanishing derivatives in directions normal to the constraint surface.

[25] The fact that there are just *four* independent invariants for the vacuum gravitational field should not be regarded as a coincidence. On the contrary, it is crucial for the purpose of point individuation and for our gauge-fixing procedure. After all, recall that in general spacetimes with matter there are fourteen invariants of this kind!

This question could be answered in the positive if a *main conjecture* advanced in Lusanna and Pauri (2004b), which is now under scrutiny, turned out to be true[26]. Then, the individuating functions of (5.7) would depend only on scalars and the distinction between Dirac and gauge observables would become *fully invariant*. In this case, one could speak of a *fully objective* (tensor-covariant) dynamical individuation of point-events. Yet, obviously, the gauge-fixing procedure would still break general covariance.

Note that the virtue of the elaborate construction described above does *not* depend on the selection of a set of physically preferred coordinates, because by modifying the functions $I^{[A]}$ of (5.2) we have the possibility of implementing *any* 'radar' coordinate transformation. So diffeomorphism invariance reappears under a different suit: we find exactly the same functional freedom of D_P in the functional freedom of the choice of the *pseudo-coordinates* $Z^{[A]}$ (i.e. of the gauge-fixing Γ_4). Thus we see that, *on shell*, both at the Hamiltonian and at the Lagrangian level, every *gauge fixing*, together with the choice of a *GLAB*, amounts to the selection of radar manifold coordinates. Yet, we can now claim that *any* 'radar' coordinatization of the manifold can be seen as embodying the physical individuation of points, because it can be implemented—locally at least—as the Komar–Bergmann intrinsic pseudo-coordinates after we choose the correct $Z^{[A]}$ and we select the proper gauge.

One crucial point concerning the hole argument still needs to be clarified.

It is clear that, for a full understanding of the role played within the Hamiltonian description by the *active* diffeomorphisms of the hole argument, it is necessary that they also be interpretable in some way as the *manifold-way* counterparts of suitable Hamiltonian gauge transformations. Here, we can only limit ourselves to state that this is actually possible by resorting to an important paper by Bergmann and Komar (1972) about the general coordinate-group symmetries of Einstein equations. In fact, it turns out (see LP1 and LP2) that active diffeomorphisms can be viewed as passive transformations on the conjunction of the spacetime manifold and the function space of the metric fields.

Thus, recalling the conclusion of §5.3.2.2 (concerning the partition of Cauchy data in gauge classes), it follows that the Hamiltonian counterpart of active diffeomorphisms—*qua* dynamical symmetries—are of *two kinds*. While those of the first kind induce mappings of the initial data on a Cauchy surface Σ_{τ_0} into gauge-equivalent Cauchy data of the same class, those of the second kind induce mappings among Cauchy data of *different* classes. The crucial point, however, is that only the first kind of diffeormorphisms can have a Lagrangian counterpart which is the identity on Σ_{τ_0}, and can thereby satisfy the assumptions of the hole argument. This entails, therefore, that *solutions of Einstein's equations that within the hole differ by*

[26] An evaluation of the degrees of freedom in connection with the Newman–Penrose formalism for tetrad gravity (Stewart 1993) tends to corroborate the conjecture. In the Newman–Penrose formalism we can define ten coordinate-independent quantities, namely the ten Weyl scalars. If we add ten further scalars built using the extrinsic curvature, we have a total of twenty scalars from which one should extract a canonical basis replacing the 4-metric and its conjugate momenta. Consequently, it should be possible to find *scalar* Dirac observables, and *scalar* gauge variables.

an active transformation on manifold points, when examined at the Hamiltonian level, turn out to be solutions simply differing by a harmless gauge transformation within the same Einstein 'universe'. On the other hand, since the active diffeomorphisms of the second kind are necessarily maps between different Einstein 'universes', they cannot be the identity on the Cauchy surface, and therefore violate the assumptions of the hole argument.

In conclusion, the difference among the solutions generated by active diffeomorphisms satisfying the conditions of the *hole argument* correspond to different 'appearances' of the intrinsic gravitational phenomena in different *GLABs*. This is what the physical content of the Leibniz equivalence boils down to. It is seen, therefore, that the hole argument has nothing to do with an alleged indeterminism and that *its philosophical bearing is dissolved.*

Finally, our procedure for the dynamical individuation of 'point-events' also shows why Stachel's *original proposal* (Stachel 1993) of a fully covariant exploitation of the Bergman–Komar invariants $I^{[A]}$ cannot work.

Here, we do not refer to Stachel's broader perspective about the significance and the possibility of generalizations of the hole story (see Stachel and Iftime 2005) which, among other things, is intended to *block the hole argument*. Within our context, we limit ourselves to stress the following: (a) spacetime does exist if a metric field is defined; (b) this metric field must be a *solution* of Einstein's equations; (c) the active diffeomorphisms that purportedly *keep the physical identity of point-events* by carrying them along in the fully covariant view are *dynamical symmetries* of Einstein's equations.

Now, how can we be sure that the functional dependence of the quantities $I^{[A]}[w_T[g(x), \partial g(x)]$ is *concretely characterized* as relating to actual solutions of Einstein's equations corresponding to given initial data? Since in the actual case we know that these quantities depend upon four Dirac observables and eight *arbitrary gauge variables*, it follows that this arbitrariness, if not properly taken into account, is unavoidably transferred into the individuation procedure and *leaves it incomplete.* Speaking of general covariance in an abstract way hides the necessity of getting rid of the above arbitrariness by a gauge fixing that, in turn, necessarily breaks general covariance. In other words, a definite physical individuation entails a *concrete* characterization of a *GLAB*, which is precisely what we do. The result is, in particular, exactly what Stachel's original suggestion intended to achieve, for our *intrinsic gauge* shows that the Hamiltonian counterparts of *active diffeomorphisms* of the first kind do map individuations of point-events into *physically* equivalent individuations. Indeed, the *on-shell* Hamiltonian gauge transformation, connecting two different gauges, is the passive counterpart of an *active* diffeomorphism D_A, which determines the *drag-along coordinate transformation* connecting the coordinates defined by the two gauges, i.e. the so-called *dual view* of the active diffeomorphism. While the active diffeomorphism *carries along* the identity of points by assumption, the passive view attributes *different physically individuated radar coordinates* to the 'same' (mathematical) point. Consider, for example, *on shell*, two world-lines intersecting at a point-event p within the hole. They define a 'point-coincidence' which is traditionally interpreted as a typical *objective spacetime occurrence*. Now, D_A maps these world-lines into different world-lines intersecting at the point p' which is just the point defined by applying D_A to p, i.e. the 'same' physical point-event. From our Hamiltonian point of view, the 'same' mathematical point gets different gauge-related individuations while the *form* of the transformed world-lines is different because their description is made in different *GLABs*, characterized by different *inertial potentials*. It is seen, therefore, that for any point-event, a given individuation by means of the Dirac observables is mapped into a physically equivalent, *GLAB*-dependent individuation.

Finally, it must be stressed that the main reason why we succeeded in carrying a concrete realization of Stachel's original suggestion to its natural completion lies in the possibility that the Hamiltonian method offers, of working *off shell*. In fact, the D_A, *qua* dynamical symmetries of Einstein's equations, must act on solutions at every stage of the procedure and fail to display the arbitrary part of the scalar invariants. On the other hand, the Hamiltonian separation of the gauge variables (characterizing the *GLAB* and ruling the generalized *inertial effects*) from the Dirac observables (characterizing the *tidal* effects) is an *off-shell* procedure. As such, it recovers the sought-after *metrical fingerprint* by working independently of the initial value problem. Once again, this mechanism is a typical consequence of the special role played by the gauge variables in GR[27].

5.3.2.4 Gravitational-field-cum-matter and the spacetime holistic texture

In conclusion, what is relevant to our discussion is that *there is a remarkable class of gauge fixings*, (5.6), that is instrumental both to the solution of the Cauchy problem and to the *physical individuation of spacetime point-events*. We propose to call this gauge the *intrinsic individuating gauge*. As we have seen above, each of the point-events of spacetime is endowed with its own physical individuation (the right *metrical fingerprint*!) as the value of the four scalar functionals of the *Dirac observables* (just four!), which describe the dynamical degrees of freedom of the gravitational field. It is important to stress that, due to the independence of the *pseudo-coordinates* from the *lapse* and *shift* functions, these degrees of freedom *are inextricably entangled with the structure of the whole 3-metric and 3-curvature in a way that is strongly gauge dependent*. This result appears prima facie as an instantiation of three-dimensional *holism*. Since, however, the extrinsic curvature has to do with the embedding of the hypersurface in M^4, the Dirac observables do *involve geometrical elements external to the Cauchy hypersurface* itself. Furthermore, since (5.7) is four-dimensional and includes the temporal gauge (fixed by the scalar $Z^{[0]}$), as soon as the Einstein–Hamilton equations are solved and the evolution in τ of the *Dirac observables* fully determined, a remarkable instantiation of a four-dimensional

[27] As already noted, if the *main conjecture* advanced by Lusanna and Pauri (2004b) is true, a canonical basis should exist having an explicit *scalar* character. Consequently, it should be possible to find *scalar* Dirac observables and *scalar* gauge variables. Then, the individuating functions of (5.7) would depend on scalars only, and the distinction between Dirac and gauge observables would become *fully invariant*.

stratified holism is recovered which, however—unless the *main conjecture* advanced in LP2 is verified—is relative to the *GLAB*. At this point we could even say that the existence of physical *point-events* in our models of general relativity appears to be synonymous with the existence of the Dirac observables for the gravitational field, and advance the *ontological* claim that, physically, *a vacuum Einstein spacetime is literally identifiable with the autonomous degrees of freedom of such a structural field*, while the specific (gauge-dependent) functional form of the *intrinsic pseudo-coordinates* maps such coordinates into the manifold's points. The intrinsic gravitational degrees of freedom are, as it were, *fully absorbed in the individuation of point-events*. Thus, in this way, *point-events also keep a special kind of property*.

Let us now briefly look at the most general case of ADM models of GR with *matter fields*, taking proper notice of the fact that we are still working with globally hyperbolic pseudo-Riemannian 4-manifolds $\mathcal{M}_4$ which are asymptotically flat at spatial infinity. The introduction of matter has the effect of modifying the Riemann and Weyl tensors, namely the curvature of the four-dimensional substratum, and to allow a measure of the gravitational field in a geometric way (for instance through effects like the geodesic deviation equation). In the presence of matter, we have *Dirac observables for the gravitational field* and *Dirac observables for the matter fields* which satisfy the coupled Einstein–Hamilton equations. As it is to be expected, however, even the functional form of gravitational observables *is modified* (relative to the vacuum case) by the presence of matter. Since the *gravitational* Dirac observables will still provide the individuating fields for point-events according to the conceptual procedure presented in this chapter, *matter will come to influence the very physical individuation of spacetime point-events*. Yet, *the ontological conclusions reached above are not altered at all.*

Finally, even in the case with matter, time evolution is still ruled by the weak ADM energy rather than by the simple canonical Hamiltonian. Therefore, the temporal variation corresponds to a *real change* and not merely to a harmless gauge transformation as in other models of GR[28].

5.4 DEVELOPING HINTS FOR THE QUANTUM GRAVITY PROGRAMME

Let us close our analysis with some hints for the quantum gravity programme that are suggested by the above results (see LP2 for more details). As is well known, there are today two inequivalent approaches: (i) the perturbative background-dependent *string* formulation, on a Fock space containing elementary particles; (ii) the non-perturbative background-independent *loop* quantum gravity formulation, based on the non-Fock so-called *polymer* Hilbert space. The latter approach still fails to accommodate elementary particles, although Ashtekar has advanced some

[28] These latter include, for instance, the spatially compact spacetime without boundary (or simply closed models) which are exploited by Earman in his *Thoroughly Modern McTaggart* (2002).

suggestions to define a *coarse-grained structure* as a bridge between standard *coherent states* in the Fock space and some *shadow states* of the discrete quantum geometry associated with the *polymer* Hilbert space.

Now, let us point out that (5.7) is a numerical identity that has a built-in *non-commutative structure*, deriving from the Dirac–Poisson structure on its right-hand side. The individuation procedure we have proposed transfers, as it were, the non-commutative Poisson–Dirac structure of the Dirac observables onto the individuated point-events, even if, of course, the coordinates on the left-hand side of the identity (5.7) are c-number quantities. One could guess that such a feature might deserve some attention in view of quantization, for instance by maintaining that the identity (5.7) could still play some role at the quantum level. We will assume here, for the sake of argument, that the *main conjecture* is verified, so that all the quantities we consider are manifestly covariant.

Let us first lay down some qualitative premisses concerning the status of Minkowski spacetime in relativistic quantum field theory (RQFT): call it *micro-spacetime* (see Pauri 2000). Such a status is indeed quite remarkable. Since it is introduced into the theory through the group-theoretical requirement of the relativistic invariance of the *statistical* results of measurements with respect to the choice of *macroscopic reference frames*, the *micro-spacetime* is therefore *anchored* to the macroscopic, medium-sized objects that asymptotically define the experimental conditions in the laboratory.[29] Thus, the spatio-temporal properties of the *micro* Minkowski manifold, including its basic causal structure, are, as it were, projected onto it *from outside*.

In classical field theories spacetime points play the role of individuals and we have seen how point-events can be individuated dynamically in a richer and holistic way. No such possibility, however, is consistently left open in a non-metaphorical way in RQFT. From this point of view, Minkowski's *micro-spacetime* in RQFT is in a worse position than classical general relativistic spacetime: it lacks the existence of Riemannian *intrinsic pseudo-coordinates*, as well as of all the non-dynamical (better, operational, and pragmatic) additional macroscopic elements that are used for the individuation of its points, like rigid rods and clocks in rigid and non-accelerated motion, or various combinations of *genidentical* world-lines of free test particles, light rays, clocks, and other test devices.

Summarizing, Minkowski's *micro-spacetime* seems to be essentially functioning like an instrumental but *external translator* of the symbolic structure of quantum theory into the *causal* language of the macroscopic, irreversible traces constituting the experimental findings within *macro*-spacetime. Such an *external translator* should be regarded as an *epistemic precondition* for the formulation of RQFT in the sense of Bohr, independently of one's attitude towards the measurement problem in quantum mechanics.

Thus, barring macroscopic Schrödinger's cat-like states of the would-be quantum spacetime, any conceivable formulation of a quantum theory of gravity would have to respect, at the *operational* level, the *epistemic priority* of a classical spatio-temporal

[29] It is just in this asymptotic sense that a physical meaning is attributed to the classical spatio-temporal *coordinates* upon which the quantum fields' operators depend as *parameters*.

continuum. In fact, the possibility of referring *directly* to 'the quantum structure of spacetime' faces at the very least a serious conceptual difficulty, concerning the *localization* of the gravitational field: what does it mean to talk about the *values* of the gravitational field *at a point* if the metric field itself is subject to quantum fluctuations? How could we identify point-events? In this case, we could no longer tell whether the separation between two points is spacelike, null, or timelike since quantum fluctuation of the metric could exchange past and future.

Accordingly, in order to give physical and operational meaning to the spatio-temporal language, we would need some sort of instrumental background, mathematically represented by a manifold structure, which, at the quantum level, should play more or less the role of a Wittgensteinian staircase. It is likely, therefore, that in order to attribute meaning to the individuality of points at some spatio-temporal scale—so as to build the basic structure of standard quantum theory—one should split, as it were, the individuation procedure of point-events from the true quantum properties, i.e. from the fluctuations of the gravitational field and the micro-causal structure. Our canonical analysis tends to prefigure a new approach to quantization, having in view a Fock space formulation which, unlike the loop quantum gravity, could even lead to a background-independent incorporation of the standard model of elementary particles (provided the Cauchy surfaces admit Fourier transforms). For a quantization programme respecting relativistic causality, two options seem available (see the discussion given in LP2):

1. Our individuation procedure suggests quantizing *only* the *gravitational* Dirac observables (assumed now as scalars in force of the *main conjecture* (see LP2)) of each Hamiltonian gauge, as well as all the *matter* Dirac observables, and then exploit the *weak ADM energy* of that gauge as the Hamiltonian for the functional Schrödinger equation (of course there might be ordering problems). This quantization would yield as many Hilbert spaces as gauge fixings, which would likely be grouped in unitary equivalence classes (we leave aside the question of what could be the meaning of inequivalent classes, were there any). In each Hilbert space the Dirac quantum operators would be distribution-valued quantum fields on a *mathematical micro-spacetime* parametrized by the 4-coordinates $(\tau, \vec{\sigma})$ associated to the chosen gauge. Strictly speaking, due to the non-commutativity of the operators $\hat{Z}^A$ associated to the classical constraint $\sigma^A - Z^A \approx 0$ defining that gauge, there would be *no spacetime manifold of point-events* to be mathematically identified by one coordinate chart over the *micro-spacetime* but only a *gauge-dependent non-commutative structure*, which is likely to lack any underlying topological structure. However, for each Hilbert space, a *coarse-grained* spacetime of point-events $(\overline{\Sigma})^A(\tau, \vec{\sigma})$, superimposed on the mathematical manifold $\mathcal{M}_4$, might be associated to each solution of the functional Schrödinger equation, via the expectation values of the operators $\hat{Z}^A$:

$$(\overline{\Sigma})^A(\tau, \vec{\sigma}) \equiv \langle \Psi \left| \hat{Z}^A_\Gamma[\mathbf{Q}^r(\tau, \vec{\sigma}), \mathbf{P}_s(\tau, \vec{\sigma})] \right| \Psi \rangle, \quad A = 0, 1, 2, 3; \quad r, s = 1, 2; \quad (5.8)$$

where $\mathbf{Q}^r(\tau, \vec{\sigma})$ and $\mathbf{P}_s(\tau, \vec{\sigma})$ are now Dirac *scalar field operators*.

Let us stress that, by means of (5.8), the *non-locality* of the *classical* individuation of point-events would be directly transferred to the basis of the ordinary, *quantum* non-locality. Also, one could evaluate in principle the expectation values of the operators corresponding to the lapse and shift functions of that gauge. Since we are considering a quantization of the 3-geometry (like in loop quantum gravity), evaluating the expectation values of the quantum 3-metric, the quantum lapse and the shift function could permit to reconstruct a coarse-grained foliation with coarse-grained so-called Wigner–Sen–Witten hypersurfaces[30].

2. In order to avoid inequivalent Hilbert spaces, we could quantize *before* adding any gauge fixing (i.e. independently of the choice of the 4-coordinates and the physical individuation of point-events). For example, using the following rule of quantization, which complies with relativistic causality: in a given scalar canonical basis, quantize the two pairs of (scalar) gravitational Dirac observables and matter Dirac observables, but leave the eight gauge variables as *c-number classical fields*. As in Schrödinger's theory with a time-dependent Hamiltonian, the momenta conjugate to the gauge variables would be represented by functional derivatives. Assuming that in the chosen canonical basis, seven among the eight constraints are gauge momenta, we would thereby get seven Schrödinger equations. Then, as suggested in (LP2), both the *super-Hamiltonian* and the *weak ADM energy* would become operators and, if an ordering existed such that the eight 'quantum constraints' satisfied a closed algebra of the form $[\hat{\phi}_\alpha, \hat{\phi}_\beta] = \hat{C}_{\alpha\beta\gamma}\,\hat{\phi}_\gamma$ and $[\hat{E}_{ADM}, \hat{\phi}_\alpha] = \hat{B}_{\alpha\beta}\,\hat{\phi}_\beta$, $(\alpha, \beta, \gamma = 1, \ldots, 8)$ (with the quantum structure functions $\hat{C}_{\alpha\beta\gamma}, \hat{B}_{\alpha\beta}$ tending to the classical counterparts for $\hbar \mapsto 0$), we might quantize by imposing nine integrable coupled functional Schrödinger equations, with the associated usual scalar product $\langle\Psi|\Psi\rangle$ being independent of τ and of the gauge variables.

Again, we would have a *mathematical micro-spacetime* and a *coarse-grained spacetime of 'point-events'*. At this point, by going to *coherent states*, we could try to recover classical gravitational fields. The 3-geometry (volumes, areas, lengths) would be quantized, perhaps in a way that agrees with the results of loop quantum gravity.

It is important to stress that, according to both suggestions, *only the Dirac observables would be quantized*. The upshot is that fluctuations in the gravitational field (better, in the Dirac observables) would entail *fluctuations of the point texture* that lends itself to the basic spacetime scheme of standard RQFT: such fluctuating texture, however, could be recovered as a coarse-grained structure. This would induce fluctuations in the coarse-grained metric relations, and thereby in the causal structure, both of which would tend to disappear in a semiclassical approximation. Such a situation should be conceptually tolerable, and even philosophically appealing, especially if compared with the impossibility of defining a causal structure within all of the attempts grounded upon a quantization of the full 4-geometry. In this connection, it would be interesting to see whether the fluctuations of the point-events metrical texture could have any relevance to the macro-objectification issue of quantum theory (see Károlyházy et al. 1985 and Penrose 1985).

[30] This foliation is called the Wigner-Sen-Witten foliation due to its properties at spatial infinity (see Lusanna 2001).

Finally, in spacetimes with matter, this procedure would entail *quantizing the generalized tidal effects and the action-at-a-distance potentials between matter elements, but not the inertial aspects of the gravitational field.* As we have seen, the latter aspects are connected with gauge variables whose variations reproduce all the possible viewpoints of local accelerated timelike observers. Quantizing also the gauge variables would be tantamount to quantizing the metric *together* with the *passive observers* and their *reference frames*, a fact that is empirically meaningless.[31]

5.5 STRUCTURAL SPACETIME REALISM

The discussion in the previous sections is substantially grounded upon the fact that GR is a gauge theory. Henneaux and Teitelboim (1992) gave a very general definition of gauge theories:

> These are theories in which the physical system being dealt with is described by more variables than there are physically independent degrees of freedom. The physically meaningful degrees of freedom then re-emerge as being those invariant under a transformation connecting the variables (gauge transformation). Thus, one introduces extra variables to make the description more transparent, and brings in at the same time a gauge symmetry to extract the physically relevant content.

The relevant fact is that, while from the point of view of the constrained Hamiltonian *mathematical* formalism general relativity is a gauge theory like any other (e.g. electromagnetism and Yang–Mills theories), from the *physical* point of view it is radically different. For, in addition to creating the distinction between what is observable and what is not, the gauge freedom of GR is unavoidably entangled with the constitution of the very *stage*, spacetime, where the *play* of physics is enacted: a stage, however, which also takes an active part in the play. In other words, the gauge mechanism has the dual role of making the dynamics unique (as in all gauge theories), and of fixing the spatio-temporal, dynamical background (the *GLAB*). It is only after a complete gauge fixing (i.e. after the individuation of a *GLAB*), and after having found the solution of Einstein's equations, that the mathematical manifold $\mathcal{M}_4$ gets a *physical individuation.*

Unlike theories such as electromagnetism (or even Yang–Mills), in GR we cannot rely from the beginning on empirically validated, gauge-invariant dynamical equations for the *local* fields. In order to get equations for *local* fields we must pay the price of general covariance which, by ruling out any background structure at the outset, conceals at the same time the intrinsic properties of point-events. With reference to the definition of Henneaux and Teitelboim, we could say, therefore, that the introduction of extra variables does indeed make the mathematical description of general relativity more transparent, but it also makes its physical interpretation more obscure and intriguing, at least prima facie. Actually, our analysis discloses a

[31] Of course, such observers have nothing to do with *dynamical measuring objects*, which should be realized in terms of the Dirac observables of matter.

deeper distinction of philosophical import. For it highlights a remarkable ontological and functional split of the metric tensor that can be briefly described as follows. On the one hand, the Dirac observables *holistically* specify the *ontic* structure of spacetime. On the other, we have seen that the gauge variables specify, as it were, the in-built *epistemic* component of the metric structure. Actually, in completing the structural properties of the general-relativistic spacetime, they play multiple roles: first of all, their fixing is necessary in order to solve Einstein's equations and to reconstruct the four-dimensional chrono-geometry emerging from the *Dirac observables*: they are essential to get a manifestly covariant and *local* metric field as a ten-dimensional tensor (the *transparency* of Henneaux and Teitelboim); but their fixing is also necessary in order to allow empirical access to the theory through the definition of a spatio-temporal *laboratory*.

The isolation of the *epistemic component of the metric* hidden behind Leibniz equivalence[32], which surfaces in the physical individuation of point-events, renders even more glaring the ontological diversity and prominence of the gravitational field with respect to all other fields, as well as the difficulty of reconciling the nature of the gravitational field with the standard approach of theories based on a *background spacetime* (to wit, string theory and perturbative quantum gravity in general). Any attempt at linearizing such theories unavoidably leads to looking at gravity from the perspective of a spin-2 theory in which the graviton stands on the same ontological level as other quanta: in the standard approach of background-dependent theories of gravity, photons, gluons, and gravitons all live on the stage on an equal footing. From the point of view gained in this chapter, however, *non-linear gravitons* are at the same time both the stage and the actors within the causal play of photons, gluons, as well as of other 'material characters' like electrons and quarks.

We can, therefore, say that general covariance represents the horizon of a priori possibilities for the physical constitution of spacetime, possibilities that must be actualized within any given solution of the dynamical equations.

We believe in conclusion that these results cast some light over the *intrinsic structure* of general relativistic spacetime that had disappeared behind the Leibniz equivalence. While Leibniz could exploit the principle of sufficient reason since for him space was *uniform*, in GR the upshot is that space (spacetime) is not *uniform* at all and shows a rich *structure*. In a way, in the context of GR, the Leibniz equivalence ends up hiding the very nature of spacetime, instead of disclosing it.

5.5.1 The Nature of Point-Events and Overcoming the Substantivalism/Relationism Debate

In 1972, Bergmann and Komar wrote:

in general relativity the identity of a world point is not preserved under the theory's widest invariance group. This assertion forms the basis for the conjecture that some physical theory

[32] As distinct from what could be called its *ontic* component, corresponding to the Dirac observables.

of the future may teach us how to dispense with world points as the ultimate constituents of spacetime altogether.

Indeed, would it be possible to build a fundamental theory that is grounded in the reduced phase space parametrized by the Dirac observables? This would be an abstract and highly non-local theory of classical gravitation but, transparency aside, it would lack all the epistemic machinery (the gauge freedom) which is indispensable for the application of the theory. Therefore, we see that, even in the context of classical gravitational theory, the spatio-temporal continuum is an *epistemic precondition* playing a role which is not too dissimilar from that enacted by the Minkowski *micro-spacetime* in RQFT. We find here much more than a clear instantiation of the relationship between canonical structure and locality that pervades contemporary theoretical physics throughout.

Can this basic freedom in the choice of the *local realizations* be equated with 'taking away from space and time the last remnant of physical objectivity', as Einstein suggested? We believe that, discounting Einstein's 'spatial worry' with *realism as locality (and separability)*, a significant kind of spatio-temporal objectivity survives. It is true that—*if* the *main conjecture* of LP2 is not verified—the *functional form* of the Dirac observables as well as the stratified holistic structure discussed above depend upon the particular choice of the *GLAB*; yet, there is no a priori *physical* individuation of the manifold points independently of the metric field. As a consequence, there is a sense in which it is not legitimate to say that the individuation procedures corresponding to different gauges individuate *different* point-events. Given the conventional nature of the primary mathematical individuation of manifold points through n-tuples of real numbers, we could say, instead, that the *identity of point-events* is *constituted* by the non-local values of gravitational degrees of freedom, while the underlying point structure of the mathematical manifold may be changed at will. A *really different* physical individuation should only be attributed to different initial conditions for the *Dirac observables* (i.e. to a different 'universe').

Taking into account our results as a whole, we want to spend a few words about their implications for the traditional debate on the absolutist/relationalist dichotomy as well as for some issues surrounding structural realism in general.

First of all, let us recall that, in remarkable contrast with respect to the traditional historical presentation of Newton's absolutism vis-à-vis Leibniz's relationism, Newton had a much deeper understanding of the nature of space and time. In a well-known passage of *De Gravitatione* (see Janiak 2004: 21, 25), he expounds what could be defined as a *structuralist view* of space and time[33]. He writes:

Perhaps now it is maybe expected that I should define extension as substance or accident or else nothing at all. But by no means, for it has its own manner of existence which fits neither substance nor accidents... so the parts of space are individuated by their positions, so that if any two could change their positions, they would change their individuality at the same time and each would be converted numerically into the other *qua* individuals. The parts of duration

[33] This reading has been clearly prefigured by DiSalle (1994).

and space are only understood to be the same as they really are only because of their mutual order and position (*propter solum ordinem et positiones inter se*); nor do they have any *principle of individuation* apart from that order and position, which consequently cannot be altered.

We have just disclosed the fact that the points of general-relativistic spacetimes, quite unlike the points of the homogeneous Newtonian space, are endowed with a remarkably rich *non-pointlike* and *holistic* structure furnished by the metric field. Therefore, the general-relativistic metric field itself or, better, its independent degrees of freedom, have the capacity of characterizing the 'mutual order and positions' of points *dynamically*, and in fact much more than this, since such mutual order is altered by the presence of matter[34].

In conclusion, we agree with Earman and Norton that the hole argument is a decisive blow against strict *manifold substantivalism*. However, the isolation of the intrinsic structure hidden behind the Leibniz equivalence—leading to our *point structuralism*—does not support the *standard* relationist view either.

Indeed, a new kind of *holistic* and *structuralistic* conception of spacetime emerges from our analysis, including elements common to the tradition of both *substantivalism* (spacetime has an autonomous existence independently of other bodies or matter fields) *and relationism* (the physical meaning of spacetime depends upon the relations between bodies or, in modern language, the specific reality of spacetime depends (also) upon the (matter) fields it contains). Indeed, even though the metric field does *not* embody the traditional notion of *substance* (rather than being 'wholly present', it has 'temporal parts'), it *exists* and plays a role for the individuation of point-events. On the other hand, each point-event itself, though holistically individuated by the metric field, has—to paraphrase Newton—'its own manner of existence', since it 'is' the 'values' of the intrinsic degrees of freedom of the gravitational field. Finally, in presence of matter, such values become dependent also on the values of the Dirac observables of matter fields and in the sense specified above they are 'relational'.

More precisely, by referring to Earman's third criterion (R_3) for relationism (see 1989: 14), 'No irreducible, monadic, spatiotemporal properties, like "is located at spacetime point p" appears in a correct analysis of the spatiotemporal idiom', we observe the following. If by 'spacetime point' we mean our *physically individuated point-event* instead of a point on the naked manifold, then—because of the *autonomous* existence of the intrinsic degrees of freedom of the gravitational field (an essential ingredient of GR)—there is a sense in which the above-mentioned spatio-temporal property should be admitted in our spatio-temporal idiom. In another sense, however, our results seem to imply a *different sort of relationism*, since the values of the 'radar' coordinates of any point-event that one obtains after the gauge-fixing procedure are irreducibly and holistically dependent upon the choice of the *GLAB*. Since in a different *GLAB* we obtain different values, such values are not intrinsically possessed. In conclusion, the point-events (the *relata*) exist (entity realism), but their

[34] Let us stress that our results concerning the holistic structure of point-events do not depend upon the *specific* gauge-fixing methodology. If we are interested in 'seeing' that structure of point-events in a given Einstein 'universe', there is no other way but relying on the technique of the intrinsic individuating gauge.

nature (their magnitudes, or properties they instantiate) are *extrinsic* or relational and not intrinsic or *monadic*. This form of relationism, however, unlike the traditional form of spacetime relationism, involves the relational and holistic nature of the properties of point-events.

These remarks show how the structural texture of spacetime in classical GR does not force us to abandon the typical entity realist attitude toward both the metric field and its points. As our case study seems to indicate, we must reject the ontic structural realist claim that (the metrical) *relations* can exist without their *relata* (the points). At the same time, we can distance ourselves from the epistemic structural realist's prudence in denying existence to entities (in our case, point-events): despite their holistic texture, the identity of point-events is sufficiently well characterized by the distinct values of the Dirac observables they exemplify. However, in view of the distinctions introduced at the end of §5.2.1, the fact that the properties of point-events are relationally dependent on the choice of a *GLAB* entails a somewhat *weaker* form of entity realism. In sum, we can use structural realism to defend both (i) a moderate form of theory realism about the approximate truth of Einstein's field equations (within the limits fixed by their domain of application) and (ii) a full-blown realism about spacetime in GR. As far as Quantum Gravity is concerned, the fluctuations of the Dirac observables do not eliminate the structuralist and holistic nature of the coarse-grained texture of 'quantum spacetime'. However, it would be difficult to claim that some kind of intrinsic individuality survives for 'point-events'.

We acknowledge that the validity of our results is restricted to the class of models of GR we worked with. Yet, we were interested in an instantiation of a question of principle, and we wanted to argue that there is a basic class of models of GR embodying both a *real notion of temporal change* and a *new structuralistic and holistic view of spacetime.*

ACKNOWLEDGMENTS

We thank our friend Luca Lusanna for many stimulating discussions and the referee for the comments we received on a previous version of the manuscript.

REFERENCES

Arnowitt, R., S. Deser, and C. W. Misner (1962) "The Dynamics of General Relativity". In L. Witten (ed.), *Gravitation: An Introduction to Current Research*. New York: Wiley, (pp. 227–65).

Belot, G., and J. Earman (1999) "From Metaphysics to Physics". In J. Butterfield and C. Pagonis (eds.), *From Physics to Philosophy*. Cambridge: Cambridge University Press (pp. 167–86).

—— —— (2001) "Pre-Socratic Quantum Gravity". In C. Callender and N. Huggett (eds.), *Physics Meets Philosophy at the Planck Scale: Contemporary Theories in Quantum Gravity*. Cambridge: Cambridge University Press (pp. 213–55).

Bergmann, P. G. (1960) "Observables in General Relativity". *Rev. Mod. Phys.* 33: 510–14.

—— (1962) "The General Theory of Relativity". In S. Flugge (ed.), *Handbuch der Physik*, iv. *Principles of Electrodynamics and Relativity*. Berlin: Springer-Verlag (pp. 247–72).

____ (1977) "Geometry and Observables". In J. S. Earman, C. N. Glymour, and J. Stachel (eds.), *Foundations of Spacetime Theories.* Minnesota Studies in the Philosophy of Science, VIII. Minneapolis: University of Minnesota (pp. 275–80).

____ and A. Komar (1960) "Poisson Brackets between Locally Defined Observables in General Relativity". *Physical Review Letters,* 4: 432–3.

____ ____ (1972) "The Coordinate Group Symmetries of General Relativity". *Int. J. Theor. Phys.* 5: 15–28.

Bueno, O., S. French, and J. Ladyman (2002) "On Representing the Relationship Between the Mathematical and the Empirical." *Philosophy of Science,* 69: 452–73.

Butterfield, J. (1989) "The Hole Truth". *British Journal for the Philosophy of Science,* 40: 1–28.

Cao, T. Y. (2003a) "Structural Realism and the Interpretation of Quantum Field Theory". *Synthese,* 136: 3–24.

____ (2003b) "Can we Dissolve Physical Entities into Mathematical Structures?" *Synthese,* 136: 51–71.

____ (2003c) "What is Ontological Synthesis? A Reply to Simon Saunders". *Synthese,* 136: 107–26.

Chakravartty, A. (1998) "Semirealism". *Studies in History and Philosophy of Science,* 29: 391–408.

Christodoulou, D., and S. Klainermann (1999) *The Global Nonlinear Stability of the Minkowski Space.* Princeton: Princeton University Press.

De Pietri R, L. Lusanna, L. Martucci, and S. Russo (2002) "Dirac's Observables for the Rest-Frame Instant Form of Tetrad Gravity in a Completely Fixed 3-Orthogonal Gauge". *General Relativity and Gravitation,* 34: 877–922.

DiSalle, R. (1994) "On Dynamics, Indiscernibility, and Spacetime Ontology". *British Journal for the Philosophy of Science,* 54: 265–87.

Dorato, M. (2000) "Substantivalism, Relationism, and Structural Spacetime Realism". *Foundations of Physics,* 30: 1605–28.

Earman J. (1989) *World Enough and Spacetime.* Cambridge, MA: MIT Press.

____ (2002) "Thoroughly Modern McTaggart or what McTaggart would have said if He had Read the General Theory of Relativity". *Philosophers' Imprint,* 2 (3), **www.philosophersimprint.org/002003/.**

____ and J. Norton (1987) "What Price Spacetime Substantivalism? The Hole Story". *British Journal for the Philosophy of Science,* 38: 515–25.

Einstein, A. (1914) "Die formale Grundlage der allgemeinen Relativitätstheorie". *Preuss. Akad. der Wiss. Sitz.* (pp. 1030–85).

____ (1916) "Die Grundlage der allgemeinen Relativitätstheorie". *Annalen der Physik,* 49: 769–822; translation by W. Perrett and G. B. Jeffrey (1952) "The Foundation of the General Theory of Relativity". In *The Principle of Relativity.* New York: Dover (pp. 117–18).

French, S., and J. Ladyman (2003) "Remodeling Structural Realism". *Synthese,* 136: 31–56.

Friedman, M. (1984) 'Critical Review of Roberto Torretti: "Relativity and Geometry"'. *Noûs,* 18: 653–64.

____ (2001) "Geometry as a Branch of Physics: Background and Context for Einstein's 'Geometry and Experience'". In D. Malament (ed.), *Reading Natural Philosophy: Essays in the History and Philosophy of Science and Mathematics to Honor Howard Stein on his 70th Birthday.* Chicago: Open Court.

Friedrich, H., and A. Rendall (2000) "The Cauchy Problem for Einstein Equations". In B. G. Schmidt (ed.), *Einstein's Field Equations and their Physical Interpretation.* Berlin: Springer. ArXiv:gr-qc/0002074.

Hacking, I. (1983) *Representing and Intervening*. Cambridge: Cambridge University Press.
Henneaux, M., and C. Teitelboim (1992) *Quantization of Gauge Systems*. Princeton: Princeton University Press.
Hilbert, D. (1917) "Die Grundlagen der Physik. (Zweite Mitteilung)". *Nachrichten von der Königlichen Gesellschaft der Wissenschaften zu Göttingen, Mathematisch-physikalische Klasse* (pp. 53–76).
Janiak, A. (2004) *Newton's Philosophical Writings*. Cambridge: Cambridge University Press.
Károlyházy, F., A. Frenkel, and B. Lukács (1985) "On the Possible Role of Gravity in the Reduction of the Wave Function". In R. Penrose and C. J. Isham (eds.), *Quantum Concepts in Space and Time*. Oxford: Clarendon Press (pp. 109–28).
Komar, A. (1955) "Degenerate Scalar Invariants and the Groups of Motion of a Riemann Space". *Proc. Natl. Acad. Sci.* 41: 758–62.
—— (1958) "Construction of a Complete Set of Independent Observables in the General Theory of Relativity". *Physical Review*, 111: 1182–7.
Ladyman, J. (1998) "What is Structural Realism?" *Studies in the History and Philosophy of Science*, 29: 409–24.
Lusanna, L. (2000) "Towards a Unified Description of the Four Interactions in Terms of Dirac–Bergmann Observables". In A. N. Mitra (ed.), *Quantum Field Theory*. New Delhi Indian National Science Academy (pp. 490–518).
—— (2001) "The Rest-Frame Instant Form of Metric Gravity". *General Relativity Gravitation*, 33: 1579–696.
—— and M. Pauri (2004a) "The Physical Role of Gravitational and Gauge Degrees of Freedom in General Relativity - I: Dynamical Synchronization and Generalized Inertial Effects". ArXiv:gr-qc/040308 v2; forthcoming in *General Relativity and Gravitation*, February 2006.
—— —— (2004b) "The Physical Role of Gravitational and Gauge Degrees of Freedom in General Relativity - II: Dirac versus Bergmann Observables and the Objectivity of Spacetime". ArXiv:gr-qc/0407007; forthcoming in *General Relativity and Gravitation*, February 2006.
Massimi, M. (2004) "Non-defensible Middle Ground for Experimental Realism: Why we are Justified to Believe in Coloured Quarks". *Philosophy of Science*, 71: 36–60.
Maudlin, T. (1990) "Substances and Spacetimes: What Aristotle would have Said to Einstein". *Studies in the History and Philosophy of Science*, 21: 531–61.
Morganti, M. (2004) "On the Preferability of Epistemic Structural Realism". *Synthese*, 142: 81–108.
Norton, J. (1987) "Einstein, the Hole Argument and the Reality of Space". In J. Forge (ed.), *Measurement, Realism and Objectivity*. Dordrecht: Reidel (pp. 153–88).
—— (1992) "The Physical Content of General Covariance". In J. Eisenstaedt and A. Kox (eds.), *Studies in the History of General Relativity*. Einstein Studies 3. Boston: Birkhäuser (pp. 281–315).
—— (1993) "General Covariance and the Foundations of General Relativity: Eight Decades of Dispute". *Rep. Prog. Phys.* 56: 791–858.
Pauri, M. (1996) "Realtà e oggettività". In *L'Oggettività nella conoscenza scientifica* Brescia: F. Angeli (pp. 79–112).
—— (2000) "Leibniz, Kant and the *Quantum*: A Provocative Point of View about Observation, Spacetime, and the Mind-Body Issue". In E. Agazzi and M. Pauri (eds.), *The Reality of the Unobservable*. Boston Studies in the Philosophy of Science, 215. Dordrecht: Kluwer Academic Publishers (pp. 257–82).
—— and M. Vallisneri (2002) "Ephemeral Point-Events: Is there a Last Remnant of Physical Objectivity?" *Dialogos*, 79: 263–303.

Penrose, R. (1985) "Gravity and State Vector Reduction". In R. Penrose and C. J. Isham (eds.), *Quantum Concepts in Space and Time.* Oxford: Clarendon Press (pp. 129–46).
Poincaré, H. (1905) *La Science et l'hypothèse.* Paris: Flammarion.
Rynasiewicz, R. (1996) "Absolute versus Relational Spacetime: An Outmoded Debate?" *Journal of Philosophy*, 43: 279–306.
Shanmugadhasan, S. (1973) "Canonical Formalism for Degenerate Lagrangians". *Journ. Math. Phys.* 14: 677–87.
Stachel, J. (1980) "Einstein's Search for General Covariance, 1912–1915". In D. Howard and J. Stachel (eds.), *Einstein and the History of General Relativity: Einstein Studies*, vol. i. Boston: Birkhäuser, 1986 (pp. 63–100).
____(1986) "What can a Physicist Learn from the Discovery of General Relativity?" In R. Ruffini (ed.), *Proceedings of the Fourth Marcel Grossmann Meeting on Recent Developments in General Relativity, Rome, 1–21 June, 1985.* Amsterdam: North-Holland.
____(1993) "The Meaning of General Covariance: The Hole Story". In J. Earman, I. Janis, G. J. Massey, and N. Rescher (eds.), *Philosophical Problems of the Internal and External Worlds: Essays on the Philosophy of Adolf Grünbaum.* Pittsburgh: University of Pittsburgh Press (pp. 129–60).
____and M. Iftime (2005) "Fibered Manifolds, Natural Bundles, Structured Sets, G-Sets and all that: The Hole Story from Space Time to Elementary Particles". ArXiv:gr-qc/0505138 v2, 28 May 2005.
Stewart, J. (1993). *Advanced General Relativity.* Cambridge: Cambridge University Press.
Wald, R. M. (1984) *General Relativity.* Chicago: University of Chicago Press.
Weyl, H. (1946) "Groups, Klein's Erlangen Program. Quantities". Ch. I, §4 of *The Classical Groups, their Invariants and Representations*, 2nd edn. Princeton: Princeton University Press (pp. 13–23).
Worrall, J. (1989) "Structural Realism: The Best of all Possible Worlds?" *Dialectica*, 43: 99–124.

6

Time and Structure in Canonical Gravity

Dean Rickles

ABSTRACT

In this chapter I wish to make some headway on understanding what *kind* of problem the 'problem of time' is, and offer a possible resolution—or, rather, a new way of understanding an old resolution.[1] The response I give is a variation on a theme of Rovelli's *evolving constants of motion* strategy (more generally: 'correlation' strategies). I argue that by giving correlation strategies a *structuralist* basis, a number of objections to the standard account can be blunted. Moreover, I show that the account I offer provides a suitable ontology for time (and space) in both classical and quantum canonical general relativity.

6.1 INTRODUCTION

Interpreting modern-day fundamental physical theories is hard. Our four best theories—three quantum field theories (describing the strong, electro-weak, and electromagnetic forces) and one classical field theory describing gravity—are *gauge* theories.[2] Interpreting these theories is complicated by the presence of a special class of symmetries (gauge symmetries) whose action does not 'disturb' any 'qualitative' properties and relations; only non-observable, non-qualitative features of a

[1] I am referring to the problem of time that appears in *canonical* formulations of both classical and quantum GR, and also in certain diffeomorphism-invariant covariant quantum field theories (e.g. topological quantum field theories: see Baez (this volume) for a clear and elementary account). More generally, though I cannot demonstrate the fact here, *any* theory that is independent of a fixed metric (or connection) on space or spacetime will be subject to the problems considered here. Since it is likely that the 'final' theory of quantum gravity will be of this form, the problem of time will almost inevitably be a problem for that theory, or, at least, will play a role in its development and eventual formulation.

[2] I should point out that this claim is not entirely uncontentious. Weinstein (2001) has argued that certain features of general relativity—namely, the fact that the gauge groups of the first three theories are Lie groups and can be viewed as acting at spacetime points whereas in general relativity the candidate for the gauge group (the diffeomorphism group) acts on the points themselves and is not a Lie group—debar it from being classified as a gauge theory proper. See Earman (2003a) for a defence of the contrary view based on the Hamiltonian formulation of general relativity.

theory (or family of models) are affected.[3] This leads to empirically superfluous elements—'surplus structure' in Redhead's sense (Redhead 1975); 'gauge freedom' in physicists' jargon—in the description of such theories that must be be dealt with in some way, either by 'elimination' or 'accommodation'. While classically inert, the decision regarding how to deal with the gauge freedom can lead to non-trivial differences at the quantum level (i.e. inequivalent quantizations). The root cause of interpretative headaches in the context of gauge theories is, then, the gauge freedom; the problem facing philosophers (and physicists!) is to explicate and provide some account of both the gauge symmetries and the elements that are acted upon by those symmetries.

The interpretative problems of gauge theory take on what is arguably their most pathological form in the context of the problem of space (better known as the 'hole argument') and the problem of time.[4] I will argue that the latter problem is essentially just a recapitulation of the former, although focused upon the Hamiltonian rather than the diffeomorphism constraint. Therefore, I think that one should respond to the problems in the same way: I favour a non-reductive gauge-invariant conception of observables coupled with a kind of structuralism. My main aims in this chapter are as follows: (1) to explain the problem of time in a way that is accessible to philosophers; (2) to provide a critique of the usual responses; (3) to disentangle the debate between substantivalists and relationalists from the problem of time; and (4) to defend a structuralist resolution of the problem of time.

6.2 CONSTRAINTS, GAUGE, AND HOLES

In their recent survey of the problem of time in quantum gravity, Belot and Earman note that there is a 'sentiment—which is widespread among physicists working on canonical quantum gravity—that there is a tight connection between the interpretive problems of general relativity and the technical and conceptual problems of quantum gravity' (2001: 214). Belot and Earman share this sentiment, and go even further in claiming that certain proposals for understanding the general covariance of general relativity *underwrite* specific proposals for quantizing gravity. These proposals are then seen as being linked to 'interpretive views concerning the ontological status of spacetime' (ibid.). I agree with their former claim but strongly disagree with the latter: such proposals cannot be seen as linked with stances concerning the ontological

[3] Belot (2003) offers a detailed philosophical survey of gauge theories; I refer the reader unacquainted with the basic details of the concept of 'gauge' to this insightful article. Redhead (2003) is an exceptionally clear, and more elementary, guide to the interpretation of gauge theories. Earman (2003a) examines the concepts of gauge theory from the perspective of the constrained Hamiltonian formalism—indeed, Earman (2003a: 153) speaks of the constrained Hamiltonian formalism as an 'apparatus ... used to detect gauge freedom'.

[4] The best places to learn about the problem of time are (still) Isham (1994) and Kuchař (1992). Belot and Earman (1999, 2001) give two excellent philosophical examinations of the problem; the latter is more comprehensive and technically demanding than the former. I am much indebted to this quadruplet of articles.

status of spacetime vis-à-vis relationalism vs. substantivalism (for reasons that will be discussed in what follows).

The crucial claim they make, for the purposes of this chapter, is that the *gauge invariance* reading of the general covariance of general relativity 'seems to force us to accept that change is not a fundamental reality in classical and quantum gravity' (ibid.). I agree with Belot and Earman that, like the hole argument, the problem of time is an aspect of the more general problem of interpreting gauge theories. I also agree with Earman's claim that the problems do not only have teeth in the quantum context, but bite in the classical context too (see Earman 2002: 6)—indeed, I don't find all that much to distinguish the two cases. In order to fully appreciate this problem, we need to take a brief detour to introduce a variety of concepts: gauge and constraints; phase spaces and possible worlds; and the interpretative problems and and options in gauge theory, including the hole argument.

6.2.1 Hamiltonian Systems: Constraints and Gauge

In this section I introduce the Hamiltonian formalism of theories, and show how the constraints arise in systems whose description possesses surplus structure.[5] I relate the presence of a certain class of constraints (those that are first class) to the presence of gauge freedom. Finally, I outline, in broad strokes, how one tackles the problem of interpreting the theories considered. This brief primer should provide enough of the technical apparatus required to understand the classical and quantum problems of time and change.

A Hamiltonian system is represented by a triple $\langle \Gamma, \omega, \mathsf{H} \rangle$ consisting of a manifold Γ (the cotangent bundle $\mathsf{T}^*\mathcal{Q}$, where $\mathcal{Q}$ is the configuration space of a system), a tensor ω (a symplectic, closed, non-degenerate 2-form), and a function H (the Hamiltonian $\mathsf{H} : \Gamma \to \mathbb{R}$). These elements interact to give the kinematical and dynamical structure of a classical theory. The manifold inherits its structure from the tensor, making it into a phase space with a symplectic geometry. The points of this space are taken to represent physically possible states of some classical system (i.e. set of particles, a system of fields, a fluid, etc.). Finally the Hamiltonian function selects a class of curves from the phase space that are taken to represent physically possible histories of the system (given the symplectic structure of the space). Any system represented by such a triple will be deterministic in the sense that knowing which phase point represents the state of the system at an initial time, there will be a *unique* curve through that point whose points represent the past and future states of the system.[6] The physical interpretation of this framework is as follows. Recall that

[5] The presentation I give here relies heavily upon Dirac (1964), Henneaux and Teitelboim (1992), and the articles in Ehlers and Friedrich (1994).

[6] In a little more detail: Hamilton's equations determine a map $\mathsf{f} \to \mathsf{X}_\mathsf{f}$ between smooth functions f on Γ and vector fields X_f on Γ. Integrating a vector field X_f associated to the smooth function f gives a unique curve through each point of Γ. The symplectic structure gives the set $C^\infty(\Gamma)$ of smooth function on Γ the structure of a Poisson algebra by means of the Poisson bracket $\{\mathsf{f}, \mathsf{g}\}$ between pairs of functions f and g. $\{\mathsf{f}, \mathsf{g}\}$ is interpreted as giving the rate of change of g with respect to the set of curves generated by f such that g is constant along the curves generated by f just

the phase space is given by the cotangent bundle of the configuration space, where points of the configuration space represent possible instantaneous configurations of some system (relative to an inertial frame). The cotangent bundle is the set of pairs (q, p), where q is an element of the configuration space and p is a covector at q. Thinking of q as representing the position of a system leads to the view that p represents that system's momentum. The value of the Hamiltonian at a point of phase space is the energy of the system whose state is represented by that point. The *physically measurable* properties of a Hamiltonian system are described by functions $A(q, p) : \Gamma \rightarrow \mathbb{R}$ in terms of a canonical basis (a set of canonical variables), with position q_i and momenta p_i, satisfying Poisson bracket relations:

$$\{q_i, p_j\} = \delta_{ij} \tag{6.1}$$

Systems described in such terms are rather simple to interpret: each point, (p, q), in the phase space represents a distinct physically possible world. Furthermore, since there is a unique curve through each point of phase space, one can interpret the phase space as *directly* representing the physically possible states of a system, and the curves as *directly* representing the physically possible histories of a system. A simple one-to-one understanding of the representation relation is possible that does not lead to indeterminism or underdetermination as regards the canonical variables, the possibilities, or the possible worlds.

Weakening the geometry of the phase space, and moving to gauge systems, however, puts pressure on this simple direct interpretation,[7] precisely because indeterminism breaks down and the canonical variables are underdetermined. When one considers systems with redundant variables and symmetries—such as Maxwell's theory and general relativity—the formulation contains constraints, where the constraints are relations of the form $\phi_m(q_i, p_i) = 0$ $(i = 1, \ldots, m)$ holding between the canonical variables. Such constraints are a by product of the Legendre transform taking one from a Lagrangian to a Hamiltonian description of a system.[8] These are known as *primary constraints*. If these constraints should be preserved by evolution a new set of constraints is generated to carry out this job. These are called *secondary constraints*.

in case $\{f, g\} = 0$. For any observable A (a function of the canonical variables), the time evolution is given by $\dot{A} = \{A, H\}$.

[7] Note that I don't say that such an interpretation isn't possible. It is, provided one either accepts the consequence of indeterminism and underdetermination, or else finds another way to deal with them.

[8] The idea of gauge freedom manifests itself at the level of the Lagrangian formalism too. The action principle $\delta \int \mathcal{L}(q, \dot{q})dt = 0$ allows us to derive Euler–Lagrange equations. Sometimes—in general relativity, for example—these equations will be non-hyperbolic, they can't be solved for all accelerations. This results in a *singular* Lagrangian, revealing itself in the singularity of the Hessian $\partial^2\mathcal{L}/\partial\dot{q}^k\partial\dot{q}^h$. This implies that when we Legendre transform to the Hamiltonian formulation, the canonical momenta are not independent, but will satisfy a set of relations called primary constraints, related to the identities of the Lagrange formalism. As I mention below, preserving these under evolution may require the imposition of higher-order constraints. Once one has a situation where all the constraints are preserved by the motion, one will have defined a submanifold where all of the constraints are satisfied—this is the 'constraint surface' $\mathcal{C}$. See Earman (2003a: 144–5) for a clear explanation of these constraints and their relation to the singularity of the Hessian.

One may wish to repeat the procedure on these, resulting in *tertiary constraints*, and so on.

The first change to note in the shift from a Hamiltonian system to a constrained Hamiltonian system is that the symplectic form is replaced by a *presymplectic form* σ, so that the phase space $\mathcal{C}$ of a gauge system inherits its geometrical structure from this. The presymplectic form induces a partitioning of the phase space into subspaces (not necessarily manifolds) known as *gauge orbits*, such that each point x in the phase space lies in exactly one orbit $[x]$. Once again we choose a Hamiltonian function on phase space, such that the value at a phase point represents the energy. However, in this case, given the weaker geometrical structure induced by the presymplectic form, the Hamiltonian is not able to determine a unique curve through the phase points. Instead, there are infinitely many curves through the points. However, the presymplectic form *does* supply the phase space with sufficient structure to determine which gauge orbit a point representing the past or future state will lie in. Hence, for two curves $t \rightarrow x(t)$ and $t \rightarrow x'(t)$ intersecting the same initial phase point $x(0)$, we find that the gauge orbit containing $x(t)$ is the same as that containing $x'(t)$: i.e., $[x(t)] = [x'(t)]$.

In a constrained system, each classical observable is represented by a function $P : \mathcal{C} \rightarrow \mathbb{R}$ on the phase space. But given that the future phase points of an initial phase point are underdetermined, it will be impossible to uniquely predict the future value of the observables. Hence, there appears to be a breakdown of determinism; the initial-value problem does not appear to be well posed, as it is for standard Hamiltonian systems. The reason is clear enough: there is a unique curve through each phase point in a Hamiltonian system but infinitely many curves through the phase points of a gauge system.

Yet there are many theories that are gauge theories and that are evidently *not* indeterministic in any pathological sense. The trick for restoring determinism and recovering a well-posed initial-value problem is to be *restrictive* about what one takes the observables to be. Rather than allowing *any* real-valued functions on the phase space to represent physical observables, one simply chooses those that are *constant* on gauge orbits, such that if $[x] = [y]$ then $f(x) = f(y)$. Such quantities are said to be *gauge invariant*. The initial-value problem is well posed for such quantities since for an initial state $x_{t=0}$, and curves $x(t)$ and $x'(t)$ through $x_{t=0}$, $f[x(t_1)] = f[x'(t_1)]$.

Another important distinction—perhaps the most important as far as the problem of time goes—between constraints is that holding between *first class* and *second class constraints*. A constraint ϕ_k is said to be first class if its Poisson bracket with any other constraints is given as a linear combination of the constraints:

$$\{\phi_k, \phi_i\} = C^j_{ki}\phi_j, \ \forall i. \tag{6.2}$$

Any constraint not satisfying these relations is second class. Our sole concern is with the first class constraints. The appearance of such constraints in a theory implies that the dynamics is restricted to a submanifold $\mathcal{C}$ of the full phase space Γ;—i.e. the *constraint surface*. Dynamical evolution on $\mathcal{C}$ has a representation in terms of an infinite family of physically equivalent trajectories. This is how the appearance of gauge freedom is represented in the constrained Hamiltonian formalism. Projecting out from $\mathcal{C}$ to Γ results in ambiguity, for any quantities that differ only by a combination of constraints come out as equal on $\mathcal{C}$. This ambiguity is a formal counterpart of the 'many-one' problems encountered in both electrodynamics formulated in terms of the vector potential and the hole argument (touched upon below); it can be seen, as such, as the origin of one kind of surplus structure; namely, that associated with gauge freedom.

A dynamical variable P (a function of the ps and qs: $P(q, p)$) is first class iff it has *weakly vanishing* Poisson bracket with all of the constraints:[9]

$$\{P, \phi_j\} \approx 0, \; j = 1, \ldots, j. \tag{6.3}$$

These quantities comprise the observables of the classical theory. They are defined by their invariance under the symmetries generated by the constraints. These symmetries are the gauge symmetries of the theory; thus, in a gauge theory the observables are defined by gauge invariance.

The constraints occurring in general relativity are all first class, implying that they generate gauge transformations. Crucially, the constraints also make up the Hamiltonian of general relativity: it is a sum of first class constraints. In a constrained Hamiltonian system, the observables must commute with the Hamiltonian since it is a constraint (or, rather, a linear combination of such)—in a gauge theory this translates into the condition that the observables must be gauge invariant. As always, the Hamiltonian generates motion via Poisson brackets of observables with the Hamiltonian. In this case, since the Hamiltonian vanishes on $\mathcal{C}$, this implies that motion is 'pure gauge'. Already we see a potential problem for the evolution of the theory's observables if the observables are defined to be the gauge-invariant quantities. The problem is this: the constraints of the theory pick out a submanifold (the constraint surface) on which observables must have vanishing Poisson bracket with the constraints. In the case of the Hamiltonian constraint (on which more below), the different points of this manifold correspond to states of the system at different times (indexed by parameter time τ). Since the constraints generate gauge transformations (i.e. along a gauge orbit) this implies that time evolution is itself a gauge transformation! This, in capsule form, is the problem of the frozen formalism of the classical theory. Let me say a little more about the kinds of constraints that appear in general relativity and how the concept of gauge freedom arises in this context.

6.2.2 Constraints and Gauge in General Relativity

The Lagrangian for general relativity contains a number of variables appearing without their corresponding velocities.[10] This implies that when we define the canonical momenta $p_i = \partial\mathcal{L}/\partial\dot{q}_i$ of the Hamiltonian formulation, we find that they vanish.

[9] The condition of weak vanishing refers to equality on the constraint surface embedded in the phase space. I say more about this in §6.2.2.

[10] Such terms become Lagrange multipliers in the Hamiltonian formulation. There are two types: the lapse function N and the shift vector N^i. These two expressions tell us how much a slice Σ is to be 'pushed forwards in time': the former acts *normally* and the latter *tangentially*.

This is a sure sign that the Hamiltonian formulation will possess constraints. Two families of constraints are picked up when we perform the Legendre transform from the Lagrangian to the Hamiltonian formulation of general relativity: diffeomorphism constraints and Hamiltonian constraints—three diffeomorphism constraints per space point and one Hamiltonian constraint per space point.[11] The diffeomorphism constraints generate infinitesimal transformations (three-dimensional diffeomorphisms) of Σ onto itself; they have the effect of 'sliding' Cauchy data along Σ in the direction of the shift vector N^i. The Hamiltonian constraints generate infinitesimal transformations of Σ onto some slice Σ' displaced normally to Σ in $\mathcal{M}$; hence, data is 'pushed' orthogonal to Σ in the direction of the lapse function N. The Hamiltonian of general relativity is a sum of these constraints such that setting lapse to zero gives a Hamiltonian that is identical to the diffeomorphism constraint and setting the shift to zero gives a Hamiltonian that is identical to the Hamiltonian constraint.

Recall that in geometrodynamics (cf. Arnowitt et al. 1962) the points in the phase space of GR are given by pairs (q, p) —where q is a Riemannian metric on a 3-manifold Σ and p is related to the extrinsic curvature K of Σ describing the way it is embedded in a four-dimensional Lorentzian manifold. In GR, the pair must satisfy the four constraint equations, and this condition picks out a surface in the phase space called the constraint surface. The observables of the theory are those quantities that have vanishing Poisson bracket with all of the constraints.[12] According to the geometrodynamical programme, each point on the constraint surface represents a physically possible (i.e. by the lights of general relativity) spacelike hypersurface of a general relativistic spacetime. Points lying on the complement of this surface are also 3-manifolds, but they do not represent physically possible spacetimes; they have metric and extrinsic curvature tensors that are incompatible with those needed to qualify as a 3-space imbedded in a general relativistic spacetime: if anything, they represent physically *impossible* states.

The constraint surface comes equipped with a set of transformations $\mathcal{C} \rightarrow \mathcal{C}$ that partition the surface into subspaces known as 'gauge orbits' (the transformations are the gauge transformations). The natural interpretation of the gauge orbits is

[11] In the connection formalism a further constraint is picked up, namely the Gauss constraint. This generates infinitesimal (global) gauge transformations. It is the only constraint that Yang–Mills theories possess, and, since these are taken to be gauge theories *par excellence*, this might provide further motivation for gauge theoretical interpretations of general relativity.

[12] Much has been made of the fact that the Poisson bracket algebra of the constraints does not close, and, therefore, does not form a Lie algebra. Steven Weinstein, for one, argues that this feature mitigates against viewing general relativity as a gauge theory. This leads him to the view that diffeomorphisms should not be viewed as gauge transformations (cf. Weinstein 1999). In fact, a more general structure called a Dirac algebra is formed that has the group of spatial diffeomorphisms, Diff(Σ), as a subgroup. This has been interpreted as implying that general relativity is not, properly speaking, a gauge theory, since it lacks a feature of Yang–Mills theories—the term 'gauge theory' commonly being reserved for Yang–Mills theories (cf. Earman 2003a: 151). I agree with Earman that this is largely 'label mongering' (2003a: 151-2). We can use 'gauge' to refer to Yang–Mills theories or we can use it to refer to theories containing arbitrary functions of time. We might even use the term more generally to refer to theories containing 'redundancy' of a certain specified type. However, it might still be instructive to see what feature is missing from GR that supposedly robs it of gauge theory status.

as representing equivalence classes of diffeomorphic models of general relativistic spacetimes. We face the problem we faced in interpreting electrodynamics: do we take the points of the orbits to represent the same state of affairs or does each point represent a distinct possibility? This leads us into the general problem of interpreting gauge theories (and, in particular, *gauge freedom*). In the case of general relativity the gauge freedom concerns the points of the spatial manifold and how the metric field (and other fields) are to be spread out over them: the intrinsic geometry of the metric is indifferent as to which points play which role in the overall relational structure determined by the fields. Satisfaction of the constraints by a solution gives a class of 'spreadings' that are compatible with Einstein's equation and some—those related by gauge transformations—may differ *only* in how the fields are spread about over the points. The hole argument uses general covariance (active diffeomorphism invariance) to demonstrate that a manifold substantivalist conception of spacetime—i.e. the view that spacetime points are real and have their identities fixed independently of any fields defined with respect to them—implies that general relativity is indeterministic. The conclusion follows by applying a diffeomorphism to any dynamical fields to the future of an initial slice through spacetime; general covariance implies that the resulting pair of diffeomorphic models (differing in how the metric is distributed over the points) solve Einstein's equation; therefore, if the points are real then the equations of motion cannot determine how the metric will evolve into the future. This procedure is essentially reapplied in the case of the problem of time: since the data on an initial slice is gauge equivalent to that on a later slice (i.e. time evolution is a gauge transformation—a diffeomorphism) they must describe a *qualitatively* identical state of affairs, differing only in which points lie under which bits of the fields. However, a substantivalist will, on the above view, have to keep them apart, giving a peculiar indeterministic world in which nothing observable (qualitative) changes! However, the prospects are no better for a relationalist, who will generally have to identify gauge-equivalent states, for the time-evolved slices will have to be identified, thus freezing out *any* kind of evolution and eradicating change.

6.2.3 Interpreting Gauge Theories

From what I have said so far we can see that there are two competing interpretations of a gauge theory: on the one hand there is a one-to-one interpretation of the phase points where we are now viewing the constraint surface as phase space, such that each point (curve) represents a distinct possible state (history) of a system; on the other hand there is a many-to-one interpretation according to which many phase points (namely, those within the same gauge orbit) represent a single possible state of a system.[13] The former leads to indeterminism and (if not supplemented

[13] This option is available because the phase points lying within the same gauge orbit are related by a gauge transformation: if they represent real possibilities then they represent qualitatively indistinguishable possibilities differing solely with respect to which individuals get which properties. Hence, the one-to-one interpretation of the representation relation if interpreted simplistically will lead to haecceitistic differences between the worlds represented by the solutions.

by a gauge-invariant account of the observables) an ill-posed initial-value problem, while the latter involves surplus structure that can be eradicated, but only in a way that violates such things as locality and (manifest) covariance.[14] Hence, though the interpretations will be empirically equivalent (at least, at the classical level) the choice is, ontologically speaking, a non-trivial matter.

The key problem in trying to interpret gauge theories is knowing what to do with the gauge freedom, the surplus that results from the equivalence of the points within the same gauge orbits (ontologically: the indistinguishability of the worlds represented by such points). There are multiple options, and hence, multiple ways of interpreting gauge theories. Let us call an interpretation that takes each phase point as representing a distinct physically possible state of a system a *direct* interpretation. Hence, each point x^i in a gauge orbit $[x]$ represents a distinct possibility. However, such a direct interpretation leads to a form of indeterminism for the reasons outlined in §6.2.1. But, since each of the phase points represents a distinct physical possibility, there is (strictly speaking) no surplus structure according to such an interpretation perhaps with the exception of the 'impossible' states: each bit of the formalism plays a role in representing reality. Recall also that the indeterminism is of a very peculiar kind: the multiple futures that were compatible with an initial state were physically (read 'qualitatively') indistinguishable, for they are represented by points lying within the same gauge orbit. Hence, the indeterminism concerns haecceitistic differences. However, for realists the indeterminism will still constitute a problem, though it is not insurmountable. As Belot notes (1998: 538):

> if we supplement this account of the ontology of the theory with an account of measurement which implies that its observable quantities are gauge-invariant, then the indeterminism will not interfere with our ability to derive deterministic predictions from the theory.

Using this method one can help oneself to gauge invariance at the level of observable ontology and remain neutral about the rest (spacetime points, quantum particles, shifted worlds, vector potentials, etc.).

Let us call an interpretation that takes many phase points (from within the same gauge orbit) as representing a single physically possible state of a system an *indirect* interpretation. There are two ways of achieving such an interpretation. The first method simply takes the representation relation between phase points from within the same gauge orbit and physically possible states to be many-to-one. Since the points of a gauge orbit represent physically indistinguishable possibilities, there is no indeterminism on this approach. Redhead suggests that 'the "physical" degrees of freedom [i.e. the fields] at [a future] time t are being multiply represented by points on the gauge orbit ... in terms of the "unphysical" degrees of freedom' (2003: 130).[15]

[14] Locality is lost since the points of gauge orbits represent states that differ in how a catalogue of properties gets distributed over a domain of points; since such points are identified in many-to-one accounts, the notion of properties attaching to points is lost—though this has been contested (quite rightly, in my opinion) on the grounds that the properties can be seen as (dynamically) 'individuating' the points (cf. Pooley, this volume). Covariance is seen to be put under pressure by the fact that the original symmetry is removed in some many-to-one accounts.

[15] Redhead's analysis seems to suggest that this is the *only* way to interpret the direct formulation (speaking in terms of vector potentials)—though he mentions that a gauge-invariant or gauge-fixing

The gauge freedom is simply an artefact of the formalism. There are superficial similarities between this approach and the modified direct approach mentioned by Belot above. However, the stance taken on this approach is that not all of the phase points represent distinct possibilities. Even on the modified direct approach this is false. The latter approach simply says that the question of whether or not all of the phase points represent distinct possibilities is irrelevant to the observable content of the theory; the observables are indifferent as to what state underlies them provided the states are gauge-related.

The second method involves treating the gauge orbits rather than phase points as the fundamental objects of one's theory. By taking the set of gauge orbits as the points of a new space, and endowing this set with a symplectic structure, one can construct a phase space for a Hamiltonian system—this new space is known as the *reduced phase space*,[16] and the original is the enlarged phase space.[17] Hence, the procedure amounts to giving a *direct* interpretation of the reduced phase space—i.e. one that takes each gauge orbit as representing a distinct physically possible state—but an *indirect* interpretation of the enlarged phase space. The resulting system is deterministic since real-valued functions on the reduced space correspond to gauge-invariant functions on the enlarged space. In effect, the structure of the reduced space *encodes* all of the gauge-invariant information of the enlarged space even though no gauge symmetry remains (i.e. there is no gauge freedom). Note, however, that complications can arise in reduced space methods: the reduced space might not have the structure of a manifold, and so will not be able to play the role of a phase space; or some phenomena might arise that requires the gauge freedom to be retained, such as the Aharonov–Bohm effect (cf. Earman 2003a: 158–9 and Redhead 2003: 132). If these complications do arise, one can nonetheless stick to the claim that complete gauge orbits represent single possible worlds, as per the above method.[18]

account can resolve the indeterminism. But clearly, it is open to us to give a direct interpretation and accept the qualitatively indistinguishable worlds that are represented by the isomorphic futures (points within the gauge orbit).

[16] In order to distinguish this approach from the previous one, let us call it a *reductive* interpretation. Note that this matches Leibniz's form of relationalism since it can be seen as enforcing the Principle of Identity of Indiscernibles ($\forall F \forall xy : Fx \equiv Fy \rightarrow x = y$) on phase points within the same gauge orbit. Thus, to complete the analogy, an enlarged phase space Γ would correspond to that containing phase points related by the symmetries associated with $\mathcal{G}_N$ (the Galilean group of Newtonian mechanics representing indistinguishable shifted, rotated, and boosted worlds) and the reduced phase space $\tilde{\Gamma}$ would correspond to the space with the symmetries removed: $\tilde{\Gamma} = \Gamma/\mathcal{G}_N$.

[17] Thus the points of the reduced space correspond to gauge orbits of the original enlarged space. Curves in the reduced space contain information about which gauge orbits the system (as represented by the enlarged space) passes through.

[18] One can even help oneself to haecceitistic notions on this interpretation by utilizing Lewis's idea of 'cheap quasi-haecceitism' (1983: 395): as long as one distinguishes between *possibilities* and *possible worlds* one can view each gauge orbit as the sum total of possibilities compatible with a single world. On the reduced account this option is not available: hence, the desire to accommodate certain modal talk and concepts may be called upon to play a role in the choice of representational geometric space.

There is another method that involves taking only a single phase point from each gauge orbit as representing a physically possible state of a system. To do this one must introduce *gauge-fixing* conditions that pick out a subset of phase points (a *gauge slice*) such that each element of this subset is a unique representative from each gauge orbit (cf. Govaerts 2001: 63). Gauge fixing thus 'freezes out' the gauge freedom of the enlarged phase space.[19] This method leads to an interpretation that is neither direct nor indirect; I shall call it a *selective* interpretation. There is a serious problem—known as a *Gribov obstruction* (ibid. 64)—facing certain gauge-fixing procedures, for some lead to different coverings of the space of gauge orbits that, while being gauge invariant, are not physically equivalent. The obstruction implies that the gauge conditions do not result in a unique 'slicing' of phase space, but may result in the selection of two or more points from within the same gauge orbit.[20]

Each of these interpretative options is seen to be applicable in both general relativity and quantum gravity; indeed, they are seen to play a crucial role in both their technical and philosophical foundations, though not, I say, to the extent that Belot and Earman suggest. Recall that the hole argument is based upon a direct, local interpretation of the models of general relativity. The argument is connected to the nature of spacetime since the gauge freedom is given by (active) diffeomorphisms of spacetime points (or by 'drag-alongs' of fields over spacetime points). What we appear to have in the hole argument is an expression of the old Leibniz shift argument couched in the language of the models of general relativity (*qua* gauge theory), with diffeomorphisms playing the role of the translations. Earman and Norton (1987) see a direct, local interpretation as being implied by spacetime (manifold) substantivalism (i.e. the view that spacetime points, as represented by a differentiable manifold, exist independently of material objects). Clearly, this view is then going to be analogous to the interpretation of Maxwell's theory that takes the vector potential as a physically real field. Such an interpretation is indeterministic: the time evolution of the potential can only be specified up to a gauge transformation. Earman and Norton extract a similar indeterminism from the direct interpretation in the spacetime case, and use this conclusion to argue against substantivalism. The 'problem of time' applies the reasoning of the hole argument (as broadly catalogued in my direct, indirect, reductive, and selective interpretations) to the evolution of data off an initial spatial slice. One's interpretation of the gauge freedom then has an impact on the question of whether or not time and change exist! However, the problems will remain in some form on *any* account that views the diffeomorphism invariance of general relativity as a gauge freedom in the theory.

[19] With reference to the hole argument, the present interpretative move would correspond to imposing a condition such that exactly one localization of the metric field relative to the points was chosen. However, in this case, it is difficult to see what could be gained by such a move; there is no symmetry or geometrical structure available to explain the various invariance principles and conservation laws.

[20] As Redhead notes (2003: 132), in the case of non-Abelian gauge theories, the application of the gauge-fixing method leads to a breakdown of unitarity (in perturbative field theory) that has to be dealt with by the ad hoc introduction of 'fictitious' ghost fields—thus replacing one type of surplus structure with another.

6.3 WHAT IS THE PROBLEM OF TIME?

There are two ways of understanding the problem of time: (1) in terms of states and (2) in terms of observables. These lead to quite distinct conceptual problems: the former leads to a problem of time and the latter leads to a problem of change.[21] The first problem concerns the fact that distinct Cauchy surfaces of the same model will be connected by the Hamiltonian constraint, and therefore will be gauge related. The gauge-invariant view demands that we view them as representing the same state of affairs. The second problem concerns the observables: no gauge-invariant quantity will distinguish between Cauchy surfaces of the above sort. Together, these problems constitute the *frozen formalism* problem of classical general relativity. Each of these classical problems transforms into a quantum version.

Let us fix some formalism so we can see how these two problems arise. We are working in the Hamiltonian formulation so we start by splitting spacetime into a space part and a time part. Thus, the spacetime manifold $\mathcal{M}$ is a background structure with the topological structure $\mathcal{M} = \mathbb{R} \times \Sigma$, with Σ a spatially compact 3-manifold. We begin with a phase space Γ, which we shall take to be the cotangent bundle defined over the space of Riemannian metrics on Σ.[22] Points in phase space are then given by pairs (q_{ab}, p^{ab}), with q_{ab} a 3-metric on Σ and p^{ab} a symmetric tensor on Σ. The physical (instantaneous) states of the gravitational field are given by points $x \in \tilde{\Gamma} \subset \Gamma$, where $\tilde{\Gamma}$ is the constraint surface consisting of points that satisfy the diffeomorphism (vector) and Hamiltonian (scalar) constraints: $\mathcal{H}_a = \mathcal{H}_\perp = 0$. These two constraints allow data to be evolved by taking the Poisson bracket of the latter with the former; thus $\{O, \mathcal{H}_a\}$ changes $O \in C^\infty\Gamma$ by a Lie derivative tangent to Σ and is generated by a spatial diffeomorphism, while $\{O, \mathcal{H}_\perp\}$ changes O in the direction normal to Σ. The Hamiltonian for the theory is given by $H = \int_\Sigma d^3x\, N^a\mathcal{H}_a + N\mathcal{H}_\perp$, where N^a and N are Lagrange multipliers called the shift vector and lapse function respectively. The dynamics are thus entirely generated

[21] If one believes that change is a necessary condition for time then the second problem will naturally pose a problem of time too, and vice versa. The necessity of time for 'real' (i.e. non-illusory) change is fairly obvious, but the (Aristotelian) converse, that time requires change, has been questioned in the philosophical literature (e.g. Shoemaker 1969).

[22] I follow 'standard procedure' of couching my discussion in terms of the metric variables. However, I should point out that the canonical approach based on these variables is now defunct and has been replaced by the connection (Ashtekar variables: cf. Ashtekar 1986) and loop representations (a nice introduction is Ashtekar and Rovelli 1992). These result in simpler expressions for the constraints and solutions for the Hamiltonian constraint (none were known for the metric variables!). The justification for sticking with the metric variables is simply that the problem of time afflicts any canonical approach and takes on much the same form regardless of which variables one coordinatizes the phase space with. Generally, one can simply imagine replacing any expression involving functionals of the metric with functionals of these other variables. I should also note that the relation between the connection and metric representations of general relativity is one of a canonical transformation on the phase space. The idea is that we 'change basis' from one set of variables to a new set of variables such that the Poisson bracket relations are preserved by these new variables. It can happen that a new set of variables simplifies certain situations, and can even help with conceptual problems. This is just what happened in the 'connection-variable turn'.

by (first class) constraints.[23] The implication is that the evolution of states (i.e. motion) is pure gauge!

What I have described above is general relativity as a constrained Hamiltonian system. The observables $\mathcal{O}_i$ for such $\mathbf{H} = 0$ systems are defined as follows:

$$O \in \mathcal{O}_i \text{ iff } \{O, \mathbf{H}\} \approx 0. \tag{6.4}$$

This condition states that observables must have weakly vanishing Poisson brackets with all of the constraints; i.e. they must vanish on the constraint surface. From this vantage point, the observables argument is well nigh ineluctable. I mentioned above that the dynamics is generated by constraints; or, in other words, the dynamics takes place on the constraint surface, and evolution is along the Hamiltonian vector fields X_H generated by the constraints on this surface (i.e. along the gauge orbits). Therefore, the observables are constants of the motion: $\frac{dO}{dt}(q(t), p(t)) = 0$ (where t is associated to some foliation given by a choice of lapse and shift). This much gives us our two problems in the classical context. As Earman sums it up: 'the Hamiltonian constraints generate the motion, motion is pure gauge, and the observables of the theory are constants of the motion in the sense that they are constant along the gauge orbits' (2003a: 152). Now to the quantum problems.

Depending upon one's interpretative strategy with regard to the constraints at the classical level, there will be distinct quantization methods for the classical theory, and these correspond to different strategies for tackling the problem of time.[24] Quantization along such lines splits into two types: one can either quantize on the extended phase space or on the reduced phase space. The former method, 'constrained quantization', is due to Dirac (1964): classical constraints are imposed as operator constraints on the physical states of the quantum theory. The latter method reduces the number of degrees of freedom of the extended phase space by factoring out the action of the symmetries generated by the constraints. Hence, the reduced space is the space of orbits of the extended space; it is a (quotient) manifold and inherits a symplectic structure (see Marsden and Weinstein 1974): gauge invariance is automatic on the reduced phase space, for observables on the reduced space will correspond to gauge-invariant functions on the unreduced space. The extended and reduced phase spaces are equivalent on a classical level, but generally they will be inequivalent on a quantum level (cf. Gotay 1984), so the choice is non-trivial.

In brief, and papering over a number of technical subtleties, the constrained (extended phase space) quantization method runs as follows:

- Choose quantum states (representation space $\mathcal{F}$):

$$\psi[q] \in L^2(\text{Riem}(\Sigma, \mu)) \tag{6.5}$$

[23] Dirac's 'conjecture' for such constraints is that they generate gauge transformations: 'transformations ... corresponding to no change in the physical state, are transformations for which the generating function is a first class constraint' (Dirac 1964: 23).

[24] Since they associate methods of dealing with the constraints (to eliminate the gauge freedom or not) with particular interpretational stances on spacetime ontology, it is in just this way that Belot and Earman claim that quantization methods are linked to the substantivalism/relationalism debate and, therefore, that quantum gravity is also implicated in the grand old debate.

- Represent the canonical variables q_{ab}, p^{ab} on $\mathcal{F}$ as:

$$\hat{q}_{ab}(x)\psi[q] = q_{ab}\psi[q] \tag{6.6}$$

$$\hat{p}^{ab}(x)\psi[q] = i(\frac{\partial}{\partial q_{ab}})\psi[q] \tag{6.7}$$

- Impose the diffeomorphism and Hamiltonian constraints:

$$\hat{\mathcal{H}}_a\psi[q] = {}^3\nabla_b\hat{p}^b_a\psi[q] = 0 \tag{6.8}$$

$$\hat{\mathcal{H}}_\perp\psi[q] = \mathcal{G}_{abcd}\frac{\partial^2}{\partial q_{ac}\partial q_{bd}}\psi[q] -{}^3 R(q)\psi[q] = 0^{25} \tag{6.9}$$

- Find a representation of a subset of classical variables on the physical state space, such that the operators commute with all of the quantum constraints.[26]

The classical observables argument filters through into this quantum set-up since, by analogy with the classical observables, the quantum observables $\hat{\mathcal{O}}_i$ are defined as follows:

$$\hat{O} \in \hat{\mathcal{O}}_i \text{ iff } [\hat{O}, \hat{H}] \approx 0. \tag{6.10}$$

Note that the weak equality '$\approx$' is now defined on the solution space of the quantum constraints; i.e. $\mathcal{F}_0 = \{\Psi : \hat{H}\Psi = 0\}$. Clearly, if eq. 6.10 did not hold, then there could be possible observables whose measurement would 'knock' a state Ψ out of $\mathcal{F}_0$. The state version of the problem then follows simply from the fact that the quantum Hamiltonian annihilates physical states: $\hat{H}\Psi = 0$. What motivates this view is the idea common to gauge theories that if a pair of classical configurations q and q' are gauge related then, for any observable O you could care to choose, $O(q) = O(q')$; so we should impose gauge invariance at the level of quantum states too: thus, $\psi(q) = \psi(q')$. The diffeomorphism constraint, eq. 6.8, is particularly easy to comprehend along such lines; it simply says that for any diffeomorphism $d : \Sigma \to \Sigma$, and state $\Psi[q]$, $\Psi[q] = \Psi[d^*q]$—in other words, no quantum state should be able to distinguish between gauge-related 3-metrics. Were this not the case, one could use the quantum theory to distinguish between classically indistinguishable states. The Hamiltonian constraint is more problematic, for it generates changes in data 'flowing off' Σ, and is seen as generating evolution. If we forbid quantum states to distinguish between states related by the Hamiltonian constraint, then there is clearly no evolution, for we must identify the 'evolved' slices Σ_0 and Σ_{t+d^*t} because evolution is a gauge motion (a diffeomorphism).

According to the alternative method, reduced phase space quantization, the constraints are solved for *prior* to quantization (i.e. at the classical level). To solve the constraints, one divides $\tilde{\Gamma}$ by its gauge orbits $[x]^i_\Gamma$. This yields a space $\tilde{\Gamma}_{red}$

[25] $\mathcal{G}_{abcd}$ is the DeWitt supermetric defined by $[|\det q|^{1/2}[(q_{ab}q_{cd} - \frac{1}{2}q_{ac}q_{bd})]$, and ${}^3R(q)$ is the scalar curvature of q on an initial hypersurface Σ. The equation (6.9) is known as the 'Wheeler–DeWitt equation'.

[26] One must also find an inner product making these self-adjoint—no easy matter when there is no background metric or connection!

equipped with a symplectic form $\tilde{\omega}$. The resulting symplectic geometry $(\tilde{\Gamma}_{\text{red}}, \tilde{\omega})$ is the reduced phase space, and in the case of general relativity corresponds to the space of non-isometric (vacuum) spacetimes. Thus, the symmetries generated by the constraints are factored out and one is left with an *intrinsic* geometrical structure of standard Hamiltonian form. In this form the canonical quantization is carried out as usual, and the observables are automatically gauge invariant when considered as functions on the enlarged space. However, since one of the constraints (the Hamiltonian constraint) was associated with time evolution, in factoring its action out the dynamics is eliminated, since time evolution unfolded along a gauge orbit (i.e. instants of time correspond to the points 'parametrizing' a gauge orbit). Thus, on this approach, states of general relativity are given by points in the reduced phase space, as opposed to the enlarged phase space used in constrained quantization approach.[27]

Of course, one can *completely* remove the ambiguity associated with gauge freedom by imposing the appropriate gauge conditions, thus allowing for an unproblematic direct interpretation. However, in the case of general relativity (and other non-Abelian gauge theories) the geometrical structure of the constraint surface and the gauge orbits can prohibit the implementation of gauge conditions, so that some gauge slices will intersect some gauge orbits more than once, or not at all. If the former occurs then some states will be multiply represented (i.e. surplus remains); if the latter occurs, some genuine possibilities will not be represented in the phase space and, therefore, will not be deemed possible. One frequently finds that the reduced phase space method is mixed with gauge-fixation methods, so that one has a partially reduced space, with the remaining gauge freedom frozen by imposing gauge conditions. Such an approach is used by a number of *internal time* responses to the problem of time. The idea is that one first solves the diffeomorphism constraint and then imposes gauge conditions on the gauge freedom generated by the Hamiltonian constraint. This is essentially the position of Kuchař (see below), and *constant mean curvature* approaches (see Carlip 1998 for a clear and thorough review). Before we consider the technical proposals for dealing with the problem of time, let us first review some of the philosophical debate concerning the nature of the problem.

6.4 A SNAPSHOT OF THE PHILOSOPHICAL DEBATE

The philosophical debate on the problem of time (what little there is of it!) has, I think, tended to misunderstand the kind of problem it is; often taking it to be

[27] Little is known about the structure of the space of 3-geometries; the (Wilson) loop variables offer the best hope of carrying out the proposed reduction, or, rather, coordinatizing the reduced space. The diffeomorphism constraint is solved by stipulating that the quantum states be knot invariants. The Gauss constraint that is picked up in the loop representation is easily solved since the Wilson loops are gauge invariant. However, the Hamiltonian constraint is still problematic, though at least *some* solutions can be found. See Brügmann (1994) for more details on these points. Thiemann has done more than anyone to make the Hamiltonian constraint respectable (see e.g. Thiemann 1996). However, there are problems even with his version.

nothing more than a result of eradicating indeterminism by applying the quotienting procedure for dealing with gauge freedom. This point of view can be seen quite clearly in action in a recent 'mini-debate' between John Earman (2002) and Tim Maudlin (2002), where both authors see the restoration of determinism via hole argument type considerations as playing a central role. Thus, Earman writes that '[i]n a constrained Hamiltonian system the intrinsic dynamics ... is obtained by passing to the reduced phase space by quotienting out the gauge orbits. When this is done for a theory in which motion is pure gauge, there is an "elimination of time" in that the dynamics on the reduced phase space is frozen' (2002: 14).[28] Before I outline some of the 'standard' responses, and my own response, it will prove instructive to examine Maudlin's views and his criticism of Earman's account. I will argue that Maudlin seriously misunderstands the nature of the problem of time.[29] Let us begin with Earman's account of the problem, and highlight its relation to other conceptions of time and change.

6.4.1 Time Series from A to D

Before we consider Maudlin's assessment of the problem of time, and of Earman's account of it, we had better have a grip on what is at stake, on what exactly the problem is saying about time and change (at least, according to Earman). To do this it will be useful to compare and contrast the various ways in which time and change have been understood, to see what the problem rules out. We introduce Earman's preferred account, based on his notion of a 'D-series', and show how it compares with the A-, B-, and C-series accounts in the philosophical literature on the philosophy of time.

According to McTaggart (1927: §§305–6), 'positions in time ... as time appears to us *prima facie*, are distinguished in two ways': first, 'each position is Earlier than some and Later than some of the others'; secondly, 'each position is either Past, Present or Future'. The distinctions encoded in the first category are permanent, while the latter category are not: 'If M is ever earlier than N, it is always earlier. But an event, which is now present, was future, and will be past.' The 'movement' or 'flow' of time is understood as 'later and later terms [passing] into the present', or, equivalently, 'as presentness [passing] into later and later terms'. The first way of understanding temporal flow corresponds to sliding the B-series 'backwards' over

[28] However, it is not entirely clear from the text whether Earman endorses the view that it is *only* when reduction is carried out that there is a problem of time. In any case, this is wrong since the problem remains whether or not one reduces the phase space; the problem concerns the gauge equivalence of states that are supposed to represent different instants of time: how can there be time and change if time evolution is along a gauge orbit, if it is a gauge transformation?

[29] As I just mentioned, Earman too appears to agree with the claim that it is quotienting in a bid to restore determinism that leads to the eradication of time evolution. This is false, as I have said, and as I shall argue in more detail below; however, I think the resolution Earman gives is along the right lines (as I explain in §6.5.2). I should point out that both Earman and Maudlin do, however, give the correct presentation of the observables argument as a problem of *change*; indeed, as I shall explain below, Earman and Maudlin appear to converge at this point, though they claim to fundamentally differ.

a *fixed* A-series (a fixed present); the second way corresponds to the opposite, the sliding of the A-series 'forwards' over a fixed B-series. McTaggart then famously argues from this basis to the unreality of time. First, he argues—from the premiss that time involves change—that the B-series depends upon the A-series, so that the only way that events can change is with respect to their A-determinations, not their B-relations. The event 'death of Queen Anne' does not change *per se*; it changes by becoming ever more past, having been future. A-series (tensed) propositions, such as 'the Battle of Waterloo is past', are true at some times (those *after* the battle), but not at others (those *before* the battle). Replacing this with a B-series version differs in this respect; the proposition 'the Battle of Waterloo is earlier than this judgement' is either always true or always false, it does not change its value as a result of the permanency of the B-series. On these grounds, McTaggart concludes that time requires the A-series: if time requires change, and events change only in terms of their A-series determinations, then time requires the A-series. But then McTaggart argues that the A-series, and therefore time, is inherently contradictory, for the A-determinations are mutually exclusive, and yet any and 'every event has them all' (1927: §329). A single event is present, will be past, and has been future. However, there is, as McTaggart realizes, no contradiction here: no event has these simultaneously. But these tensed verbs, *is* present, *will be* past, and *has been* future, need cashing out. McTaggart (cf. ibid., §330) suggests that 'X has been Y' is tantamount to 'X is Y at $t < t_0$'; 'X will be Y' is tantamount to 'X is Y at $t > t_0$'; and 'X (temporally) is Y' is tantamount to 'X is Y at $t = t_0$'. In other words, the analysis requires there to be 'moments' of past, present, and future time. But, McTaggart asks, what are these moments? The A-determinations cannot fix them once and for all, for the same reasons as with events. If we attempt to say that the moments do not have their A-determinations simultaneously, then the analysis must be reapplied: a moment M is future and will be present and then past, which we then rewrite as above, thus courting the same problem, producing 'higher-order' moments, *ad infinitum*.

So, without the A-series there is no change; the B-series alone is not sufficient for time, because time involves change. Moreover, the B-series depends on the A-series, since the former is essentially temporal (McTaggart 1908: 461): the distinctions it marks out are temporal, and yet without the A-series there is no time; therefore, there is no B-series! So much for the A- and B-series; what is left to put in their place? McTaggart suggests that an ordering remains, the *C-series*, but it cannot be temporal, for it does not involve change. The C-series consists of an ordering of events themselves.[30] That we have a string of events, X, Y, Z, implies that there is any change no more than the ordering of the letters of the alphabet implies change. When the A-series is superimposed on the C-series, however, then the C-series becomes a B-series.

[30] For a relationalist about time this is simply what time *really* is. There is little sense in saying that the relationalist is an eliminativist about time; he simply *reduces* it, or at least re*defines* time in terms of material happenings (non-temporal entities). (There are, I think, analogies to reductive theories of modality here; especially Lewis's definition of modal notions in terms of a (non-modal) plurality of concrete worlds (Lewis 1986)—this analogy is stretched further when we discuss Julian Barbour's solution to the problem of time below in §6.5.2.)

McTaggart's analysis relies on the notion that change applies to events; the argument is grounded in times and events. What of objects? It seems that objects change their properties. Indeed, this is what most people mean by change. Change in events, if it is of the kind suggested by McTaggart, is somewhat spurious; it does not *exhaust* what is meant by change. But, in any case, McTaggart believes that *any* kind of change, including the changes that objects undergo, requires an A-series. Modern philosophers of time are divided on this point, and the schools of thought can be split according to whether they agree with McTaggart or not about the necessity of the A-series for change. The nay-sayers are grouped into the category of B-theorists or 'detensers', and the yea-sayers are grouped into the cateory of A-theorists or 'tensers'. The A-theorists will say that the B-theorists cannot properly accommodate the notion of the *passage* of time—that, following McTaggart, they claim is essential—and can, at best, allow that it is an illusion. The B-theorist denies that passage is necessary for time and change, and is happy to see it done away with. Both sides claim support from physics: B-theorists generally wield spacetime theories (special relativity) and the A-theorists wield mechanical theories such as quantum mechanics.

The A-theorists' and B-theorists' theories are often said to underwrite a 'dynamic' and 'static' conception of time respectively. The static conception represents the moments of time as an 'eternally' existing line, such that each individual moment is equally as real as any other. No fundamental ontological distinction is to be made regarding any 'elements' or 'sections' of the line. This is not the case with the dynamic conception according to which the different times are assigned different ontological status: Broad's 'growing block' theory, for example, views the future as unreal, and the past and present as real; the presentist denies reality to any times other than the present.

There are many and varied ways of responding to McTaggart (token-reflexive or indexical analyses of tenses;[31] presentism, non-property-based becoming, etc.); however, the most important for my purposes is the class of responses that attempt to ground a notion of real time and change within the B-series alone (see, for example, Mellor 1998). Both Earman and I agree that the B-series is sufficient for change in the sense that different properties and relations are instantiated at different times, such that if those times were equal we would have a contradiction—in Wheeler's words 'time is what stops everything happening at once'! But strip the dynamical A-determinations from the world and one is left with a *static* block of events ordered earlier to later; indeed, it isn't clear that 'earlier' and 'later' can be anything other than arbitrary directions, for it is the dynamical *flow* that gives direction to the B-series ordering a direction, and this belongs to the A-series.

Earman introduces a character called 'Modern McTaggart', who attempts to revive the conclusions of old McTaggart by utilizing a gauge-theoretic interpretation of

[31] This line of response grounds the moments of time mentioned above contextually by supplying a Now, and combining this with the B-series (see Russell 1940 and Reichenbach 1947). Thus, one provides a notion of presentness with, say, the time of utterance of a sentence, S_t, and then analyses the tenses in terms of this 'present' and a string of earlier and later times. 'The Battle of Waterloo is Past' is parsed as 'The Battle of Waterloo is earlier than S_t'.

general relativity. Earman is dismissive of A-theories; he claims that they are not part of the scientific image, though he does at least pay lip-service to Shimony's attempt to account for 'transience'. Earman sets up the problem of time as a McTaggartesque consequence of the Hamiltonian formulation of general relativity. The problem targets B-series change—different properties at different times—in that 'no genuine physical magnitude takes on different values at different times' (2002: 2–3). The problem is that given the gauge-theoretic conception of observables as gauge invariants, and given that time evolution is a gauge transformation (being generated by a first class constraint), the observables mustn't change from one time to the next.[32] On the assumption that time requires change, and assuming that the observables exhaust what might undergo change (that is, assuming that the set is complete), it appears that a version of McTaggart's conclusion follows: time is unreal according to Hamiltonian general relativity! If we deny that time requires change, then though there might be temporal evolution, because it is along a gauge orbit, there will be no qualitative, B-series change; just fixed values. Earman's response is to argue that general relativity is nonetheless compatible with change, though in neither the B- nor the A-series senses. Instead he introduces a 'D-series' ontology consisting of a 'time ordered series of occurrences or events, with different occurrences or events occupying different positions in the series' (2002: 3). These are events formed from the coincidence quantities familiar from Einstein's 'point-coincidence argument'. Earman writes that '[t]he occurrence or non-occurrence of a coincidence event is an observable matter [in the technical sense of *observable*] ... and that one such event occurs earlier than another such event is also an observable matter ... Change now consists in the fact that different positions in the D-series are occupied by different coincidence events' (ibid. 14). Thus, Earman maintains that his D-series is temporally ordered. But this is simply McTaggart's C-series; and, according to McTaggart, that ordering was not temporal. Earman owes us, but does not give us—claiming that those who demand that Becoming is required for change 'are stuck in the manifest image' (ibid. 5)—an explanation of how this *is* a temporal series, and in what sense change can be said to occur. He does tell us in what sense it does *not* occur, for according to Earman, 'common sense B-series property change is not to be found in physical events themselves but only in the mode in which we represent these events to ourselves' (ibid.). However, as I later demonstrate, Earman cannot make do with a single D-series; if he must have it, then he must have many. Let us next see what Maudlin makes of Earman's account.

[32] Again, I feel that Earman limits the problem too much by focusing on the *removal* of the gauge freedom as the source of the problems, rather than the gauge freedom itself. Thus, he writes that for that 'class of gauge theories where the very dynamics is implemented by a gauge transformation ... [w]hat such a theory describes when the gauge freedom ... is killed is a world without B-series change' (2002: 7). I say the B-series change is ruled out regardless of whether the gauge freedom is removed. The difference is subtle: when the gauge freedom is removed, time itself is removed, for time evolution is along a gauge orbit; when the gauge freedom is retained, there is time evolution of a sort, but it is gauge, therefore there is no B-series change. Earman is clearly clinging to the idea that without change there is no time; but the formalism does not force this.

6.4.2 Maudlin *versus* Earman

Maudlin is responding to the aforementioned paper of Earman, wherein the latter upholds the seriousness of the problem of frozen dynamics, and defends a response to the problem based on the idea that there is, at a fundamental level, no B-series type change according to general relativity. As we saw, Earman argues that changes in the magnitudes of things are, at best, an artefact of the local representations (a particular chart, for example) we might choose to use to describe the world. What is real is a series of events, a D-series. Earman is very much taken with the Hamiltonian formalism, and believes that the frozen dynamics is something that must be accommodated by any sound interpretation of canonical general relativity. I agree. Maudlin does not; rather, he thinks the frozen formalism involves 'some Alice-in-Wonderland logic' (ibid. 13)! Maudlin distinguishes two separate arguments in Earman's paper that appear to lead to the frozen formalism: the 'Hamiltonian Argument' and the 'Observables Argument'—corresponding, more or less, to my 'states' and 'observables' arguments. He takes the crux of the Hamiltonian Argument to consist in the following observation:

Applying this standard method ['quotienting out'] to the GTR does indeed restore the determinism of the theory—but at a price. The price is that the dynamics of the theory becomes 'pure gauge'; that is, states of the mathematical model which we had originally taken to represent physically different conditions occurring at different times are now deemed equivalent since they are related by a 'gauge transformation'. We find that what we took to be an 'earlier' state of the universe is 'gauge equivalent' to what we took to be a 'later' state. If gauge equivalent states are taken to be physically equivalent, it follows that *there is no physical difference between the 'earlier' and the 'later' states*: there is no real physical change. (ibid. 2)

Maudlin's claim is that 'the key to the Hamiltonian Argument' is based 'in the freedom to foliate' (ibid. 7). A specific foliation is an essential ingredient of any Hamiltonian formulation, for we need an initial data set on a hypersurface. However, in relativistic theories there are many ways to slice up the spacetime manifold $\mathcal{M}$. Given an arbitrary foliation, a phase space can be constructed so that points of this space represent instantaneous states (in this case 3-geometries). The complete four-dimensional solution (i.e. a model of general relativity) is given by a trajectory through the phase space. One and the same solution can be represented by many different trajectories depending upon the foliation that one chooses. He then claims that this yields an indeterminism of the kind that the quotienting procedure is used to eradicate; one can make foliations that agree up to some point, and then diverge thereafter. But, he claims that it is a *faux* indeterminism. The quotienting is unnecessary, and not only is it unnecessary it leads to 'silly' claims such as 'change is not real, but merely apparent' (ibid. 11). Claims, says Maudlin, that Earman thinks are revealed about the deep structure of general relativity by the constrained Hamiltonian formalism. For Maudlin, any such interpretation is absurd. As he explains:

Any interpretation which claims that the deep structure of the theory says that there is no change at all—and that leaves completely mysterious why there *seems* to be change and why the merely apparent changes are correctly predicted by the theory—so separates our experience from physical reality as to render meaningless the evidence that constitutes our grounds for believing the theory. So the only real question is not *that* the constrained Hamiltonian formalism is yielding nonsense in this case, but *why* it is yielding nonsense. And the freedom to foliate provides the perfectly comprehensible answer. (ibid. 12)

Maudlin's opening line here is facetious. First, the canonical approach is a formal framework *not* an interpretation. Prima facie, on a surface reading of the formalism there appears to be no scope for change, therefore, given the apparent existence of change, something is wrong. However, there *is* scope for interpreting the formalism so as to introduce change, as I show in §6.5.2. We can, on these interpretations, say why the surface reading is 'yielding nonsense'. The answer is related to the gauge-invariant response to the hole argument (the response that is supposed to cause the problems of change in the first place): only change with respect to the manifold is ruled out; if we focus on those quantities that are independent of the manifold we can restore change by considering the 'evolving' relationships between these quantities.[33] As regards the observables argument, he rather oddly simply regurgitates what is the gauge-theoretical lesson of general relativity, that local quantities cannot be observables:

the Observables Argument gets any traction only by considering candidates for observables (values at points of the bare manifold) which are neither the sorts of things one actually uses the GTR to predict nor the sorts of things one would expect—quite apart from diffeomorphism invariance—to be observables. (ibid. 18)

Thus, Maudlin has in fact simply accepted the gauge-invariance interpretation without realizing it; he mistakenly thinks that the gauge interpretation goes hand in hand with the quotienting procedure. That values at the points of the bare manifold are not the things one predicts cannot be separated from the issues of diffeomorphism invariance, for it is precisely this that results in the problems for local field quantities that we have seen in the hole argument. Thus, we can agree, and Earman will agree, that the observables argument gets off the ground by considering the 'wrong' type of observables, but this is to adopt a substantive response that buys into the gauge-theoretical interpretation! (I return to this point below, for it backfires on his account of the Hamiltonian argument.)

Maudlin concludes from this 'double debunking' that the frozen formalism problem is simply a result of either a 'bad choice of formalism or a bad choice of logical form of an observable' (ibid. 18). I proceed to attack Maudlin's account of the problem in two stages.

First, putting aside his analysis of the source of the indeterminism that requires the framework of gauge theory (which I consider below), Maudlin's responses to the indeterminism are: (1) ignore it; (2) gauge-fix it: and (3) quotient it. He thinks that the first two are 'viable solutions' to the problem, but that the third rests on some kind of confusion (it is being applied in a domain where it should not be). Not

[33] In §6.6 I suggest that we should view the correlations themselves as observables, following Rovelli's interpretation.

so. The underdetermined local field quantities that gauge invariance is invoked to dispel cannot simply be ignored. To ignore them is to tend towards anti-realism, for it amounts to the suggestion that we should worry neither about how our theories represent nor about what they represent. The gauge-fixing response essentially sides with those who believe that the problem of change is a real problem, for it is tantamount to a resolution in terms of gauge-invariant quantities: one fixes a set of coordinates, thus breaking general covariance, and defines the quantities with respect to the points of this coordinate system. Maudlin doesn't give an example of a viable gauge fixing, but the only ones I know of will involve using either the invariants of the gravitational field or else some ideal, phenomenological dust field, or some other material objects: these are wholly unrealistic idealizations. The quotienting procedure is one way to sop up the indeterminism, but it is not the only way: nobody said that quotienting was the root of the problem, yet Maudlin appears convinced that it is. Quotienting would certainly eradicate *any* kind of evolution, since the evolution happens along a gauge orbit and the points ('temporal instants') would be identified in a quotienting strategy. But even unreduced the points represent indistinguishable states so that there will be no qualitative difference between one instant and the next: there will be no B-series (or A-series!) change. Bizarrely, Maudlin does not even consider gauge-invariant observables as a viable response! Yet the observables he suggests as the kinds of things we actually measure are of just this kind: e.g. 'the amount by which light from the sun is redshifted when it reaches the Earth' and 'the position of the perihelion of Mercury *relative* to the sun' (2002: 13, my italics). Or *almost* of the kind, for Maudlin assumes that the time of measurement is unproblematic. But of course, the freedom to foliate means that a time choice will be arbitrary: the time of measurement is far from unproblematic! Position relative to the sun *when*? How is the 'when' of the first quantity determined? To fix matters one will need to invoke a physical clock. One then considers the above observables suggested by Maudlin when the clock reads a certain figure. Maudlin even gives an example where a physical clock is invoked: 'the position of the perihelion of Mercury after some number of orbits'. Here Mercury is being used as a clock, the orbits being the 'ticks'. However, now the position will need to be defined relationally, presumably by the Earth or some other useful reference object: we don't just measure the position *simpliciter* in general relativity, we always assume a frame. If we don't want the frame to be arbitrary, we had better make it physical. In case we do use an arbitrary frame, we had better be able to demonstrate that the quantity in question is independent of the specific choice. If Maudlin disagrees with this then he is talking about something other than general relativity, for he is apparently assuming that time is a fixed background structure. If he agrees, which I'm sure he must, then the observables are just (what I shall later call) 'correlation observables', and are, more or less, the same as Earman's 'coincidence observables' that we met earlier. He gives an example of what he takes to be a good quantity for general relativity that brings this similarity to the fore: thus, he writes that '[w]hat we *can* identify by observation are the points that satisfy definite descriptions such as "the point where these geodesics which originate here meet", and against *these* sorts of [local] quantities Earman's diffeomorphism argument has exactly zero force' (ibid.). Indeed, but here Maudlin is essentially gauge-fixing spacetime points and then constructing gauge-invariant quantities by attaching them to the physically defined points—the reasoning (as Chris Isham likes to put it) that for some quantity 'ϕ', physical object 'thing', and space point x: $\phi(\text{thing})$ is gauge invariant but $\phi(x)$ is not. If Maudlin is willing to go this far, then why not allow that change is accounted for with just such observables: the evolution and change concerns the relations between things or quantities, rather than the having and losing of properties at times? One can form a chain of values for ϕ by using the values of 'thing' as the 'ticks' of a clock—this is essentially what Rovelli proposes (see §6.5.2). Moreover, all of this is perfectly possible in the context of the Hamiltonian formulation. Indeed, on the preferred choice of polarization, holonomies, with no local spacetime dependence, are used as the fundamental variables, thus taking on board the lesson that Maudlin has clearly internalized without being aware of it.[34]

Secondly, Maudlin diagnoses the Hamiltonian argument as the freedom to foliate a spacetime in general relativity. Different slicings of spacetime yield different trajectories through phase space, which are to represent four-dimensional solutions. But, says Maudlin, we can make a pair of foliations agree up to a point (so that their corresponding phase trajectories do likewise) and diverge thereafter (again, likewise for the trajectories). This results in an indeterminism to which the three options listed above apply. The quotienting option removes the indeterminism by declaring the solutions equivalent, and forming a reduced phase space out of the equivalence classes. Maudlin thinks this is absurd for the reasons given. But the source of the indeterminism is *not* the freedom to foliate a spacetime, it is the freedom to drag the dynamical fields around without generating a distinguishable scenario. The indeterminism concerns 'local' quantities that are attached to manifold points. Any local quantities will be altered by the dragging. We cannot even assume that we have a spacetime in the Hamiltonian formulation, since for that to be the case we require a solution.

Thus, Maudlin can claim that he is willing to accept the indeterminism that follows from such gauge transformations rather than quotienting if he likes,[35] but the fact that the indeterminism is unobservable is tantamount to saying that the time evolution is unobservable, which simply lets the problem in through the back door. As regards the observables argument it seems to me, as I hope to have demonstrated, that far from showing it to be 'broken-backed', Maudlin has simply taken a stance

[34] As further evidence that Maudlin misunderstands the nature of the problems of time and change, he mentions that on the basis of his arguments, the quantum gravitational problems of time and change might be 'equally chimerical' (ibid. 18). His worry is that if local observables cause a problem in the classical theory 'then we should anticipate difficulties in defining the observables in the quantum version' (ibid. 19). But no one is suggesting that we use these kinds of observables! There are *proofs* that no such observables are available in general relativity, classical or quantum (Torre 1993). These quantities are not forced upon us in the Hamiltonian formulation.

[35] Something he is willing to do on the grounds that the indeterminism is '*completely phoney*' (ibid. 9; see also p. 16).

(and a highly non-trivial one at that) with respect to the observables argument. Specifically, he opts for the view that the 'proper' observables of general relativity are relational quantities involving intersections of quantities.[36] However, what is missing from Maudlin's suggested quantities is a time of occurrence (or, in some cases, the *position* of occurrence). It is not enough to say that two things meet at a point; one must say *when* they meet, and to do this one needs a clock. Likewise, it is not enough to say when something happens, one must say *where* it happens. And the 'when' and 'where' are not given a priori; one can arbitrarily shift the points of the manifold around, so these cannot ground the where and the when. There is no background temporal or spatial structure in general relativity, so this will have to be a physical clock or a physical reference frame. The coincidence of the hand of the clock and the meeting of the geodesics is a diffeomorphism-invariant quantity that satisfies the constraints. It is a constant of the motion, so it does not change over time.[37] It will certainly be hard to write such a quantity as a phase function, but that is not of moment for what is at stake here. As I argued, the observables Maudlin mentions sound suspiciously like Earman's coincidence quantities. This is just what many physicists take to be the 'lesson' of the hole argument and the problem of time: the proper observables are independent of the manifold and, therefore, independent of time as well as space. One way of understanding the observable content of the theory is to view the points of spacetime as relationally (dynamically) individuated in the manner Maudlin suggests; this was, of course, Stachel's position too (see, for example, Stachel 1993). The problem remains: how do we reconcile this with the manifest change we seem to observe? I review some options in the next section.

6.5 CATALOGUE OF RESPONSES

Those approaches to classical and quantum gravity that attempt to understand these theories without change and time existing at a fundamental level I shall call *timeless*, and those that disagree I call *timefull*. An alternative pair of names for these views, suggested by Kuchař, are 'Parmenidean' and 'Heraclitean' respectively (1993b). But it is important to note that the debate here is not directly connected to the debate in the philosophy of time between 'A-theorists' and 'B-theorists' (or 'tensers' and 'detensers', if you prefer). Both of these latter camps agree that time exists, but disagree as to its nature. By contrast, the division between timefull and timeless interpretations concerns whether or not time (at a fundamental level) exists *simpliciter*! I begin by reviewing several timefull responses.

[36] Note that Maudlin gives no account as to the nature of the 'individual' elements participating in these intersections. The standard line is to take these elements as having some physical reality independently of the relation; but this leads to serious problems as we shall see in §6.5.2.

[37] As I intimated above, and as I will discuss in §6.5.2, by stringing a sequence of such quantities together one can get a fairly robust account of change.

6.5.1 Timefull Stratagems

Recall that the observables argument required that in order to class as kosher, the relevant observables must have vanishing Poisson brackets with *all* of the constraints. This idea filtered through into the quantum version, modified appropriately. Like Maudlin, Kuchař has been a vociferous opponent of this 'liberal' gauge-invariant approach to observables.[38] He agrees with the plan to the level of the diffeomorphism constraint, so that $\{\mathrm{O}, \mathcal{H}_a\} \approx 0$, $[\hat{\mathrm{O}}, \hat{\mathcal{H}}_a] \approx 0$ and $\mathcal{H}_a \Psi = 0$; but does not agree that we should apply the same reasoning to the Hamiltonian constraint. Thus, neither states nor observables should distinguish between metrics connected by $\mathrm{Diff}(\Sigma)$: only the 3-geometry ${}^3\mathcal{G}$ counts. But the alterations generated by the Hamiltonian constraint are a different matter says Kuchař:

> $[\mathcal{H}_\perp]$ generates the dynamical change of data from one hypersurface to another. The hypersurface itself is not directly observable, just as the points $x \in \Sigma$ are not directly observable. However, the collection of the canonical data $(q_{ab}(1), p^{ab}(1))$ on the first hypersurface is clearly distinguishable from the collection $(q_{ab}(2), p^{ab}(2))$ of the evolved data on the second hypersurface. If we could not distinguish between those two sets of data, we would never be able to observe dynamical evolution. (1993b: 20)

Ditto for states: the Wheeler–DeWitt equation does not say that an evolved state is indistinguishable from some initial state—as the diffeomorphism constraint does—rather, it 'tells us how the state evolves' (ibid. 21). More colourfully:

> I would say that the state of the people in this room now, and their state five minutes ago should not be identified. These are not merely two different descriptions of the same state. They are physically distinguishable situations. (Ashtekar and Stachel 1991: 139)

Thus, Kuchař concludes that 'if we could observe only constants of motion, we could never observe any change' (ibid.). On this basis he distinguishes between two types of variable: *observables* and *perennials*. The former class are dynamical variables that remain invariant under spatial diffeomorphisms but *do not* commute with the Hamiltonian constraint; while the latter are observables that *do* commute with the Hamiltonian constraint. Kuchař's key claim is that one can observe dynamical variables that are not perennials.[39]

In their assessment of Kuchař's proposal, Belot and Earman (1999: 183) claim that he 'endeavours to respect the spirit of general covariance of general relativity without treating it as a principle of gauge invariance'. For this reason they see his strategy as underwritten by substantivalism. I argue against the connection between

[38] Though, unlike Maudlin, he has a constructive alternative that submits to quantization. He also takes the problem of time much more seriously than Maudlin; he doesn't think that it can simply be ignored. The positions end up being very different, with Maudlin occupying a position almost identical to Earman and, as will become evident, Rovelli.

[39] He goes further than this, arguing that perennials are in fact hard to come by. I do not deal with this aspect of his argument here. In fact, I think that relational observables show that they are not at all hard to come by. How one makes a quantum theory out of these is, of course, quite another matter. The hard task is to find quantum operators that correspond to such classical observables without facing operator ordering ambiguities, and so on.

the denial of gauge invariance and substantivalism in Rickles (2005a); for now I note that Kuchař *does* treat general covariance as a principle of gauge invariance as far as the diffeomorphisms of Σ are concerned (and additionally, in the connection representation, as far as the $\mathsf{SO}(3)$ Gauss constraint goes). Observables are gauge-invariant quantities on his approach; the crucial point is simply that the Hamiltonian constraint should not be seen as a generator of gauge transformations. Viewed in this light, according to Belot and Earman's own taxonomy (ibid., §2), Kuchař's position should more properly be seen as underwritten by a *relationalist* interpretation of space coupled with a *substantivalist* interpretation of time! Let me spell out some more of the details of Kuchař's idea.

Kuchař's claim that observables should not have to commute with the Hamiltonian constraint leads almost inevitably to the conclusion that the observables do not act on the space of solutions; or, as he puts it 'if $\Psi \in \mathcal{F}_0$ and $\hat{\mathsf{F}}$ is an observable, $\hat{\mathsf{F}}\Psi \notin \mathcal{F}_0$' (1993b: 26). This, amongst other things, motivates the *internal time* strategy, where an attempt is made to construct a time variable T from the classical phase space variables. This strategy conceives of general relativity (as described by Γ) as a parametrized field theory. The idea is to find a notion of time *before* quantization hidden amongst the phase space variables so that a time-dependent Schrödinger equation can be constructed; the quantum theory's states then evolve with respect to the background time picked out at the classical level. Kuchař's method involves finding four (scalar) fields $\mathsf{X}^{\mathsf{A}} = (\mathcal{T}(\mathsf{x}; \mathsf{q}, \mathsf{p}], \mathcal{Z}^{\mathsf{a}}(\mathsf{x}; \mathsf{q}, \mathsf{p}])$ (where $\mathsf{A} = 0, 1, 2, 3$ and $\mathsf{a} = 1, 2, 3$) from the full phase space Γ that when defined on $\overline{\Gamma}$ represents a spacelike embedding $\mathsf{X}^{\mathsf{A}} : \Sigma \rightarrow \mathcal{M}$ of a hypersurface Σ in the spacetime manifold $\mathcal{M}$ (without metric). These kinematical variables are to be understood as positions in the manifold, and the dynamical variables (separated out from the former variables within the phase space) are observables evolving along the manifold. The constraints are then understood as conditions that identify the momenta P_{A} conjugate to X^{A} with the energy-momenta of the remaining degrees of freedom: they thus determine the evolution of the true gravitational degrees of freedom between hypersurfaces.

There are two broadly 'technical' ways of dealing with Kuchař's arguments. The first involves demonstrating that general relativity is not a parametrized field theory; the second involves showing that *observing* change is compatible with the view that all observables are constants of the motion. I deal with the second when I get to the timeless responses; the first I outline now. Clearly, we need to test whether or not the identification between the phase space Γ of general relativity and the phase space Υ of a parameterized field theory goes through. The proposal requires that there is a canonical transformation $\Phi : \Upsilon \rightarrow \Gamma$ such that $\Phi(\overline{\Upsilon}) = \overline{\Gamma}$. However, there can be no such transformation because $\overline{\Upsilon}$ is a manifold while $\overline{\Gamma}$ is not (cf. Torre 1993). Hence, there are serious, basic technical issues standing in the way of this approach: general relativity is not a parameterized field theory!

Along more 'philosophical' lines, one might perhaps question the line of reasoning that led Kuchař to deny that observables commute with all of the constraints in the first place. Is it an empirical input that determines the break, or is it something internal to the theory? I think that it is neither, but is instead an intuitive belief that change is a real feature of the world, and that change happens when things change by changing in the values of their observables. He takes the fact that the liberal gauge-invariance position entails that observables are constants of the motion as providing a reductio of that view, and as providing a counter-example to Dirac's conjecture that first class constraints generate gauge transformations. But there are ways of understanding change; we can understand change as the possession of incompatible properties by things at different times (ruled out by the gauge interpretation), but we can incorporate a notion of change as variation: the rug, for example, changes from blue to green as one moves across it. Or, one can get a simulacrum of change by piecing together unchanging parts, as one finds in the old-fashioned movies. All we really need to do is explain the *appearance* of change; to assume a substantial metaphysics of time and change and then base ones physical theories on this metaphysics is a dangerous move in my opinion. Intuition strongly suggests that there is a unique notion of simultaneity; physics suggests that our intuitions need to be revised. Regardless of this, if the problem of time can be resolved in a liberalistic gauge-invariant way, then we should opt for that on the grounds that violating the 'first class constraint'–'gauge transformation' connection, which has worked so well in other gauge theories, is too high a price to pay. In keeping it we can retain a unified interpretative picture of these theories.

An alternative (internal) timefull approach uses *matter* variables coupled to spacetime geometry instead of (functionals) of the gravitational variables as above. Thus, one might consider a space-filling dust field, each mote of which is considered to be a clock (i.e. the proper time of the motes gives a preferred time variable and, therefore, amounts to fixing a foliation). These variables are once again used to 'label' spacetime points. This includes an internal time variable against which systems can evolve, and which can function as the fixed background for the construction of the quantum theory. Another internal approach, *unimodular gravity*, amounts to a modification of general relativity, according to which the cosmological constant is taken to be a dynamical variable for which the conjugate is taken to be 'cosmological' time.[40] The upshot of this is that the Hamiltonian constraint is augmented by a cosmological constant term $\lambda + q^{-\frac{1}{2}}(x)$, $x \in \Sigma$, giving the super-Hamiltonian constraint $\lambda + q^{-\frac{1}{2}}(x)\mathcal{H}_{\perp}(x) = 0$. The presence of this extra term (or, rather, its conjugate τ) unfreezes the dynamics, thus allowing for a time-dependent Schrödinger equation describing dynamical evolution with respect to τ. The conceptual details of this approach are, however, more or less in line with gauge-fixation methods like that mentioned above.[41] Another popular, but now aged approach is that which takes surfaces of constant mean

[40] The idea to use unimodularity as a response to the problem of time was originally suggested by Unruh (1989). For a nice philosophical discussion of unimodular gravity see Earman (2003b)—§6 of his paper focuses the discussion on the problem of time. See also Isham (1993: 63).

[41] Isham (ibid. 62) goes so far as to say that it is in line with reference fluid methods since it amounts to the imposition of a coordinate condition (on the metric γ_{ab}): $\det \gamma_{ab}(x^i) = 0$. See ibid. 60–2 for more details on the notion of a reference fluid and how it might offer a solution to the problem of time.

curvature $\tau = q_{ab}p^{ab}/\sqrt{\det q} = \text{const.}$ as providing a time coordinate by providing a privileged foliation of spacetime.[42]

The basic idea underlying each of these approaches is to introduce some preferred *internal* time variable so that general relativity can be set up as a time-dependent system describing the evolution in time of a spatial geometry (possibly involving the extrinsic curvature and possibly coupled to matter or some reference fluid). With this background time parameter in hand, the quantization proceeds along the lines of other quantum field theories since there will be a non-zero Hamiltonian for the theory. Naturally, the selection of a preferred time coordinate breaks the general covariance of the theory, for it is tantamount to accepting that there is a preferred reference frame. One would have to demonstrate that the resulting quantum theory is independent of the choice.[43]

I do not think that these timefull approaches are the correct direction to go. Aside from the technical difficulties, they either represent a step backwards towards unphysical, ad hoc, or arbitrary background structures, or else they point to the idea that a 'robust' or 'external' notion of time is required to get a quantum theory up and running. The proposals in the next subsection show that this is simply false.

Before I leave the 'timefull' methods, I should first mention one more related approach: Hájiček's *perennial formalism* (1996, 1999), according to which the dynamics is constructed solely from the geometry of phase space, and no reference is made to spacetime. The idea is to begin with some system whose time evolution is well understood, like a Newtonian system, and transform the spacetime structure into a phase space structure so that a quantum time evolution can be reconstructed from phase space objects. Then one attempts to find similar phase spaces for systems without background spacetimes, effectively 'guessing' a theory. This approach links technically to Kuchař's scheme, but conceptually it links up to the timeless approaches—especially Rovelli's evolving constants scheme. However, questions need to be asked about the way the phase space is constructed, for it is not intrinsically done, but is parasitic on what we know of phase spaces for systems with background spacetime structure (fixed metrics and connections). If the virtue of this approach is that it retains background independence, then we would surely like the formalism to reflect this property.[44]

6.5.2 Timeless Stratagems

We come now to the timeless strategies; the most radical of which is surely Barbour's. I deal with this first, and then outline the view I favour. Butterfield (2001) has written a fine account of Barbour's timelessness as outlined in the latter's book *The End of*

[42] This approach was first suggested by York (1971). See Beig (1994) for a nice discussion.

[43] Note that Kuchař's approach escapes this objection since it quantizes the 'multi-time' formalism according to which dynamical evolution takes place along deformations of *arbitrary* hypersurfaces embedded in $\mathcal{M}$ (see Isham 1993: 46).

[44] Compare this with Earman's point that the relationalist should be able to construct his theories in relationally pure vocabulary, rather than 'piggy backing' on substantivalist formulations (1989: 135).

Time (2003); he describes the resulting position as 'a curious, but coherent, position which combines aspects of modal realism *à la* Lewis and presentism *à la* Prior' (ibid. 291). I agree that these aspects do surface; however, I disagree with his account on several key substantive points. In particular, I will argue—*contra* Butterfield—that Barbour's brand of timelessness is connected to a denial of persistence, and as such is not timeless at all; rather, it is changeless. I go further: far from denying time, Barbour has in fact *reduced* it (or, rather, the instants of time) to the points of a relative configuration space!

The central structure in Barbour's vision is the space of Riemannian 3-metrics modulo the spatial diffeomorphism group (known as 'superspace'): Riem(Σ)/Diff(Σ). Choosing this space as the configuration space of the theory amounts to solving the diffeomorphism constraint; this is Barbour's *relative configuration space* that he labels 'Platonia' (ibid. 44). The Hamiltonian constraint (i.e. the Wheeler–DeWitt equation, eq. 6.9) is then understood as giving (once solved, and 'once and for all' (Barbour 1994: 2875)) a *static* probability distribution over Platonia that assigns amplitudes to 3-geometries (Σ, q) in accordance with $|\Psi[q]|^2$. Each 3-geometry is taken to correspond to a 'possible instant of experienced time' (ibid.) This much is bullet biting and doesn't get us far as it stands; there remains the problem of accounting for the appearance of change. This he does by introducing his notion of a 'time capsule', or a 'special Now', by which he means 'any fixed pattern that creates or encodes the appearance of motion, change or history' (Barbour 2003: 30). Barbour *conjectures* that the relative probability distribution determined by the Wheeler–DeWitt equation is peaked on time capsules; as he puts it 'the timeless wavefunction of the universe concentrates the quantum mechanical probability on static configurations that are time capsules, so that the situations which have the highest probability of being experienced carry within them the appearance of time and history' (ibid.). What sense are we to make of this scheme?

Barbour's approach is indeed timeless in a certain sense: it contains no reference to a background temporal metric in either the classical or quantum theory. Rather, the metric is defined by the dynamics, in true Machian style. Butterfield mentions that Barbour's denial of time might sound (to a philosopher) like a simple denial of temporal becoming—i.e. a denial of the A-series conception of time. He rightly distances Barbour's view from this B-series conception. Strictly speaking, there is neither an A-series nor a B-series on Barbour's scheme. Barbour believes that space is fundamental, rather than spacetime.[45] This emerges from his Machian analysis of general relativity. What about Butterfield's mention of presentism and modal realism? Where do they fit in?

Presentism is the view which says that only presently existing things actually exist.[46] The view is similar in many respects to modal actualism, the view that only actually

[45] I might add that Belot writes that he does 'not know of any philosopher who entertains, let alone advocates, substantivalism about space as an interpretive option for GR' (1996: 83). I think that Barbour's proposal ends up looking like just such an interpretative option—a position recently defended by Pooley (2002).

[46] The consensus amongst philosophers seems to be that special and general relativity are incompatible with presentism (cf. Callender (2000), Savitt (2000), and Saunders (2002)). I

existing things exist *simpliciter*. Yet Butterfield claims that Barbour's view blends with modal realism. What gives? We can make sense of this apparent mismatch as follows: Barbour believes that there are *many* Nows that exist 'timelessly', even though we happen to be confined to *one*. The following passage brings the elements Butterfield mentions out to the fore:[47]

All around NOW ... are other Nows with slightly different versions of yourself. All such nows are 'other worlds' in which there exist somewhat different but still recognizable versions of yourself. (ibid. 56)

Clearly, given the multiplicity of Nows, this cannot be presentism conceived of along Priorian lines, though we can certainly see the connection to modal realism; talk of other nows being 'simultaneously present' (ibid.) surely separates this view from the Priorian presentist's thesis. That Barbour's approach is not a presentist approach is best brought out by the lack of temporal flow; there is no A-series change. Such a notion of change is generally tied to presentism. Indeed, the notion of many nows existing simultaneously sounds closer to eternalism than presentism; i.e. the view that past and future times exist with as much ontological robustness as the present time. These points also bring out analogies with the 'many-worlds' interpretation of quantum mechanics; so much so that a more appropriate characterization might be a 'many-Nows' theory.[48] Thus, I don't think that Butterfield's is an accurate diagnosis. What is the correct diagnosis?

There is a view, which has become commonplace since the advent of special relativity, that objects are four-dimensional; objects are said to 'perdure', rather than 'endure': this latter view is aligned to a three-dimensionalist account according to which objects are wholly present at each time they exist; the former view is known as 'temporal part theory'. The four-dimensionalist view is underwritten by a wide variety of concerns: for metaphysicians these concerns are to do with puzzles about change; for physics-minded philosophers they are to do with what physical theory has to say. Change over time is characterized by differences between successive temporal parts of individuals. Whichever view one chooses, the idea of *persisting* individuals plays a role; without this, the notion of change is simply incoherent, for change requires there to be a *subject* of change. Although Barbour's view is usually taken to imply a three-dimensionalist interpretation (by Butterfield for one), I think it is also perfectly compatible with a kind of temporal parts type theory. However, rather

think that special relativity allows for presentism in a certain sense—we simply need to modify what we mean by 'present' in this context, distinguishing it from what we mean in Newtonian mechanics—and that general relativity (classical and quantum) too allows for presentism in the canonical formulation (a view recently defended by Monton (2001) in the context of timefull, 'fixed foliation' strategies). But we need to distinguish the kind of presentism that classical and quantum general relativity allows for from that which special relativity allows for, and that Newtonian mechanics allows for. However, this is not the place to argue the point.

[47] Fans of Lewis's *On The Plurality of Worlds* (1986) will notice a remarkable similarity to a certain famous passage from that work. Hence the suggested link to modal realism alluded to earlier.

[48] Indeed, Barbour himself claims that his approach suggests what he calls a 'many-instants ... interpretation of quantum mechanics' (ibid.). However, it seems clear that the multiplicity of Nows is as much a classical as a quantum feature.

than the structure of time being linear (modelled by $\mathbb{R}$), it is, in a very rough sense, non-linear (modelled by relative configuration space) and the 'temporal evolution' is probabilistic (governed by a solution to the Hamiltonian constraint). We see that the parts themselves do not change or endure and they cannot perdure since they are three-dimensional items and the parts occupying distinct 3-spaces (and, indeed, the 3-spaces themselves) are not genidentical; rather, the quantum state 'jumps' around from Now to Now in accordance with the Hamiltonian constraint in such a way that the parts contain records that 'appear' to tell a story of linear evolution and persistence. Properly understood, then, Barbour's views arise from a simple thesis about identity over time, i.e. a denial of persistence:

> We think things persist in time because structures persist, and we mistake the structure for substance. But looking for enduring substance is like looking for time. It slips through your fingers. (ibid. 49)

In denying persisting individuals, Barbour has given a philosophical grounding for his alleged timelessness. However, as I mentioned earlier, the view that results might be seen as not at all timeless: the relative configuration space, consisting of Nows, can be seen as providing a *reduction* of time, in much the same way that Lewis's plurality of worlds provides a reduction of modal notions.[49] The space of Nows is given once and for all and does not alter, nor does the quantum state function defined over this space, and therefore the probability distribution is fixed too. But just as modality lives on in the structure of Lewis's plurality, so time lives on in the structure of Barbour's Platonia. However, also like Lewis's plurality, believing in Barbour's Platonia requires substantial imagination stretching. Of course, this isn't a knock-down objection; with a proposal of this kind I think we need to assess its cogency on a cost versus benefit basis. As I show below, I think that the same result (a resolution of the problem of time) can be achieved on a tighter ontological budget. However, I think there is real value in Barbour's analysis of the problem of time, and philosophers of time would do well to further consider the connections between Lewis's and Barbour's reductions, and the stand-alone quality of the view of time that results.[50]

Not quite as radical as Barbour's are those timeless views that accept the fundamental timelessness of general relativity and quantum gravity that follows from the gauge-invariant conception of observables, but attempt to introduce a *thin* notion of time and change into this picture. A standard approach along these lines is to account for time and change in terms of time-independent correlations between gauge-dependent quantities. The idea is that one never measures a gauge-dependent quantity, such as position of a particle; rather, one measures 'position at a time', where the time is defined by some *physical* clock.[51] Thus, in the general relativistic

[49] Roughly, Lewis's idea is that the notions of *necessity* and *possibility* are to be cashed out in terms of holding at all or some of a class of 'flesh and blood' worlds.

[50] I expect that the view of most philosophers of time would be that Barbour has simply outlined a variation of eternalism, albeit a peculiar one.

[51] See the exchange between DeWitt, Rovelli, Unruh, and Kuchař in Asktekar and Stachel (1991: 137–40) for a nice quick introduction to the timeless vs. timefull views: Rovelli and DeWitt

context, we might consider the spatial volume of the universe, $V = \int_{\Sigma} \sqrt{-\det g}\, d^3x$; this is gauge dependent (for compact Σ) and, therefore, is not an observable. Now suppose we wish to measure some quantity defined over Σ, say the total matter density $\rho(x)$, $\forall_i x_i \in \Sigma$. Of course, this too is a gauge-dependent quantity; but the *correlation* between V and ρ when they take on a certain value is gauge *independent*. In this way, one can define an instant of time; one can write $\tau = \rho(V)$ or $\tau = V(\rho)$. One can then use these correlations to function as a clock giving a monotonically increasing time parameter τ against which to measure some other quantities—this is admittedly a rather loose example, but gets the point across well enough. Unruh objects to this method along the following lines:

> one could [try to] define an instant of time by the correlation between Bryce DeWitt talking to Bill Unruh in front of a large crowd of people, and some event in the outside world one wished to measure. To do so however, one would have to express the sentence 'Bryce DeWitt talking to Bill Unruh in front of a large crowd of people' in terms of physical variables of the theory which is supposed to include Bryce DeWitt, Bill Unruh, and the crowd of people. However, in the type of theory we are interested in here, those physical variables are all time independent, they cannot distinguish between 'Bryce DeWitt talking to Bill Unruh in front of a large crowd of people' and 'Bryce DeWitt and Bill Unruh and the crowd having grown old and died and rotted in their graves.' ... The subtle assumption [in the correlation view] is that the individual parts of the correlation, e.g. DeWitt talking, are measurable when they are not. (1991: 267)

Belot and Earman question Unruh's interpretation of the correlation view, and suggest that it might be better understood 'as a way of explaining the illusion of change in a changeless world' (2001: 234). The basic idea is that one deals in quantities of the form 'clock-1-reads-t_1-when-and-where-clock-2-reads-t_2'. We get the illusion of change by (falsely) taking the elements of these relative (correlation) observables to be capable of being measured independently of the correlation. They suggest that Rovelli's notion of evolving constants of motion is a good way of 'fleshing out' the relative observables view.

Rovelli's *evolving constants of motion* proposal is made within the framework of a gauge-invariant interpretation. He accepts the conclusion that quantum gravity describes a fundamentally timeless reality, but argues that sense can be made of dynamics and change within such a framework. Take as a naive example of an observable m = 'the mass of the rocket'. This cannot be an observable of the theory since it changes over (coordinate) time; it fails to commute with the constraints, $\{m, H\} \neq 0$, because it does not take on the same value on each Cauchy surface. Rovelli's idea is to construct a one-parameter family of observables (constants of the motion) that can represent the sorts of changing magnitudes we observe. Instead of speaking of, say, 'the mass of the rocket' or 'the mass of the rocket at t', which are both gauge-dependent quantities (unless t is physical), one speaks instead of 'the mass of the rocket when it entered the asteroid belt', $m(0)$, and 'the mass of the rocket when it

are firmly in favour of the correlation view, while Unruh and Kuchař are firmly against it. I outline Unruh's and Kuchař's objections below.

reached Venus', $m(1)$, and so on up until $m(n)$. These quantities are gauge invariant, and, hence, constants of the motion; but, by stringing them together in an appropriate manner, we can explain the appearance of change in a property of the rocket. The change we normally observe taking place is to be described in terms of a one-parameter family of constants of motion, $\{m(t)\}_{t\in\mathbb{R}}$, an *evolving* constant of motion.[52]

A similar criticism to Unruh's comes from Kuchař (1993b: 22), specifically targeting Rovelli's approach. Kuchař takes Rovelli to be advocating the view that observing 'a changing dynamical variable, like Q [a particle's position, say], amounts to observing a one-parameter family $Q'(\tau_1) := Q' + P'\tau = Q - P(T - \tau), \tau \in R$ of perennials' (ibid. 22). By measuring $Q'(\tau)$ at τ_1 and τ_2 'one can infer the change of Q from $T = \tau_1$ to $T = \tau_2$' (ibid.). So the idea is that a changing observable can be constructed by observing correlations between two dynamical variables T and Q, so that varying τ allows one a notion of 'change of Q with respect to T'. Kuchař objects that one has no way of observing τ that doesn't smuggle in non-perennials. But this is a non sequitur; one doesn't need to observe τ independently of Q: we can simply *stipulate* that the two are a 'package deal', inseparable. In this way, I think both Unruh's and Kuchař's objections can be successfully dealt with. I outline this view further in the next section, where I attempt to strengthen the correlation solution.

Rovelli's approach has a certain appeal from a philosophical point of view. It bears similarities to four-dimensionalist views on time and persistence. The basic idea of both of these views is that a changing individual can be constructed from unchanging parts. Change over time is conceptually no different from variation over a region of space. (I think philosophers of time might perhaps profit from a comparison of Rovelli's proposal with four-dimensionalist views.) However, technically, it is hard to construct such families of constants of motion as phase functions on the phase space of general relativity. To the extent that they can be constructed at all, they result in rather complicated functions that are hard to represent at the quantum level (i.e. as quantum operators on a Hilbert space: cf. Hájiček 1996: 1369), and face the full force of the factor ordering difficulties (cf. Ashtekar and Stachel (eds.) 1991: 139).[53] For this, and other, reasons, Rovelli has recently shifted to something more like the original correlation view I outlined above (see Rovelli 2002; rather surprisingly, his earlier 1991 paper contains much the same view).

As with the evolving constants of motion programme, Rovelli believes that the observables of general relativity and quantum gravity are relative quantities expressing correlations between dynamical variables. The problem Rovelli sets himself in his *partial observables* programme, as if in answer to Unruh's complaint, is this: 'how can a

[52] Rovelli, in collaboration with Connes (1994), has argued that the 'flow' of time can be explained as a 'thermodynamical' effect, and is state-dependent. The thermal time is given by the state-dependent flow generated by the statistical state s over the algebra of observables: $\frac{dq}{dt} = -\{q, \log s\}$. Hence, the Hamiltonian is given by $-\log s$, so that the (statistical) state that a system occupies determines the Hamiltonian *and* the associated flow. Rovelli connects this idea up to his evolving constants proposal by identifying the thermal time flow with the one-parameter group of automorphisms of the algebra of observables (as given by the Tomita flow of a state).

[53] But see Montesinos et al. (1999) for a construction of such a family for a simple $SL(2, \mathbb{R})$ model.

correlation between two nonobservable [gauge-dependent] quantities be observable?' (ibid. 124013-1). He distinguishes between *partial* and *complete* observables, where the former is defined as a physical quantity to which we can associate a measurement leading to a number, and the latter is defined as a quantity whose value (or probability distribution) can be predicted by the relevant theory, i.e. a (gauge-invariant) Dirac observable. Partial observables can be measured but not predicted, and complete observables are correlations between partial observables that can be both measured and predicted. The above question can then be rephrased in these terms: 'how can a pair of partial observables make a complete observable?' (see p. 124013-5). His answer is somewhat surprising, for he argues that this question is just as applicable to classical non-relativistic theories as it is to relativistic theories. However, to make sense of the answer, there is a further distinction to be made within the class of partial observables that only holds in non-general relativistic (more generally: background dependent) theories: *dependent* and *independent*. These can be understood as follows: take two partial observables, q and t (position and time); if we can write $\mathsf{q}(\mathsf{t})$ but not $\mathsf{t}(\mathsf{q})$ then we say that q is a dependent partial observable and t is an independent partial observable. He then traces the confusion in Unruh's objection to the notion of localization in space and time and, in particular, that this makes no sense in the context of general relativistic physics. The absolute localization admitted in non-relativistic theories means that the distinction can be disregarded in such quantum theories since 'the space of observables reproduces the fixed structure of spacetime' (p. 124013-1). However, where the structure of spacetime is dynamical, t and q are partial observables for which we cannot assume that an external clock or spatial reference frame exists. Going back to Unruh's example, we see that Unruh, DeWitt, and the crowd of people are analogues of partial observables. Unruh assumes that the dependent/independent distinction must hold. However, this is just what Rovelli denies:

> A pre-GR theory is formulated in terms of variables (such as q) evolving as functions of certain distinguished variables (such as t). General relativistic systems are formulated in terms of variables ... that evolve with respect to each other. General relativity expresses relations between these, but in general we cannot solve for one as a function of the other. Partial observables are genuinely on the same footing. (Rovelli, ibid. 124013-3)

The theory describes relative evolution of (gauge-dependent) variables as functions of each other. No variable is privileged as the independent one (cf. Montesinos et al. 1999: 5).[54] How does this resolve the problem of time? The idea is that coordinate time evolution and physical evolution are entirely different beasts. To get physical evolution, all one needs is a pair $\mathcal{C}$, C consisting of an extended configuration space

[54] Earman appears to endorse this view, and claims that the events (he calls them 'Komar events') formed by such coincidences between gauge-dependent variables can be strung together to give a temporal evolution, generating a 'D-series'. However, I think that *coincidences* narrow the class of observables down too much. Moreover, I argue below that if Earman means to follow this kind of account—and I think it is clear that he does (see e.g. Earman 2002: 22)—of the evolution of observables, then the D-series cannot be formed: a unique series is incompatible with the multi-fingered time evolution that goes in tandem with the relational approach—Earman is no doubt aware of this, but, to the best of my knowledge, nowhere makes it explicit.

(coordinated by partial observables) and a function on $T^*\mathcal{C}$ giving the dynamics. The dynamics concerns the relations between elements of $\mathcal{C}$, and though the individual elements do not have a well-defined evolution, relations between them (i.e. correlations) do: they are independent of coordinate time.

Let me spell this out some more, for my own response is based on Rovelli's, albeit with an interpretative twist. Consider two non-gauge-invariant (i.e. gauge-dependent) functions α and β. These are our partial observables; we can suppose that α is the volume of a compact hypersurface and that β is the matter density defined on a compact hypersurface. Recall that neither of these quantities is predictable, for their evolution will be gauge dependent. We want to construct from this pair of partial observables a complete observable $\mathcal{E}^{\tau}_{\alpha|\beta}$ (where τ will be understood to be a 'clock' variable). To do this we consider the relational quantity that is formed by correlating the values of the two partial observables. We arbitrarily take one of the partial observables to be the 'clock', whose values will parameterize the evolution of the other. Let β be the clock. $\mathcal{E}^{\tau}_{\alpha|\beta}$ then gives the quantity that gives the value of α when, under the flow generated by the constraints, the value of β is τ. Thus, a partial observable is evolved along a gauge flow, such that the evolution is a gauge transformation, and is to be understood as a clock 'ticking' along the gauge orbit. On its own, of course, this is an expression of the problem of change since evolution along a gauge orbit is just the problem! But when we correlate another partial observable with the values at which $\beta = \tau$ we form a time-independent observable since the value of α when $\beta = \tau$ does not change. Variation in τ allows for the formation of a 1-parameter family of complete observables that correspond to empirically observable change.[55] The evolution does not occur with respect to some background time parameter, but with respect to the values of the arbitrary clock; the complete observables will enable us to *predict* the value of α at the 'time' $\beta = \tau$. More precisely, the evolution will be a map $\mathfrak{E}^{\tau} : \mathcal{E}^{\tau_0}_{\alpha|\beta} \rightarrow \mathcal{E}^{\tau+\tau_0}_{\alpha|\beta}$, taking complete observables into complete observables. The fact that the clock β is arbitrary (since it can be chosen from $C^{\infty}(\mathcal{C}) \subset C^{\infty}(\Gamma)$) implies that the theory is a multi-fingered time formalism: there are numerous (infinitely many) choices that one can make for the clocks, and so there are numerous times—though not all choices will be 'good' clocks physically speaking.

The multi-fingered time result implies that Earman's D-series of coincidence events—which I take to be of the form of Rovelli's complete observables, albeit including the four invariant components of the gravitational field (of which he speaks as if it were unique)—cannot be applicable. Earman (2002: 14) claims that '[t]he occurrence or non-occurrence of a coincidence event is an observable matter' and that when 'one such event occurs earlier than another such event' that 'is also an observable matter'; '[c]hange now consists in the fact that different positions in the D-series are occupied by different coincidence events'. This is not equivalent to the B-series, consisting of a string of events which are either earlier-than, later-than, or simultaneous with each other, because, according to Earman, that 'can be described

[55] This is why the response matches temporal parts theories: temporal parts do not change in themselves, but by forming an individual from a string of such parts a persisting, changing thing emerges.

in terms of the time independent correlations between gauge dependent quantities which change with time' (ibid. 15). B-series change is an artefact of the local representations (the elements of the equivalence class of metrics) rather than a real feature of the world, that associated with the equivalence class itself (for which his D-series is supposed to apply). This is a strange way of viewing the content of B-series time, and I have never seen any philosophers of time dabbling with such concepts before: why does the B-series depend on gauge-dependent quantities? Perhaps it is a way to understand the B-series given an ontology that sticks by the gauge-dependent quantities, but for different ontologies it needn't follow. If, for example, we adopt an ontology of events then it seems that Earman has simply constructed a B-series all over again. If we view the ontology as per Rovelli's partial/complete observables approach then the multi-fingered time makes the D-series dependent on an arbitrary choice of clock. If we are to make any sense of Earman's D-series within this framework then it will be as one series among infinitely many, each corresponding to a clock choice. But then it does nothing more than to give a name to $\mathcal{E}^{\tau_i}_{\alpha|\beta}$.

However, both Earman and Rovelli appear to want to cling to the notion that the elements of the relations (the partial observables or coinciding elements) have some independent physical reality.[56] This is most explicit in Rovelli, who takes the extended configuration space (physically impossible states and all!) to have physical significance as the space of the partial observables. I agree that, without empirical evidence to the contrary, the extended space should be retained since it gives us more conceptual elbow room; but I favour a view whereby gauge invariance itself picks out the physical parts of this space. The interpretation then follows the correlation view, but with the correlates and the correlations understood as simply different aspects of one and the same basic structure. The natural interpretation of Rovelli's view is that there is no physical distinction between gauge-dependent and independent quantities. This implies that there are physically real quantities that are not predictable, even though we can associate a measurement procedure with them; indeed, Rovelli claims that these variables 'are the quantities with the most direct physical interpretation in the theory' (ibid. 124013-7)—I discuss this, what I take to be a problematic issue with Rovelli's approach, in Rickles (2005b).

It is interesting to note how this links up to Belot and Earman's interpretative taxonomy regarding constraints and spacetime ontology. Since Belot and Earman equate the view that there are physically real quantities that do not commute with the constraints with (straightforward) substantivalism, it appears that Rovelli would have to class as a substantivalist, for his partial observables are just such quantities! Combined with the role reversal of Kuchař given earlier, this makes something of a mockery of their taxonomy, for they have Kuchař and Rovelli as the archetypical substantivalist and relationalist respectively. This, I would urge, is yet another aspect

[56] Note that Rovelli reads the gauge-fixation methods involving dust variables, curvature scalars, and the like as partial observables. What occurs in these strategies is that the partial observables are taken to be independent so that they are able to function as coordinate systems. However, as Rovelli notes, since the dependent and independent players can have their roles permuted, the distinction collapses (ibid. 124013-4).

of my claim that the relationalist/substantivalist controversy doesn't get any support from those problems with their roots in the interpretation of gauge symmetries (see Rickles 2005a). However, I think better justice can be done to Rovelli's view if we take the measurability of the gauge-dependent quantities as *derived* from the more fundamental correlations of which they are a part. I explain this structuralist gloss on Rovelli's position in the next section.

6.6 ENTER STRUCTURALISM

Rovelli, and other defenders of the correlations view,[57] are of the opinion that the observables of general relativity and quantum gravity are *relative* quantities that express correlations between dynamical, and hence gauge-dependent, variables. The problems posed to the correlation-type timeless strategies are based upon an understanding that is couched in terms of relationalism. The fact that correlations between *material* systems are required to define instants of time (and points of space) does indeed look, superficially, to entail relationalism. I suspect that this entailment is what was motivating the objections of Unruh and Kuchař. The assumption was that if it is relations doing the work, then the relata must have some physical significance independently of these relations. This is just what I deny: the distinction between material systems and space and time simply amounts to different aspects of one and the same physical structure (cf. Stein 1967). It is not that relations can be free standing; maybe they can, but in this case we have clear relata entering into the relations: DeWitt, Unruh, and a crowd of people! The question concerns the relative ontological priority of these relata over the relations. Relationalists will argue that the relations supervene upon the relata so that the relata are fundamental. Substantivalists will argue that the relata enter into their relations only in virtue of occupying a position in some underlying spatio-temporal structure that exists independently of both the relations and relata. An alternative position will see the relata as being some kind of epiphenomena or 'by-product' resulting from intersections occurring between the relations. But there is a middle way between these two extremes: neither relations nor relata have ontological priority. The relata are individuated in virtue of the relations and the relations are individuated by the relata.[58] Thus, the idea is to understand the correlation view structurally: one cannot ontologically decompose or factor the relative observables in to their relata, since the relata have no physical

[57] Others include DeWitt (see Ashtekar and Stachel 1991: 137), Marolf (1994), Page and Wooters (1983), and, on the philosophical side, Earman (2002: see below). The proposal of Page and Wooters' involves the idea that one deals with conditional probabilities for outcomes of pairs of observables. One then takes the observables as defining an instant of time (*qua* the value of a physical clock variable) at which the other observable is measured. A notion of evolution emerges in terms of the dependence of conditional probabilities on the values of the (internally defined) clock variables.

[58] Thus, though admittedly similar, this should be distinguished from Teller's brand of *relational holism* (see his 1991). Teller argues that in some cases—entanglement is the example he focuses on—we should view relations as being primitive (non-supervenient).

significance 'outside' (i.e. independently of) the correlations. But one need not imbue the relations themselves with ontological primacy either. Thus, one can evade the objection that gauge-dependent quantities are independently measurable by taking the correlations and correlates to be interdependent.

I shall call the overall structure formed from such correlations a *correlational network*, and the correlates I shall call *correlata*. It is important to note that the correlata need not be material objects, and we can find suitable items from the vacuum case. One is able to use (any) four invariants of the metric tensor to provide an intrinsic coordinate system that one can use to set up the necessary correlational network.[59] Thus, this approach does not imply relationalism; but it does not imply substantivalism either (neither sophisticated nor straightforward). The reason is, of course, that those interpretations require a stance to be taken with regard to the primacy of some category of object (points, fields, or whatever). Each of these other positions is problematic in the context of the problem of time since they both require that some set of objects take the ontological burden to function as a clock or a field of clocks.

Earman too seems to defend a version of the correlation view. His account is based on his notion of *coincidence events*; thus, quoting a passage already quoted, he writes:

> The occurrence or non-occurrence of a coincidence event is an observable matter ... and that one such event occurs earlier than another such event is also an observable matter. ... Call this series of coincidence events the *D-series* ... Change now consists in the fact that different positions in the D-series are occupied by different coincidence events. (2002: 14)

Earman claims that the coincidence event (represented by the functional relationship $g^{\mu\nu}(\phi^{\lambda})$: 'the *Komar state*') 'floats free of the points of $\mathcal{M}$' and 'captures the intrinsic, gauge-independent state of the gravitational field' (ibid.). General covariance implies that if this state is represented by one spacetime model it is also represented by any model from a diffeomorphism class of its copies. Now, Earman's interpretation of this, and his resolution of the problem of time, is to claim that the notion of spacetime points, properties localized to points, and change couched in terms of relationships between these, is to be found 'in the representations' and not 'in the world' (ibid.). This conclusion is clearly bound to the idea that in order to have any kind of change, a *subject* is required to undergo the change and *persist* under the change. In getting rid of the notion of a subject (i.e. spacetime points), Earman sees the only way out as abolishing change. The idea that change is a matter of representation is one way (not a particularly endearing one, say I) of accounting for the psychological impulse to believe that the world itself contains changing things, though I think it needs

[59] This is, of course, the method developed by Bergmann and Komar (1972). They used the four eigenvalues of the Riemann tensor. Dorato and Pauri (this volume) use this method and these 'Weyl scalars' to argue for a form of structuralism they call 'spacetime structural realism'. This is a far cry from what I have in mind since they retain fairly robust notions of independent object (the metric field) in their approach. Hans Westman, of The Perimeter Institute, has shown how one can build up an entire manifold structure (which he has named the 'point-coincidence manifold') from such invariant quantities (i.e. observables). In this case the correlational network would simply be represented by the point-coincidence manifold.

spelling out in much more detail than Earman has given us. But—quite aside from the fact that I don't think the existence of spacetime points is ruled out[60]—I don't see why Earman needs to go to this extreme; there is *variation* in the structure formed from the various correlations. True, we don't get any notion or account of time *flow* from this variation, but that is a hard enough problem outside general relativity and quantum gravity anyway (but see Connes and Rovelli 1994).

However, some other remarks of Earman's show that he doesn't have in mind the same view as mine. For instance, Earman (2002: 16–17) makes the following observations:

> [T]he gauge interpretation of diffeomorphism invariance ... calls into question the traditional choices for conceiving the subject vs. attribute distinction. The extremal choices traditionally on offer consist of taking individuals to be nothing but bundles of properties vs. taking individuals to have a 'thisness' (*haecceitas*) that is not explained by their properties. The gauge interpretation of GTR doesn't provide any grounds for *haecceitas* of spacetime points. Nor does it fit well with taking spacetime points as bundles of properties since it denies that the properties that were supposed to make up the bundle are genuine properties. The middle way between the *haecceitas* view and the bundles-of-properties view takes individuals and properties to require each other, the slogan being that neither exists independently of the states of affairs in which individuals instantiate properties.

As Earman goes on to explain, in the context of general relativity this middle way fares no better than the bundle-of-properties view since the gauge interpretation of general covariance 'implies that the state of affairs composed of spacetime points instantiating, say, metrical properties do not capture the literal truth about physical reality; rather, these states of affairs are best seen as representations of a reality ... that itself does not have this structure.' What Earman means by 'representation' in this context, is, I think, what Rovelli calls a 'local universe' (1992): a physically possible world in which properties are 'attached' to spacetime points. However, as Earman and Rovelli point out, this is not how general relativity represents the world; it does so by means of an equivalence class of such local universes, yielding a very 'non-local' description. If we extend the account Earman gives to include *relations* rather than simply properties (which clearly *do* require subjects of some sort) then we can in fact get directly at the structure Earman mentions.

Instead of the view Earman outlines, I have something more along the lines of Skyrms's 'Tractarian Nominalism' (1981). The idea here is to understand individuals, properties, and relations as 'abstractions' from the structure of the world (from facts) but not as existing independently of that structure: 'We may conceive of the world not as a world of individuals or as a world of properties and relations, but as a world of facts—with individuals and relations being equally abstractions from the facts' (p. 199). Likewise, the 'totality of facts' (the structure of the world) itself is 'composed' of such facts. As regards the question of ontological priority, then, we see that relations and relata share the same status: 'the Tractarian Nominalist ... takes both objects and relations quite seriously, and puts them on par. Neither is reduced

[60] For example, Simon Saunders's (2003) account of identity allows that spacetime points exist as *individual* objects while respecting diffeomorphism symmetry.

to the other' (p. 202). Armstrong too defends a similar account, and it is perhaps even more applicable to Unruh's decomposition problem. Thus, speaking in terms of 'states of affairs' rather than 'facts', he writes that 'while by an act of selective attention they [individuals, properties, and relations] may be *considered* apart from states of affairs in which they figure, they have no existence outside states of affairs' (1987: 578). Likewise, the correlations are the fundamental things; they are things that can be measured and predicted. The correlata are measurable only in virtue of their position in the correlation, and have no independence outside this. However, the correlata are our epistemic 'access point' to the correlations, and this is why, I think, Rovelli imbues his partial observables with fundamental significance. If his position is to escape the interpretative troubles highlighted by Unruh and Kuchař, however, the primacy needs to be reversed and shifted to the complete observables. By taking these seriously, as ontologically basic, those difficulties are easily resolvable.

This *structuralist* way of understanding the correlation view avoids Unruh's and Kuchař's objections, and it sidesteps Earman's worry. Not only does it resolve these objections, and the problem of time, it also provides a suitable ontological framework for classical and quantum gravity, according to which there are neither primitive points nor objects to be individuated. Rather, one has a correlational network that fluctuates quantum mechanically as a whole. This, I suggest, is a safe and sane ontological basis from which to view time and space in both classical and quantum (canonical) gravity.

6.7 CONCLUSION

There are two aspects to the problem of time, concerning both the states and observables. These each have both a classical and a quantum variant. I argued that the problems struck whether or not one reduced the phase space to the space of physical states (contra Maudlin). The responses split into two broad categories: timefull and timeless. I argued against the former strategies on the ground that they either break covariance by introducing external background structure, or else assume that a time variable is required for the construction of a quantum theory. The timeless approaches show that this is false. I defended the correlation view from objections by Unruh and Kuchař by utilizing the notion of non-decomposability of the correlations. I claimed that this latter notion is best understood structurally.

ACKNOWLEDGEMENTS

I would like to thank Steven French, Carl Hoefer, and Joseph Melia for a very useful discussion of the matters contained in and around this article. I should also like to thank Oliver Pooley for stimulating chats about background independence and the status of space and time in quantum gravity. Finally, I thank Luca Lusanna, J. Brian Pitts, and the anonymous referees of this book for their helpful comments and suggestions.

REFERENCES

Armstrong, D. M. (1987) "The Nature of Possibility." *Canadian Journal of Philosophy*, 16: 575–94.

Arnowitt, R., S. Deser, and C. W. Misner (1962) "The Dynamics of General Relativity". In L. Witten (ed.), *Gravitation: An Introduction to Current Research*. New York: Wiley & Sons (pp. 227–65).

Ashtekar, A. (1986) "New Variables for Classical and Quantum Gravity." *Physical Review Letters*, 57: 2244–7.

—— and C. Rovelli (1992) "Connections, Loops, and Quantum General Relativity'. *Classical and Quantum Gravity*, 9: 3–12.

—— and J. Stachel (eds.) (1991) *Conceptual Problems of Quantum Gravity*. Boston: Birkhäuser.

Barbour, J. (1994) "The Timelessness of Quantum Gravity: I. The Evidence from the Classical Theory". *Classical and Quantum Gravity*, 11: 2853–73.

—— (2003) *The End of Time*. London: Weidenfeld & Nicholson.

Beig, R. (1994) "The Classical Theory of Canonical General Relativity". In J. Ehlers and H. Friedrich (eds.), *Canonical Relativity: Classical and Quantum*. Berlin: Springer-Verlag (pp. 59–80).

Belot, G. (1996) 'Whatever is Never and Nowhere is Not: Space, Time, and Ontology in Classical and Quantum Gravity'. Unpublished Ph.D. thesis, University of Pittsburgh: **www.pitt.edu/~gbelot/Papers/dissertation.pdf**.

—— (1998) "Understanding Electromagnetism". *British Journal for the Philosophy of Science*, 49: 531–55.

—— (2003) "Symmetry and Gauge Freedom". *Studies in the History and Philosophy of Modern Physics*, 34: 189–225.

—— and J. Earman (1999) "From Physics to Metaphysics". In J. Butterfield and C. Pagonis (eds.), *From Physics to Metaphysics*. Cambridge: Cambridge University Press (pp. 166–86).

—— —— (2001) "PreSocratic Quantum Gravity". In C. Callender and N. Huggett (eds.), *Physics Meets Philosophy at the Planck Scale*. Cambridge: Cambridge University Press (pp. 213–55).

Bergmann, P. G., and A. Komar (1972) "The Coordinate Group Symmetries of General Relativity". *Int. J. Theor. Phys.* 5: 15–28.

Brügmann, B. (1994) "Loop Representations". In J. Ehlers and H. Friedrich (eds.), *Canonical Relativity: Classical and Quantum*. Berlin: Springer-Verlag (pp. 213–53).

Butterfield, J. (2001) "Critical Notice: *The End of Time*". *British Journal for the Philosophy of Science*, 53(2): 289–330.

Callender, C. (2000) "Shedding Light on Time". *Philosophy of Science*, 67: S587–S599.

—— and N. Huggett (eds.) (2001) *Physics Meets Philosophy at the Planck Scale*. Cambridge: Cambridge University Press.

Carlip, S. (1998) *Quantum Gravity in 2+1 Dimensions*. Cambridge: Cambridge University Press.

Connes, A., and C. Rovelli (1994) "Von Neumann Algebra Automorphisms and Time Versus Thermodynamics in Generally Covariant Quantum Theories". *Classical and Quantum Gravity*, 11: 2899–918.

Dirac, P. A. M. (1964) *Lectures on Quantum Mechanics*. Yeshiva University, Belfer Graduate School of Science Monographs Series, No. 2.

Earman, J. (1989) *World Enough and Space-Time: Absolute versus Relational Theories of Space and Time*. Cambridge, MA: MIT Press.

____ (2002) "Throughly Modern McTaggart: Or What McTaggart Would Have Said If He Had Learned the General Theory of Relativity". *Philosophers' Imprint*, 2(3).

____ (2003) "Ode to the Constrained Hamiltonian Formalism". In K. Brading and E. Castellani (eds.), *Symmetries in Physics: Philosophical Reflections*. Cambridge: Cambridge University Press (pp. 140–62).

____ (2003b) "The Cosmological Constant, the Fate of the Universe, Unimodular Gravity, and all that". *Studies in the History and Philosophy of Modern Physics*, 34, 4: 559–77.

____ and J. Norton (1987) "What Price Substantivalism? The Hole Story". *British Journal for the Philosophy of Science*, 38: 515–25.

Ehlers, J., and H. Friedrich (eds.) (1994) *Canonical Gravity: From Classical to Quantum*. Berlin: Springer-Verlag.

Gotay, M. (1984) "Constraints, Reduction, and Quantization". *Journal of Mathematical Physics*, 27(8): 2051–66.

Govaerts, J. (2001) "The Quantum Geometer's Universe: Particles, Interactions and Topology". ArXiv:hep-th/0207276.

Hájiček, P. (1996) "Time Evolution and Observables in Constrained Systems". *Classical and Quantum Gravity*, 13: 1353–75.

____ (1999) "Covariant Gauge Fixing and Kuchař Decomposition". ArXiv:gr-qc/9908051.

Henneaux, M., and C. Teitelboim (1992) *Quantization of Gauge Systems*. Princeton: Princeton University Press.

Hoefer, C. (1996) "The Metaphysics of Space-Time Substantivalism". *Journal of Philosophy*, 93: 5–27.

Isham, C. J. (1993) "Canonical Quantum Gravity and the Problem of Time". In *Integrable Systems, Quantum Groups, and Quantum Field Theories*. Dordrecht: Kluwer Academic Publishers, 1993 (pp. 157–288). Ar Xiv:gr-qc/9210011.

____ (1994) "Prima Facie Questions in Quantum Gravity". In J. Ehlers and H. Friedrich (eds.), *Canonical Relativity: Classical and Quantum*. Berlin: Springer-Verlag, 1994 (pp. 1–21).

Kuchař, K. (1992) "Time and Interpretations of Quantum Gravity". In *Proceedings of the 4th Canadian Conference on General Relativity and Relativistic Astrophysics*. Singapore: World Scientific Press (pp. 211–314).

____ (1993a) "Matter Time in Canonical Quantum Gravity". In B. Hu, M. Ryan, and C. Vishveshvara (eds.), *Directions in General Relativity*. Cambridge: Cambridge University Press (pp. 201–21).

____ (1993b) "Canonical Quantum Gravity". ArXiv:gr-qc/9304012.

Lewis, D. K. (1983) 'Individuation by Acquaintance and by Stipulation.' Reprinted in D. Lewis, *Papers in Metaphysics and Epistemology*. Cambridge: Cambridge University Press (pp. 373–402).

____ (1986) *The Plurality of Worlds*. Oxford: Basil Blackwell.

McTaggart, J. M. E. (1908) "The Unreality of Time". *Mind*, 17: 457–74.

____ (1927) *The Nature of Existence*. Cambridge: Cambridge University Press.

Marolf, D. (1994) "Almost Ideal Clock in Quantum Cosmology: A Brief Derivation of Time". ArXiv:gr-qc/9412016.

Marsden, J., and A. Weinstein (1974) "Reduction of Symplectic Manifolds with Symmetry". *Reports in Mathematical Physics*, 5: 121–30.

Maudlin, T. (2002) "Throughly Muddled McTaggart: Or How to Abuse Gauge Freedom to Generate Metaphysical Monstrosities". *Philosophers' Imprint*, 2(4).

Mellor, D. H. (1998) *Real Time II*. London: Routledge.

Montesinos, M., C. Rovelli, and T. Thiemann (1999) "An $\mathsf{SL}(2,\mathbb{R})$ model of Constrained Systems with Two Hamiltonian Constraints". ArXiv:gr-qc/9901073.

Monton, B. (2001) "Presentism and Quantum Gravity". **http://philsci-archive.pitt.edu/archive/00000591/.**

Page, D. N., and W. K. Wooters (1983) "Evolution without Evolution: Dynamics Described by Stationary Observables". *Physical Review D*, 27: 2885–92.

Pooley, O. (2002) 'The Reality of Spacetime'. Unpublished PhD thesis, Oxford University.

Redhead, M. L. G. (1975) "Symmetry in Intertheory Relations". *Synthese*, 32: 77–112.

—— (2003) "The Interpretation of Gauge Symmetry". In K. Brading and E. Castellani (eds.), *Symmetry in Physics: Philosophical Reflections.* Cambridge: Cambridge University Press (pp. 124–39).

Reichenbach, H. (1947) *Elements of Symbolic Logic*. New York: Macmillan.

Rickles, D. (2005a) "A New Spin on the Hole Argument". *Studies in the History and Philosophy of Modern Physics*, 36: 415–34.

—— (2005b) "Interpreting Quantum Gravity". *Studies in the History and Philosophy of Modern Physics*, 36: 691–715.

Rovelli, C. (1991) "Is There Incompatibility between the Ways Time is Treated in General Relativity and Standard Quantum Mechanics". In A. Ashtekar and J. Stachel (eds.), *Conceptual Problems of Quantum Gravity*. New York: Birkhaüser (pp. 126–39).

—— (1992) "What is Observable in Classical and Quantum Gravity". *Classical and Quantum Gravity*, 8: 297–316.

—— (2002) "Partial Observables". *Physical Review D*, 65: 124013–124013-8.

—— (2004) *Quantum Gravity.* Cambridge: Cambridge University Press.

Russell, B. (1940) *An Inquiry into Meaning and Truth*. New York: W. W. Norton.

Saunders, S. (2002) 'How Relativity Contradicts Presentism.' In C. Callender (ed.), *Time, Reality, and Experience*. Cambridge: Cambridge University Press (pp. 277–92).

—— (2003) 'Indiscernibles, General Covariance, and Other Symmetries: The Case for Non-eliminativist Relationalism.' In A. Ashtekar, D. Howard, J. Renn, S. Sarkar, and A. Shimony (eds.), *Revisiting the Foundations of Relativistic Physics: Festschrift in Honour of John Stachel.* Dordrecht: Kluwer (pp. 151–73).

Savitt, S. (2000) "There's No Time Like the Present (in Minkowski Spacetime)". *Philosophy of Science*, 67: S563–S574.

Shoemaker, S. (1969) "Time without Change". *Journal of Philosophy*, 66(12): 363–81.

Skyrms, B. (1981) "Tractarian Nominalism". *Philosophical Studies*, 40: 199–206.

Stachel, J. (1993) "The Meaning of General Covariance: The Hole Story". In J. Earman et al. (eds.), *Philosophical Problems of the Internal and External Worlds: Essays on the Philosophy of Adolf Grünbaum.* Pittsburgh: Pittsburgh University Press/Konstanz Universitäts Verlag (pp. 129–60).

Stein, H. (1967) "Newtonian Space-Time". *Texas Quarterly*, 10: 174–200.

Teller, P. (1991) "Relativity, Relational Holism, and the Bell Inequalities". In J. T. Cushing and E. McMullin (eds.), *Philosophical Consequences of Quantum Theory: Reflections on Bell's Theorem.* Notre Dame, Ind.: University of Notre Dame Press (pp. 208–23).

Thiemann, T. (1996) "Anomaly Free Formulation of Nonperturbative 4-Dimensional Lorentzian Quantum Gravity". *Physics Letters B*, 380: 257.

Torre, C. G. (1993) "Gravitational Observables and Local Symmetries". *Physical Review D*, 48: R2373–R2376.

Unruh, W. G. (1989) 'Unimodular Theory of Canonical Quantum Gravity.' *Phys. Rev. D* 40: 1048–52.

—— (1991) "No Time and Quantum Gravity". In R. Mann and P. Wesson (eds.), *Gravitation: A Banff Summer Institute.* Singapore: World Scientific Press (pp. 260–75).

____ and R. Wald (1989) "Time and the Interpretation of Quantum Gravity". *Physical Review D*, 40: 2598–614.

Weinstein, S. (1999) "Gravity and Gauge Theory". In D. A. Howard (ed.), *Proceedings of the 1998 Biennial Meeting of the Philosophy of Science Association: Supplement to Volume 66, No. 3, Part I, Contributed Papers*, S146–S155.

____ (2001) "Conceptual and Foundational Issues in the Quantization of Gravity". Unpublished Ph.D. dissertation, Northwestern University.

York, J. W. (1971) "Gravitational Degrees of Freedom and the Initial Value Problem". *Physical Review Letters*, 26: 1656–8.

7

The Case for Background Independence

Lee Smolin

ABSTRACT

The aim of this chapter is to explain carefully the arguments behind the assertion that the correct quantum theory of gravity must be background independent. We begin by recounting how the debate over whether quantum gravity must be background independent is a continuation of a long-standing argument in the history of physics and philosophy over whether space and time are relational or absolute. This leads to a careful statement of what physicists mean when we speak of background independence. Given this we can characterize the precise sense in which general relativity is a background-independent theory. The leading background-independent approaches to quantum gravity are then discussed, including causal set models, loop quantum gravity, and dynamical triangulations, and their main achievements are summarized along with the problems that remain open. Some first attempts to cast string/$\mathcal{M}$ theory into a background-independent formulation are also mentioned.

The relational/absolute debate has implications also for other issues such as unification and how the parameters of the standard models of physics and cosmology are to be explained. The recent issues concerning the string theory landscape are reviewed and it is argued that they can only be resolved within the context of a background-independent formulation. Finally, we review some recent proposals to make quantum theory more relational.

7.1 INTRODUCTION

During the last three decades research in theoretical physics has focused on four key problems, which, however, remain unsolved. These are

1. The problem of quantum gravity.
2. The problem of further unifying the different forces and particles, beyond the partial unification of the standard model.
3. The problem of explaining how the parameters of the standard models of particles physics and cosmology, including the cosmological constant, were chosen by nature.
4. The problem of what constitutes the dark matter and energy, or whether the evidence for them is to be explained by modifications in the laws of physics at very large scales.

One can also mention a fifth unsolved problem, that of resolving the controversies concerning the foundations of quantum mechanics.

All five problems have remained unsolved, despite decades of determined effort by thousands of extremely talented people. While a number of approaches have been studied, most expectations have been put on string theory as it appears to provide a uniquely compelling unification of physics. Given that the correct perturbative dynamics for gauge fields, fermions, and gravitons emerges from a simple action expressed in terms of worldsheets and that, in addition, there are strong indications[1] that the quantum corrections to these processes are finite to each order of string perturbation theory (Berkovits 2004; D'Hoker and Phong 2002a, 2002b), it is hard not to take string theory seriously as a hypothesis about the next step in the unification of physics. At the same time, there remain open problems.

Despite knowing a great deal about the different perturbative string theories and the dualities that relate them, it is widely believed that a more fundamental formulation exists. This would give us a set of equations, solutions to which would give rise to the different perturbative string theories. While there is a lot of evidence for the existence of this more fundamental formulation, in the dualities that relate the different string perturbation theories, there is as yet no agreed-upon proposal as to either the principles or the equations of this formulation.

It is also unfortunately the case that the theory makes, as yet, no falsifiable predictions for doable experiments, by which the applicability of the theory to nature could be checked. This is because of the landscape of discrete vacua which have been uncovered in the last few years. Powerful effective field theory[2] arguments have made it plausible that the theory comes in an infinite number of versions (DeWolfe et al. 2005). These appear to correspond to an infinite number of possible universes and low-energy phenomenologies. Even if one imposes the minimal phenomenological constraints of a positive vacuum energy and broken supersymmetry, there are argued to be still a vast ($> 10^{500}$) number of theories (Kachru et al. 2003). There thus appears to be no uniqueness and no predictability so far as observable parameters are concerned, for example, one can get any gauge group and many different spectra of Higgs and fermions.

Of course, these two issues are related. It seems very likely that the challenge posed by the landscape would be resolved if we had a more fundamental formulation of string theory. This would enable us to establish which of the vacua described by effective field theory are truly solutions to the exact theory. It would also allow us to study the dynamics of transformations between different vacua.

Another striking feature of the present situation is that we have no unique predictions for the post-standard model physics which will be explored in upcoming experiments at the LHC. This is true in spite of the fact that we have had three decades since the formulation of the standard model to discover a convincing theory that would give us unique predictions for these experiments. The theory many of

[1] But no proof.

[2] These are ordinary classical field theories which are argued to describe low-energy limits of string theories, whose existence is conjectured, but not demonstrated.

our colleagues believe, the supersymmetric extension of the standard model, has too many parameters to yield unique predictions for those experiments.

It is beyond doubt that research in string theory has nonetheless led to a large number of impressive results and conjectures, some of great mathematical beauty. Several mathematical conjectures have been suggested by work in string theory that turned out to be provable by more rigorous means. A number of interesting conjectures and results have been found for the behaviour of supersymmetric gauge theories. All of this suggests that string theory has been worth pursuing. At the same time, the present situation is very far from what was expected when people enthusiastically embraced string theory twenty years ago.

If so much effort has not been rewarded with success, it might be reasonable to ask whether some wrong assumption was made somewhere in the course of the development of the theory. The purpose of this chapter is to propose such a hypothesis. This hypothesis is made with an open-minded spirit, with the hopes of stimulating discussion.

To motivate my hypothesis, we can start by observing that theorists' choices of how to approach the key issues in fundamental physics are largely determined by their views on three crucial questions.

- *Must a quantum theory of gravity be background independent, or can there be a sensible and successful background-dependent approach?*
- *How are the parameters of the standard models of physics and cosmology to be determined?*
- *Can a cosmological theory be formulated in the same language we use for descriptions of subsystems of the universe, or does the extension of physics from local to cosmological require new principles or a new formulation of quantum theory?*

It is the first issue that divides most string theorists from those who pursue alternative approaches to quantum gravity.

The most basic argument for background independence is simply that the experimental success of general relativity proves that the geometry of spacetime is dynamical. The geometry evolves according to equations of motion just like any other field in physics. Hence, in any quantum theory of spacetime the spacetime geometry should be treated as a quantum dynamical degree of freedom. It seems impossible that this most basic insight of Einstein's theory should be reversed and physics go back to a formulation in which fields, particles, or strings move on a fixed, non-dynamical background. But that is how string theory has been formulated, up till now.

It is sometimes asserted that string theory incorporates general relativity, because the Einstein equations (up to higher-order corrections) are a necessary condition that must be satisfied by a spacetime on which a string is to propagate consistently. This is true, but it is not the only necessary condition that must be satisfied. All string theories so far formulated explicitly in terms of the dynamics of the string worldsheets require supersymmetry or an equivalent constraint, to cancel an instability which manifests itself as the presence of a tachyon. In all known cases this requires that the spacetime have a timelike or null killing field[3]. This reduces the possible cases to the measure zero subset of solutions of Einstein's equations in which the geometry is stationary and hence non-dynamical.

While there have been attempts to construct string theory on time-dependent backgrounds, unfortunately, no explicit construction of amplitudes for string propagation have ever been written down for such backgrounds. This is why the evidence for the existence of the vast numbers of string theories making up the landscape is restricted to effective field theory.

This brings us to the second issue, which determines the attitude different people take to the landscape. There are, roughly, three possible approaches: (1) a unique theory leading to unique predictions. (2) Anthropic approaches, according to which our universe may be very different from a typical member of an ensemble or landscape of theories[4]. (3) Dynamical, or evolutionary approaches, according to which the dynamics of reproduction of universes results in our universe being a typical member of the ensemble (Smolin 1997a). The first has been, traditionally, the basis of the hopes for a unified theory, but the recent results suggest that unification leads not to a single, unique theory, but to a multitude of possible theories. This leaves the other two options.

The third issue has been long appreciated by those who have attempted to formulate a sensible quantum theory of cosmology, but it has recently been raised in the contexts of attempts to resolve the problems of the landscape in terms of cosmological theories and hypotheses.

In this chapter I would like to make two observations and a hypothesis about these issues.

- These three debates are closely related and they are unlikely to be resolved separately.
- These three debates are aspects of a much older debate, which has been central to thinking about the nature of space and time, going back to the beginning of physics. This is the *debate between relational and absolute theories of space and time.*

In particular, as I will explain below, background-dependent attempts at quantum gravity and anthropic approaches to the landscape are the contemporary manifestations of the absolute side of the old debate. Similarly, background-independent approaches to quantum gravity and dynamical or evolutionary approaches to the landscape are firmly within the relational tradition.

Now here is my thesis, which it is the task of this chapter to support:

The reason that we do not have a fundamental formulation of string theory, from which it might be possible to resolve the challenge posed by the landscape, is that it has been so far developed as a background-dependent theory. This is despite there being compelling arguments that a fundamental theory must be background independent. Whether string

[3] This is because the algebra of supersymmetry transformations closes on the Hamiltonian, which generates a symmetry in time.

[4] A critique of the attempts to resolve the landscape problem through the anthropic principle is given in Smolin (2004a).

theory turns out to describe nature or not, there are now few alternatives but to approach the problems of unification and quantum gravity from a background-independent perspective.

This essay is written with the hope that perhaps some who have avoided thinking about background-independent theories might consider doing so now. To aid those who might be so inclined, in the next section I give a sketch of how the absolute/relational debate has shaped the history of physics since before the time of Newton. Then, in §7.3, I explain precisely what is meant by relational and absolute theories. §7.4 asks whether general relativity is a relational theory and explains why the answer is: partly. We then describe, in §7.5, several relational approaches to quantum gravity. There have been some remarkable successes, which show that it is possible to get highly non-trivial results from background-independent approaches to quantum gravity (Smolin 2004b). At the same time, there remain open problems and challenges. Both the successes and open problems yield lessons for any future attempt to make a background-independent formulation of string theory or any other quantum theory of gravity.

§§7.6 to 7.8 discuss what relationalism has to offer for the problems in particle physics such as unification and predictability. It is argued that the apparent lack of predictability emerging from studies of the string theory landscape is a symptom of relying on background-dependent methodologies in a regime where they cannot offer sensible answers. To support this, I show that relationalism suggests methodologies by which multiverse theories may nevertheless make falsifiable predictions.

Many theorists have asserted that no approach to quantum gravity should be taken seriously if it does not offer a solution to the cosmological constant problem. In §7.9 I show that relational theories do offer new possibilities for how that most recalcitrant of issues may be resolved.

§7.10 explores another application of relationalism, which is to the problem of how to extend quantum theory to cosmology. I review several approaches which have been called 'relational quantum theory'. These lead to formulations of the holographic principle suitable for quantum gravity and cosmology.

7.2 A BRIEF HISTORY OF RELATIONAL TIME

The debate about whether space and time are relational is central to the history of physics. Here is a cartoon sketch of the story[5].

Debate about the meaning of motion goes back to the Greeks. But the issues of interest for us came into focus when Newton proposed his form of dynamics in his book *Principia Mathematica*, published in 1687. Several of his rough contemporaries, such as Descartes, Huygens, and Leibniz, espoused *relational* notions of space and time, according to which space and time are to be defined only in terms of relationships among real objects or events. Newton broke with his contemporaries

[5] A full historical treatment of the relational/absolute debate is in Barbour's book (Barbour 1989, 2001).

to espouse an *absolute* notion of space and time, according to which the geometry of space and time provided a fixed, immutable, and eternal background, with respect to which particles moved. Leibniz responded by proposing arguments for a relational view that remain influential to this day [6].

Leibniz's argument for relationalism was based on two principles, which have been the focus of many books and papers by philosophers to the present day. The *principle of sufficient reason* states that it must be possible to give a rational justification for every choice made in the description of nature. I will refer the interested reader to the original texts (Leibniz 1698; Francks and Woolhouse 1999; Alexander 1956) for the arguments given for it, but it is not hard to see the relevance of this principle for contemporary theoretical physics. A theory that begins with the choice of a background geometry, among many equally consistent choices, violates this principle. So does a theory that allows some parameters to be freely specified, and allows no mechanism or rational argument why one value is observed in nature.

One circumstance that the principle of sufficient reason may be applied to is spacetimes with global symmetries. Most distributions of matter in such a space will not be invariant under the symmetries. One can then always ask, why is the universe where it is, rather than ten feet to the left, or rotated 30 degrees? Or, why did the universe not start five minutes later? This is sometimes called the problem of underdetermination: nothing in the laws of physics answers the question of why the whole universe is where it is, rather than translated or rotated.

As there can be no rational answer why the universe is where it is, and not ten feet to the left, the principle of sufficient reason says this question should not arise in the right theory. One response is to demand a better theory in which there is no background spacetime. If all there is to space is an emergent description of relations between particles, questions about whether the whole universe can be translated in space or time cannot arise. Hence, the principle of sufficient reason motivates us to eliminate fixed background spacetimes from the formulation of physical law.

Conversely, if one believes that the geometry of space is going to have an absolute character, fixed in advance, by some a priori principles, you are going to be led to posit a homogeneous geometry. For what, other than particular states of matter, would be responsible for inhomogeneities in the geometry of space? But then spacetime will have symmetries which leave you prey to the argument just given. So from the other side also, we see that the principle of sufficient reason is hard to square with any idea that spacetime has a fixed, absolute character.

One way to formulate the argument against background spacetime is through a second principle of Leibniz, *the identity of the indiscernible*. This states that any two entities which share the same properties are to be identified. Leibniz argues that were this not the case, the first principle would be violated, as there would be a distinction between two entities in nature without a rational basis. If there is no experiment that could tell the difference between the state in which the universe is here, and the

[6] Some essential texts, accessible to physicists, are Leibniz (1698); Francks and Woolhouse (1999); Alexander (1956).

state in which it is translated ten feet to the left, they cannot be distinguished. The principle says that they must then be identified. In modern terms, this is something like saying that a cosmological theory should not have global symmetries, for they generate motions and charges that could only be measured by an observer at infinity, who is hence not part of the universe. In fact, when we impose the condition that the universe is spatially compact without boundary, general relativity tells us there are no global spacetime symmetries and no non-zero global conserved charges[7].

But it took physics a long time to catch up to Leibniz's thinking. Even if philosophers were convinced that Leibniz had the better argument, Newton's view was easier to develop, and took off, whereby Leibniz's remained philosophy. This is easy to understand: a physics where space and time are absolute can be developed one particle at a time, while a relational view requires that the properties of any one particle are determined self-consistently by the whole universe.

Leibniz's criticisms of Newton's physics were sharpened by several thinkers, the most influential of which was Mach (1893), who in the late nineteenth century gave an influential critique of Newtonian physics on the basis of its treatment of acceleration as absolute.

Einstein was among those whose thinking was changed by Mach. There is a certain historical complication, because what Einstein called 'Mach's principle' was not exactly what Mach wrote. But that need not concern us here. The key idea that Einstein got from, or read into, Mach, was that acceleration should be defined relative to a frame of reference that is dynamically determined by the configuration of the whole universe, rather than being fixed absolutely, as in Newton's theory.

In Newton's mechanics, the distinction between who is accelerating and who is moving uniformly is a property of an absolute background spacetime geometry, that is fixed independently of the history or configuration of matter. Mach proposed, in essence, eliminating absolute space as a cause of the distinction between accelerated and non-accelerated motion, and replacing it with a dynamically determined distinction. This resolves the problem of underdetermination, by replacing an a priori background with a dynamical mechanism. By doing this Mach showed us that a physics that respects Leibniz's principle of sufficient reason is more predictive, because it replaces an arbitrary fact with a dynamically caused and observationally falsifiable relationship between the local inertial frames and the distribution of matter in the universe. This for the first time made it possible to see how, in a theory without a fixed background, properties of local physics, thought previously to be absolute, might be genuinely explained, self-consistently, in terms of the whole universe.

There is a debate about whether general relativity is 'Machian', which is partly due to confusion over exactly how the term is to be applied. But there is no doubt that general relativity can be characterized as a partly relational theory, in a precise sense that I will explain below.

[7] That is, special solutions may have symmetries. But, as we will discuss in §7.4, there are no symmetries acting on the space of physical solutions of the theory, once these have been identified with equivalence classes under diffeomorphisms (Kuchař 1976, 1982).

To one schooled in the history of the relational/absolute debate[8], it is easy to understand the different choices made by different theorists as reflecting different expectations and understandings of that debate (Earman 1989; Norton 1987; Smolin 2001a; Rovelli 1991; Barbour and Pfister 1996). The same can be said about the debates about the merits of the Anthropic Principle as a solution to the very puzzling situation that string theory has found itself in recently (Smolin 2004a). To explain why, we need some precise definitions.

7.3 WHAT PHYSICISTS MEAN WHEN WE TALK ABOUT RELATIONAL SPACE AND TIME

While many physicists have been content to work with background-dependent theories, from the earliest attempts at quantum gravity there has been a community of those who shared the view that any approach must be background independent. Among them, there has been a fair amount of discussion and reflection concerning the roots of the notion of background independence in older relational views of space and time. From this has emerged a rough consensus as to what may be called the physicists' relational conception of space and time[9].

Any theory postulates that the world is made up of a very large collection of elementary entities (whether particles, fields, events, or processes). Indeed, the fact that the world has many things in it is essential for these considerations—it means that the theory of the world may be expected to differ in important aspects from models that describe the motion of a single particle, or a few particles in interaction with each other.

The basic form of a physical theory is framed by how these many entities acquire properties. In an absolute framework the properties of any entity are defined with respect to a single entity—which is presumed to be unchanging. An example is the absolute space and time of Newton, according to which positions and motions are defined with respect to this unchanging entity. Thus, in Newtonian physics the background is three-dimensional space, and the fundamental properties are a list of the positions of particles in absolute space as a function of absolute time: $x_i^a(t)$. Another example of an absolute background is a regular lattice, which is often used in the formulation of quantum field theories. Particles and fields have the property of being at different nodes in the lattice, but the lattice does not change in time.

The entities that play this role may be called the background for the description of physics. The background consists of presumed entities that do not change in

[8] The understanding that working physicists like myself have of the relevance of the relational/absolute debate to the physical interpretation of general relativity and contemporary efforts towards quantum gravity is due mainly to the writings and conference talks of a few physicists—primarily John Stachel (1989) and Julian Barbour (1984, 2000). Also important were the efforts of philosophers who, beginning in the early 1990s were kind enough to come to conferences on quantum gravity and engage us in discussion.

[9] Philosophers distinguish several versions of relationalism (Saunders 2003), among which, what is described here is what some philosophers call *eliminative relationalism*.

time, but which are necessary for the definition of the kinematical quantities and dynamical laws.

The most basic statement of the relational view is that

R1 *There is no background.*

How then do we understand the properties of elementary particles and fields? The relational view presumes that

R2 *The fundamental properties of the elementary entities consist entirely in relationships between those elementary entities.*

Dynamics is then concerned with how these relationships change in time.

An example of a purely relational kinematics is a graph. The entities are the nodes. The properties are the connections between the nodes. The state of the system is just which nodes are connected and which are not. The dynamics is given by a rule which changes the connectivity of the graph.

We may summarize this as

R3 *The relationships are not fixed, but evolve according to law. Time is nothing but changes in the relationships, and consists of nothing but their ordering.*

Thus, we often take background independent and relational as synonymous. The debate between philosophers that used to be phrased in terms of absolute versus relational theories of space and time is continued in a debate between physicists who argue about background-dependent versus background-independent theories.

It should also be said that for physicists relationalism is a strategy. As we shall see, theories may be partly relational, i.e. they can have varying amounts of background structure. One can then advise that progress is achieved by adopting the

Relational strategy: *Seek to make progress by identifying the background structure in our theories and removing it, replacing it with relations which evolve subject to dynamical law.*

Mach's principle is the paradigm for this strategic view of relationalism. As discussed above, Mach's suggestion was that replacing absolute space as the basis for distinguishing acceleration from uniform motion with the actual distribution of matter would result in a theory that is more explanatory, and more falsifiable. Einstein took up Mach's challenge, and the resulting success of general relativity can be taken to vindicate both Mach's principle and the general strategy of making theories more relational.

7.4 GENERAL RELATIVITY AS A PARTLY RELATIONAL THEORY

Even if quantum gravity is not a quantization of general relativity, the right quantum theory of gravity must have general relativity as an appropriate limit. By this we mean that we must recover the basic principles that underlie how general relativity describes nature, as well as all of its solutions. As a result, it is not possible to have a useful discussion about how space and time are to be understood in quantum gravity unless we have a clear understanding of the physical interpretation of classical general relativity. This is the task of this section.

As I will describe, general relativity can be characterized as a partly relational theory. As such, it serves as a good example of the power of the relational strategy.

There is one clarification that should be stated at the outset: the issue of whether general relativity is Machian or relational is only interesting if we take general relativity as a possible cosmological theory. This means that we take the spatial topology to be compact, without boundary. In some models of subsystems of the universe, one does not do this. In these cases space has a boundary and one has to impose conditions on the metric and fields at the boundary. These boundary conditions become part of the background, as they indicate that there is a region of spacetime outside the dynamical system which is being modelled.

There is of course nothing wrong with modelling subsystems of the universe with boundaries on which we impose boundary conditions. One way to do this is to assume that the system under study is isolated, so that as one moves away from it the spacetime satisfies asymptotic conditions. But the boundary or asymptotic conditions can only be justified by the assertion that the system modelled is a subsystem of the universe. No fundamental theory could be formulated in terms that require the specification of boundary or asymptotic conditions because those conditions imply that there is a part of the universe outside the region being modelled. Thus, one cannot assert that a theory defined only with the presence of such conditions can be fundamental.

But at the same time, the fact that asymptotic conditions can be imposed does not mean general relativity is not fundamental, since it can also be formulated for cosmologies by making the universe compact without boundary. It does mean that it is only interesting to ask if general relativity is a relational theory in the cosmological case[10].

General relativity is a complicated theory and there has been a lot of confusion about it. However, I will show now why it is considered to be mainly, but not purely, a relational theory. One reason it is complicated is that there are several layers of structure.

- Dimension
- Topology
- Differential structure
- Signature
- Metric and fields

We denote a spacetime by (M, g_{ab}, f), where M refers to the first four properties, g_{ab} is the metric, and f stands for all the other fields.

10 But it is worth asking whether the fact that GR allows models with boundary conditions means that it is incomplete, as a fundamental theory.

It is true that in general relativity the dimension, topology, differential structure, and signature are fixed. They can be varied from model to model, but they are arbitrary and not subject to law. These do constitute a background[11].

Then why do we say the theory is relational? Given this background, we can define an equivalence relation called a diffeomorphism. A diffeomorphism ϕ is a smooth, invertible map from a manifold to itself[12]

$$\phi(M, g_{ab}, f) \rightarrow (M', g'_{ab}, f') \tag{7.1}$$

which takes a point p to another point $\phi \cdot p$, and drags the fields along with it by

$$(\phi \cdot f)(p) = f(\phi^{-1} \cdot p) \tag{7.2}$$

The diffeomorphisms of a manifold constitute a group, *Diff*(M), called the group of diffeomorphisms of the manifold. The basic postulate which makes GR a relational theory is

R4 *A physical spacetimes is defined to correspond, not to a single* (M, g_{ab}, f), *but to an equivalence class of manifolds, metrics, and fields under all actions of Diff*(M). This equivalence class may be denoted $\{M, g_{ab}, f\}$.

The important question for physics is, what information is coded inside an equivalence class $\{M, g_{ab}, f\}$, apart from the information that is put into the specification of M?

The key point is that the points and open sets that define the manifold are not preserved under *Diff*(M), because any diffeomorphism except the identity takes points to other points. Thus, the information coded in the equivalence classes cannot be described simply as the values of fields at points.

The answer is that

1. Dimension and topology are coded in $\{M, g_{ab}, f\}$
2. Apart from those, all that there is, is a system of relationships between events. Events are not points of a manifold, they are identifiable only by coincidences between the values of fields preserved by the actions of diffeomorphisms.

The relations between events are of two kinds:

(a) *causal order* (i.e. which events causally precede which, given by the lightcone structure).

(b) *measure* (the spacetime volumes of sets defined by the causal order).

It can be shown that the information in a spacetime $\{M, g_{ab}, f\}$ is completely characterized by the causal structure and the measure (Malament to appear). Intuitively, this is because the conformal metric[13] determines, and is determined by,

[11] A very interesting question is whether the restriction to fixed dimension and topology is essential or may be eliminated by a deeper theory.

[12] More generally to another manifold.

[13] Defined as the equivalence class of metrics related by local conformal transformations $g_{ab} \rightarrow g'_{ab} = \phi^2 g_{ab}$, where ϕ is a function.

the light cones and hence the causal structure. The remaining conformal factor then determines the volume element.

The problem of underdetermination raised in §7.2 is solved by the identification, in **R4**, of physical histories with equivalence classes. For the spatially compact case, once we have modded out by the diffeomorphisms, there remain no symmetries on the space of solutions (Kuchař 1976, 1982). But why should we mod out by diffeomorphisms? As Einstein intuited in his famous 'hole argument', and Dirac codified, one must mod out by diffeomorphisms if one is to have deterministic evolution from initial data (Stachel 1989; Earman 1989; Norton 1987; Smolin 2001a; Rovelli 1991, 2004).

This establishes that, apart from the specification of topology, differential structure, and dimension, general relativity is a relational physical theory.

7.4.1 The Problem of Time and Related Issues

As I emphasized at the beginning of this section, a truly fundamental theory cannot be formulated in terms of boundaries or asymptotic conditions. This, together with diffeomorphism invariance, implies that the Hamiltonian is a linear combination of constraints[14]. This is no problem for defining and solving the evolution equations, but it does lead to subtleties in the question of what is an observable. One important consequence is that one cannot define the physical observables of the theory without solving the dynamics. In other words, as Stachel emphasizes, *there is no kinematics without dynamics*. This is because all observables are relational, in that they describe relations between physical degrees of freedom. You cannot just ask what is happening at a manifold point, or an event, labelled by some coordinate, and assume you are asking a physically meaningful question. The problem is that because of diffeomorphism invariance, points are not physically meaningful without a specification of how a point or event is to be identified by the values of some physical degrees of freedom. As a result, even observables that refer to local points or regions of physical spacetime are non-local in the sense that as functions of initial data they depend on data in the whole initial slice.

As a result, the physical interpretation of classical general relativity is more subtle than is usually appreciated. In fact, most of what we think we understand naively about how to interpret classical GR applies only to special solutions with symmetries, where we use the symmetries to define special coordinates. These methods do not apply to generic solutions, which have no symmetries. It is possible to give a physical interpretation to the generic solutions of the theory, but only by taking into account the issues raised by the facts that all physical observables must be diffeomorphism invariant, and the related fact that the Hamiltonian is a sum of constraints (Rovelli 2004).

We see here a reflection of Leibniz's principles, in that the interpretation that must be given to generic solutions, without symmetries, is completely different from that given to the measure zero of solutions with symmetries.

[14] This is reviewed in Rovelli (2004); Thiemann (2001); Smolin (2001a).

One can actually argue something stronger (Smolin 2001a): Suppose that one could transform general relativity into a form in which one expressed the dynamics directly in terms of physical observables. That is, observables which commute with all the constraints, but still measure local degrees of freedom. Then the solutions with symmetries might just disappear. This is because, being diffeomorphism invariant, such observables can distinguish points only by their having different values of fields. Such observables must degenerate when one attempts to apply them to solutions with symmetries. Thus, expressed in terms of generic physical observables, there may be no symmetric solutions. If this is true this would be a direct realization of the identity of the indiscernible in classical general relativity.

Thus, even at the classical level, there is a distinction between background-independent and background-dependent approaches to the physical interpretation. If one is interested only in observables for particles moving within a given spacetime, one can use a construction that regards that spacetime as fixed. But if one wants to discuss observables of the gravitational field itself, one cannot use background-dependent methods, for those depend on fixing the gravitational degrees of freedom to one solution. To discuss how observables vary as we vary the solution to the Einstein equations we need functions of the phase space variables that make sense for all solutions. Then one must work on the full space of solutions, in either configuration space or phase space.

One can see this with the issue of time. If by time you mean time experienced by observers following world-lines in a given spacetime, then we can work within that spacetime. For example, in a given spacetime time can be defined in terms of the causal structure. But if one wants to discuss time in the context in which the gravitational degrees of freedom are evolving, then one cannot work within a given spacetime. One constructs instead a notion of time on the infinite dimensional phase or configuration space of the gravitational field itself. Thus, at the classical level, there are clear solutions to the problems of what is time and what is an observable in general relativity.

Any quantum theory of gravity must address the same issues. Unfortunately, background-dependent approaches to quantization evade these issues, because they take for granted that one can use the special symmetries of the non-dynamical backgrounds to define physical observables. To usefully address issues such as the problem of time, or the construction of physical observables, in a context that includes the quantum dynamics of the spacetime itself, one must work in a background-independent formulation.

However, while the problem of time has been addressed in the context of background-independent approaches to quantum gravity, the problem has not been definitively solved. The issue is controversial and there is strong disagreement among experts. Some believe the problem is solved, at least in principle, by the application of the same insights that lead to its solution in classical general relativity (Rovelli 2004). Others believe that new ideas are needed (Smolin 2001a). While I will not dwell on it here, the reader should be aware that the problem of time is a key challenge that any complete background-independent quantum theory of gravity must solve.

7.5 RELATIONALISM AND THE SEARCH FOR THE QUANTUM THEORY OF GRAVITY

We next consider how the debate between relational and absolute approaches to spacetime reappears in the search for a quantum theory of gravity.

Let us begin by noting that conventional quantum theories are background-dependent theories. The background structures for a quantum theory include space and time, either Newtonian or, in the case of quantum field theory (QFT), some fixed, background spacetime. There are additional background structures connected with quantum mechanics, such as the inner product. It is also significant that the background structures in quantum mechanics are connected to the background space and time. For example, the inner product codes probability conservation, in a given background time coordinate.

Thus, when we attempt to unify quantum theory with general relativity we have to face the question of whether the resulting theory is to be background dependent or not. There are two kinds of approaches, which take the two possible answers—yes and no. These are called background-independent and background-dependent approaches.

Background-dependent approaches study quantum theory on a background of a fixed classical spacetime. These can be quantum theories of gravity in a limited sense in which they study the quantization of gravitational waves defined as moving (to some order of approximation) on a fixed background spacetime. One splits the metric into two pieces

$$g_{ab} = b_{ab} + h_{ab} \tag{7.3}$$

where b_{ab} is the background metric, a fixed solution to the Einstein equations, and h_{ab} is a perturbation of that solution. In a background-dependent approach one quantizes h_{ab} using structures that depend on the prior specification of b_{ab}, as if h_{ab} were an ordinary quantum field, or some substitute such as a string.

Background-dependent approaches include

- perturbative quantum general relativity;
- string theory.

Perturbative quantum general relativity does not lead to a good theory, nor are the problems cured by modifying the theory so as to add supersymmetry or other terms to the field equations.

It is hard to imagine a set of better-motivated conjectures than those that drove interest in string theory. Had string theory succeeded as a background-dependent theory, it would have served as a counter-argument to the thesis of this essay[15]. Conversely, given that the problems string theory faces seem deeply rooted in the structure of the theory, it may be worthwhile to examine the alternative, which is background-independent theories.

[15] A more detailed summary of the results achieved in string theory and other approaches to quantum gravity, together with a list of problems that remain unsolved, is given in Smolin (2003).

In recent years there has been healthy development of a number of different background-independent approaches to quantum gravity. These include,

- Causal sets;
- Loop quantum gravity (or spin foam models);
- Dynamical triangulations models;
- Certain approaches to non-commutative geometry (Connes and Moscovici 2005);
- A number of approaches that posit a fundamental discrete quantum theory from which classical spacetime is conjectured to emerge at low energies (Dreyer 2004);
- Attempts to formulate string theory as a background-independent theory.

I will briefly describe the first three. These are well enough understood to illustrate both the strengths of the relational view for quantum gravity and the hard issues that any such approach must overcome.

7.5.1 The Causal Set Theory

To describe the causal set model we need the definition of a causal set.

A causal set is a partially ordered set such that the intersection of the past and future of any pair of events is a finite set. The elements of the causal set are taken to be physical events and their partial ordering is taken to code the relation of physical causation.

The basic premisses of the causal set model are Bombelli et al. (1987); Martin et al. (2001); Rideout and Sorkin (2000, 2001); Rideout (2002):

1. A history of the universe consists of nothing but a causal set. That is, the fundamental events have no properties except their mutual causal relations[16].
2. The quantum dynamics is defined by assigning to each history a complex number which is to be its quantum amplitude[17].

The motivation for the causal set hypothesis comes from the expectation that the geometry of spacetime becomes discrete at the Planck scale. This leads one to expect that, given any classical spacetime $\{M, g_{ab}\}$, one will be able to define a causal set C which approximates it. The precise sense in which this is possible is:

We say that a causal set C approximates a classical spacetime, $\{M, g_{ab}, f\}$, if, to each event e in C, there is an event e in $\{M, g_{ab}, f\}$, such that (1) the causal relations are preserved and (2) there is on average 1 event e coming from C per Planck volume of $\{M, g_{ab}, f\}$.

We note that when a causal set does approximate a classical spacetime, it does so because it is the result of a fair sampling of the relations that define the spacetime, which are the causal order and measure.

[16] The events of a causal set are sometimes called 'elements' to emphasize the principle that each corresponds to a finite element of spacetime volume.

[17] In the causal set literature the dynamics is sometimes formulated in terms of quantum measure theory, which is a variant of the consistent histories formulation of quantum mechanics.

However if the discrete quantum theory is to be more fundamental there should be a procedure to define the classical spacetime $\{M, g_{ab}, f\}$ from some kind of classical or low energy limit of the causal set theory. This has not yet been achieved. A main reason is the following problem, which we call the inverse problem for causal sets (Markopoulou and Smolin forthcoming).

The inverse problem for causal sets: Given a classical spacetime $\{M, g_{ab}, f\}$, it is easy to define a causal set C which approximates it in the sense just defined. But almost no causal set C approximates a low-dimensional manifold in this sense. Moreover, we do not have a characterization, expressed only in terms of the relations in a causal set, C, which would allow us to pick out those causal sets that do approximate spacetimes. We can only do this by first constructing classical spacetimes, and then extracting from them a causal set that approximates them. Moreover no dynamical principle has been discovered which would generate causal sets C that either directly approximate low-dimensional classical spacetimes, or have coarse grainings or approximations that do so.

This is an example of a more general class of problems, which stems from the fact that combinatorially defined discrete structures are very different from continuous manifolds.

A very general combinatorial structure is a graph. The possibility of a correspondence between a graph and a smooth geometry is based on two definitions.

Definition: The metric on a graph Γ is defined by $g(j, k)$ for two nodes k and j is the minimal number of steps to walk from j to k along the graph.

Definition: A graph Γ is said to approximate a manifold and metric $\{M, g_{ab}\}$ if there is an embedding of the nodes of Γ into points of $\{M, g_{ab}\}$ such that the graph distance $g(j, k)$ is equal to the metric distance between the images of the nodes in $\{M, g_{ab}\}$.

It is easy to see that the following issue confronts us.

Inverse problem for graphs. Given any $\{M, g_{ab}\}$ it is easy to construct a graph Γ that approximates it. But, assuming only that the dimension is much less than the number of nodes, for almost no graphs do there exist low-dimensional smooth geometries that they approximate.

Because of the inverse problem, it is fair to say that the causal set programme has unfortunately so far failed to lead to a good physical theory of quantum gravity, But it is useful to review the logic employed:

Logic of the causal set programme:

- GR is relational, and the fundamental relations are causal relations.
- But GR is continuous and it is also non-quantum mechanical.
- We expect that a quantum theory of spacetime should tell us if the set of physical events is discrete.

- Therefore a quantum spacetime history should consist of a set of events which is a discrete causal structure.
- Moreover, the causal structure is sufficient to define the physical classical spacetime, so it should be sufficient to describe a fundamental quantum history.
- But this programme so far fails because of the inverse problem.

Given the seriousness of the inverse problem, it is possible to imagine that the solution is that there are more fundamental relations, besides those of causality. It should be said that this direction is resisted by some proponents of causal sets, who are rather 'purist' in their belief that the relation of causality is sufficient to constitute all of physics. But a possible answer to this question is given by another programme, *loop quantum gravity*, where causal relations are local changes in relational structures that describe the quantum geometry of space (Markopoulou 1997, 2002).

7.5.2 Loop Quantum Gravity

Loop quantum gravity was initiated in 1986 and is by now a well-developed research programme, with on the order of 100 practitioners. There is now a long list of results, many of them rigorous. Here I will briefly summarize the key results that bear on the issue of relational space and time[18].

7.5.2.1 Basic Results of Loop Quantum Gravity

Loop quantum gravity is based on the following observation, introduced by Sen and Ashtekar for general relativity and extended to a large class of theories including general relativity and supergravity in spacetime dimensions three and higher[19].

- General relativity and supergravity, in any spacetime dimension greater than or equal to $2+1$, can be rewritten as gauge theories, such that the configuration space is the space of a connection field, A_a, on a spatial manifold Σ. The metric information is contained in the conjugate momenta. The gauge symmetry includes the diffeomorphisms of a spacetime manifold, usually taken to be $\Sigma \times R$. The dynamics takes a simple form that can be understood as a constrained topological field theory. This means that the action contains one term, which is a certain topological field theory called *BF* theory, plus another term which generates a quadratic constraint.

Consider such a classical gravitational theory, T, whose histories are described as diffeomorphism equivalence class of connections and fields, $\{M, A_a, f\}$. To define the action principle one must assume that the topology, dimension, and differential structure of spacelike surfaces, Σ , are fixed.

The following results have then been proven (Smolin 2004b):

[18] For details of the results, including those mentioned below, and references, see Smolin (2004b). Books on loop quantum gravity include Ashtekar (1988, 1991); Gambini and Pullin (1996); Rovelli (2004); Thiemann (2001) and review papers include Ashtekar and Lewandowski (2004); Rovelli (1998); Smolin (1992, 1997b); Baez (1998, 2000); Perez (2003).

[19] See for example, Smolin (2002) and (2004b) and references contained therein.

1. The quantization of T results in a unique Hilbert space, H, of diffeomorphism-invariant states. There is a recent uniqueness theorem (Lewandowski et al. 2005; Fleischhack 2004; Sahlmann 2002a, 2002b; Sahlmann and Thiemann 2003a, 2003b; Okolow and Lewandowski 2003), which guarantees that for dimension of Σ two or greater, there is a unique quantization of a gauge field such that (i) the Wilson loops are represented by operators that create normalizable states, (ii) its algebra with the operator that measures electric field flux is represented faithfully, and (iii) the diffeomorphisms of Σ are unitarily implemented without anomaly.

   This unique Hilbert space has a beautiful description. There is an orthonormal basis of H whose elements are in one-to-one correspondence with the embeddings of certain labelled graphs Γ in Σ. (The label set varies depending on the dimension, matter fields, and with supersymmetry.)

   Because H carries a unitary representation of $Diff(\Sigma)$ it is possible rigorously to mod out by the action of diffeomorphisms and construct a Hilbert space, H^{diff}, of spatially diffeomorphism-invariant states. This has a normalizable basis in one-to-one correspondence with the diffeomorphism classes of the embeddings in Σ of the labelled graphs.

   This is a very satisfactory description from the point of view of relationalism. There is no more relational structure than a graph, as two nodes are distinguished only by their pattern of connections to the rest of the graph. The labels come from the theory of representation of a group or algebra $\mathcal{A}$. The edges are labelled by representations of $\mathcal{A}$, which describe properties shared between the nodes they connect. The labels on nodes are invariants of $\mathcal{A}$, which likewise describe properties shared by the representations on edges incident on those nodes.

   Because there is a background topology, there is additional information coded in how the edges of the graph knot and link each other. Given the choice of background topology, this information is also purely relational.

2. A quantum history is defined by a series of local moves on graphs that take the initial state to the final state (Markopoulou 1997, 2002). The set of local moves in each history define a causal set.

   Hence, the events of the causal set arise from local changes in another set of relations, that which codes the quantum geometry of a spatial slice. The structure that merges the relational structure of graphs with that of causal sets is now called a *causal spin foam.*

3. The amplitudes for local moves that follow from the quantization of the Einstein equations are known in closed form. The sums over those amplitudes are known to be ultraviolet finite. Similarly, the quantum Einstein equations in the Hamiltonian form have been implemented by exact operator equations on the states.

   In the case of a spin foam model for $2 + 1$ gravity coupled to massive particles, it has been shown in detail that the theory can be re-summed, yielding an effective field theory on a non-commutative spacetime (Freidel and Livine 2005). This provides an explicit demonstration of how physics in classical spacetime can emerge from a non-trivial background-independent quantum theory of gravity. The resulting effective field theory has in addition deformed Poincaré symmetry, which confirms, in this case, the general conjecture that the low-energy limit of loop quantum gravity has deformed Poincaré symmetry (Smolin 2004b).

4. The quantum spacetime is discrete in that each node of the graph corresponds to a finite quanta of spatial volume. The operators that correspond to volumes, areas, and lengths are finite, and have discrete spectra with finite non-zero minimal values. Hence a graph with a finite number of nodes and edges defines a region of space with finite volume and area.
5. There are a number of robust predictions concerning subjects like black hole entropy. Evidence has recently been found that both cosmological and black hole singularities bounce, so the evolution of the universe continues through apparent classical cosmological singularities.
6. There are explicit constructions of semiclassical states, coarse-grained measurements of which reproduce classical geometries. Excitations of these states, with wavelengths long in Planck units, relative to those classical geometries, have been shown to reproduce the physics of quantum fields and linearized gravitational waves on those backgrounds.

7.5.2.2 Open Problems of Loop Quantum Gravity

There are of course many, in spite of the fact that the theory is well defined.

- *Classical limit problem*: Find the ground state of the theory and show that it is a semiclassical state, excitations of which are quantum field theory and classical GR.
- *Do science problem*: By studying the excitations of semiclassical states, make predictions for doable experiments that can test the theory up or down.
- *Remove the remaining background-dependence problem:* The results so far defined depend on the fact that the dimension and topology of the spatial manifold, Σ, is fixed, so that the graphs are embedded in Σ. This helps by lessening the inverse problem. Can this be removed—and the inverse problem solved—so that all the structure that was background for previous theories, including dimension and topology, is explained as following from solutions to a relational theory[20]?

  We note that in some formulations of spin foam models, the dependence on a fixed background topology is dropped, so that the states and histories are defined as pure combinatorial structures. But this makes the problem of recovering classical general relativity from the low energy limit more complicated.
- *The problem of time*: The different proposals that have been made to resolve the problem of time in quantum gravity and cosmology can all be studied in detail in loop quantum gravity and related cosmological models. While there are some interesting results, the opinion of this author is that the problem remains open.

[20] For more on the inverse problem and its implications, see Markopoulou and Smolin (in preparation).

These are hard problems, and remain unsolved, but some progress is being made on all of them.

It is important to mention that there are real possibilities for experimental tests of the theory. This is because the discrete structure of space and time implies modifications in the usual relations between energy and momenta

$$E^2 = p^2 + m^2 + l_p E^3 + \ldots \tag{7.4}$$

This turns out to have implications for experiments currently under way, having to do with ultra-high-energy cosmic rays and gamma ray bursts, amongst others[21]. Loop quantum gravity appears to make predictions for these experiments (Smolin 2005).

7.5.2.3 Lessons from Loop Quantum Gravity for the Relational Programme

So far as the relational/absolute debate is concerned, loop quantum gravity (LQG) teaches us several lessons:

- So long as we keep as background those aspects of space and time that are background for classical GR (the topology, dimension, and differential structure), we can find a quantum mechanical description of the metric and fields. Thus LQG is partly relational, in exactly the same way that GR is partly relational.
- Loop quantum gravity does give us a detailed description of quantum spatial and spacetime geometry. There are many encouraging results, such as finiteness, and the derivation of an explicit language of states, histories, and observables for general background-independent theories of quantum gravity. It is possible to do non-trivial computations to study the dynamics of quantum spacetime, and applications to physical problems such as black holes and cosmology yield results that are sensible and, in some cases, testable. It is very satisfying that the description of quantum geometry and quantum histories is formulated using beautiful relational structures such as graphs and causal sets.
- This description is flexible and can accommodate different hypotheses as to the dimension of spacetime, matter couplings, symmetries, and supersymmetries.
- There do remain hard open problems having to do with how a classical spacetime is to emerge from a purely background-independent description. A related challenge is to convincingly resolve the problem of time. Nevertheless, significant progress is being made on these problems (Freidel and Livine 2005), and it even appears to be possible to derive predictions for experiment by expanding around certain semiclassical states (Smolin 2005).
- The main barrier to making an entirely relational theory of quantum spacetime appears to be the inverse problem.

[21] See Smolin (2004b) for a brief review of this important subject, with additional references.

7.5.3 Causal Dynamical Triangulation Models

These are models for quantum gravity, based on a very simple construction (Ambjorn et al. 1992, 2004; Ambjorn 1995; Agishtein and Migdal 1992). A quantum spacetime is represented by a combinatorial structure, which consists of a large number N of d dimensional simplexes (triangles for two dimensions, tetrahedra for three, etc.) glued together to form a discrete approximation to a spacetime. Each such discrete spacetime is given an amplitude, which is acquired from a discrete approximation to the action for general relativity. Additional conditions are imposed, which guarantee that the resulting structure is the triangulation of some smooth manifold (otherwise there is a severe inverse problem.) For simplicity the edge lengths are taken to be all equal to a fundamental scale, which is considered a short distance cut-off.[22] One defines the quantum theory of gravity by a discrete form of the sum over histories path integral, in which one sums over all such discrete quantum spacetimes, each weighted by its amplitude.

These models were originally studied as an approach to Euclidean quantum gravity (that is the path integral sums over spacetimes with Euclidean signature, rather than the Lorentzian signature of physical spacetime). In these models the topology is not fixed, so one has a model of quantum gravity in which one can investigate the consequences of removing topology from the background structure and making it dynamical (Ambjorn et al. 1992, 2004; Ambjorn 1995; Agishtein and Migdal 1992).

More recently, a class of models have been studied corresponding to Lorentzian quantum gravity. In these cases additional conditions are fixed, corresponding to the existence of a global time slicing, which restricts the topology to be of the form of $\Sigma \times R$, where Σ is a fixed spatial topology (Ambjorn et al. 2000a, 2000b, 2001a, 2001b, 2002; Ambjorn and Loll 1998; Loll 2001; Dittrich and Loll 2002)[23].

Some of the results relevant for the debate on relationalism include,

- In the Euclidean case, for spacetime dimensions $d > 2$, the sum over topologies cannot be controlled. The path integral is, depending on the parameters of the action chosen, unstable to the formation of either an uncontrolled spawning of 'baby universes', or to a crunch down to degenerate triangulations. Neither converges to allow a coarse-grained approximation in terms of smooth manifolds of any dimension.
- In the Lorentzian case, when the simplices have spacetime dimension $d = 2, 3, 4$, where the topology is fixed and the formation of baby universes suppressed, there is evidence for convergence to a description of physics in manifolds whose macroscopic dimension is the same as the microscopic dimension. For the case of $d = 4$ the results are recent and highly significant (Ambjorn et al. 2000a, 2000b, 2001a, 2001b, 2002, 2004, 2005a, 2005b, 2005c; Ambjorn and Loll

[22] There is a different, but related approach, called Regge calculus, in which the triangulations are fixed while the edge lengths are varied.

[23] The condition of a fixed global time slicing can be relaxed to some extent (Markopoulou and Smolin 2004).

1998; Loll 2001; Dittrich and Loll 2002). In particular, there is now detailed numerical evidence for the emergence of $3 + 1$ dimensional classical spacetime at large distances from a background-independent quantum theory of gravity (ibid.).
- The measure of the path integral is chosen so that each triangulation corresponds to a diffeomorphism class $\{M, g_{ab}, f\}$. The physical observables such as correlation functions measured by averaging over the triangulations correspond to diffeomorphism-invariant relational observables in spacetime.

These results are highly significant for quantum gravity. It follows that earlier conjectures about the possibility of defining quantum gravity through the Euclidean path integral cannot be realized. The sum has to be done over Lorentzian spacetimes to have a hope of converging to physics that has a coarse-grained description in smooth spacetimes. Further, earlier conjectures about summing over topologies in the path integral also cannot be realized.

As far as relationalism is concerned we reach a similar conclusion to that of loop quantum gravity. There is evidence for the existence of the quantum theory when structures including topology, dimension, and signature are fixed, as part of the background structure, just as they are in classical general relativity. When this is done one has a completely relational description of the dynamics of a discrete version of metric and fields. Furthermore, in the context of each research programme there has recently been reported a detailed study showing how classical spacetime emerges from an initially discrete, background independent theory. This is an analytic result in the case of spin foam models in $2 + 1$ dimensions, with matter (Freidel and Livine 2005), and numerical results in $3 + 1$ dimensions in the causal dynamical triangulations case (Ambjorn et al. 2000a, 2000b, 2001a, 2001b, 2002, 2004, 2005a, 2005b, 2005c; Ambjorn and Loll 1998; Loll 2001; Dittrich and Loll 2002). This is very encouraging, given that the problem of how classical spacetime emerges is the most challenging problem facing background-independent approaches to quantum gravity.

7.5.4 Background-Independent Approaches to String and $\mathcal{M}$ Theory

It has been often argued that string theory requires a background-independent formulation. This is required, not just because any quantum theory of gravity must be background independent, but because there is a need to unify all the different perturbative string theories into one theory. As this must combine theories defined on different backgrounds, it must not be restricted by the choice of a particular background.

There are some claims that string theory does not need a background-independent formulation, and can be instead defined for fixed boundary or asymptotic conditions as dual to a field theory on a fixed background, as in the AdS/CFT correspondence. To respond to this, it first should be emphasized that the considerable evidence in favour of some form of an AdS/CFT correspondence falls short of a proof of actual equivalence, which would be needed to say that a full quantum theory of gravity, rather than just limits of correlation functions taken to the boundary, is coded in the dual conformal field theory (Arnsdorf and Smolin 2001; Smolin 2003). But even granting the full Maldacena conjecture it is hard to see how a theory defined only in the presence of boundary or asymptotic conditions, as interesting as that would be, could be taken as a candidate for a complete formulation of a fundamental theory of spacetime. This is because the boundary or asymptotic conditions can only be interpreted physically as standing for the presence of physical degrees of freedom outside the theory. For example, the timelike or null killing fields at the boundary stand for the reading of a clock which is not part of the physical systems. Such a formulation cannot be applied to cosmological problems, where the problem is precisely to formulate a consistent theory of the entire universe as a closed system. General relativity with spatially compact boundary conditions is such a theory. Hence, it seems reasonable to require that a quantum theory of gravity, which is supposed to reproduce general relativity, must also make sense as a theory of a whole universe, as a closed system.

Some string theorists have also claimed that string theory does not need a background-independent formulation, because the fact that string perturbation theory is, in principle, defined on many different backgrounds is sufficient. This assertion rests on exaggeration and misunderstanding. First, string perturbation theory is so far only defined on stationary backgrounds that have timelike killing fields. But this is a measure zero of solutions to the Einstein equations. It is, however, difficult to believe that a consistent string perturbation theory can be defined on generic solutions to the Einstein equations because, in the absence of timelike killing fields, one cannot have spacetime supersymmetry, without which the spectrum will typically contain a tachyon[24].

More generally, this assertion misses completely the key point that general relativity is itself a background-independent theory. Although we sometimes use Einstein's equations as if they were a machine for generating solutions, within which we then study the motion of particles of fields, this way of seeing the theory is inadequate as soon as we want to ask questions about the gravitational degrees of freedom themselves. Once we ask about the actual local dynamics of the gravitational field, we have to adopt the viewpoint which understands general relativity to be a background-independent theory within which the geometry is completely dynamical, on an equal footing with the other degrees of freedom. The correct arena for this physics is not a particular spacetime, or even the linearized perturbations of a particular spacetime. It is the infinite dimensional phase space of gravitational degrees of freedom. From this viewpoint, individual spacetimes are just trajectories in the infinite dimensional phase or configuration space; they can play no more of a role in a quantization of spacetime than a particular classical orbit can play in the quantization of an electron.

To ask for a background-independent formulation of string theory is to ask only that it conserve the fact that the dynamics of the Einstein equations does not require, indeed does not allow, the specification of a fixed background metric. For, if one

[24] There are a few exceptions, but typically non-supersymmmetric string theories have tachyons. Also, note that for none of the theories in the landscape is it known how to construct the free string worldsheet theory.

means anything at all by a quantum theory of gravity, one certainly means a theory by which the degrees of freedom of the classical theory emerge from a suitable limit of a Hilbert space description. This does not commit one to the belief that the elementary degrees of freedom are classical metrics or connections, nor does it commit one to a belief that the correct microscopic dynamics have to do with the Einstein equations. But it does imply that a quantum theory must have a limit in which it reproduces the correct formulation of general relativity as a dynamical system, which is to say in the background-independent language of the classical phase space. It would seem very unlikely that such a background-independent formulation can emerge as a classical limit of a theory defined only on individual backgrounds, which are just trajectories in the exact phase space.

In fact, there have been a few attempts to develop a background-independent approach to string and $\mathcal{M}$ theory (Smolin 2000a, 2000c, 2000d, 2001b, forthcoming; Azuma and Bagnoud 2002; Bagnoud et al. 2002; Azuma et al. 2001; Azuma 2001). These have been based on two lessons from loop quantum gravity: (i) Background-independent quantum theories of gravity can be based on matrix models, so long as their formulation depends on no background metric. Such a model can be based on matrices valued in a group, as in certain formulations of spin foam models. (ii) The dynamics of all known gravitational theories can be understood by beginning with a topological field theory and then extending the theory so as to minimally introduce local degrees of freedom. This can be extended to supergravity, including the eleven-dimensional theory (Ling and Smolin 2001; Smolin 1997c).

By combining these, a strategy was explored in which a background-independent formulation of string or $\mathcal{M}$ theory was to be made which is an extension of a matrix Chern–Simons theory (Smolin, 2000a, 2000c, 2000d, 2001b, forthcoming; Azuma and Bagnoud 2002; Bagnoud et al. 2002; Azuma et al. 2001; Azuma 2001). The Chern–Simons theory provides a starting point which may be considered a membrane dynamics, but without embedding in any background manifold. The background manifold and embedding coordinates then arise from classical solutions to the background-independent membrane model. It was then found that background-dependent matrix models of string and $\mathcal{M}$ theory emerged by expanding around these classical solutions.

A recent development in this direction is a proposal for how to quantize a certain reduction of $\mathcal{M}$ theory non-perturbatively (Smolin forthcoming).

These few, preliminary, results, indicate that it is not difficult to invent and study hypotheses for background-independent formulations of string theory.

7.6 RELATIONALISM AND REDUCTIONISM

I would now like to broaden the discussion by asking: *Does the relational view have implications broader than the nature of space and time?* I will argue that it does[25]. A starting point for explaining why is to begin with a discussion of reductionism.

[25] The arguments of this and the following sections are developed from Smolin (1997a).

To a certain degree, reductionism is common sense. When a system has parts, it makes sense to base an understanding of it on the laws that the parts satisfy, as well as on patterns that emerge from the exchanges of energy and information among the parts. In recent years we have learned that very complex patterns can emerge when simple laws act on the parts of a system, and this has led to the development of the study of complex systems. These studies have shown that there are useful principles that apply to such complex systems and these may help us to understand an array of systems from living cells to ecosystems to economic systems. But this is not in contradiction to reductionism, it is rather a deepening of it.

But there is a built-in limit to reductionism. If the properties of a complex system can be understood in terms of their parts, then we can keep going and understanding the parts in terms of their parts, and so on. We can keep looking at parts of parts until we reach particles that we believe are elementary, which means they cannot be further divided into parts. These still have properties; for example, we believe that the elementary particles have masses, positions, momenta, spin, and charges.

When we reach this point we have to ask what methodology we can follow to explain the properties of the elementary particles. As they have no parts, reductionism will not help us. At this point we need a new methodology.

Most thinking about elementary particle physics has taken place in the context of quantum field theory and its descendants such as string theory, which are background-dependent theories. Let us start by asking how well these background-dependent theories have done resolving the problem of how to attribute properties to particles thought to be elementary. After this we will see if background-independent theories can do better.

In a background-dependent theory, the properties of the elementary entities have to do with their relationships to the background. This is clear in ordinary quantum field theory, where we understand particles to be representations of the Poincaré group and other externally imposed symmetries and gauge invariances. In these theories the particle states are labelled precisely by how they transform under symmetries of the background. The specification of the gauge and symmetry groups are indeed part of the background, because they are fixed for all time, satisfy no dynamical principles, and do not evolve.

The search for an explanation of the properties of the elementary particles within quantum field theory and string theory has been based on three hypotheses:

Unification: *All the forces and particles are different quantum states of some elementary entities.*

This elementary entity was at first thought, by Einstein and his friends, to be a field, giving rise to the once maligned subject of unified field theory. In more recent times it is thought to be a string. These are not so far apart, for a low-energy approximation to a string theory is a unified field theory. So most actual calculations in string theory involve classical calculations in unified field theories that are descendants of the theories Einstein and friends such as Kaluza and Klein studied many years ago.

But is unification enough of a criterion to pick out the right theory of nature? By itself it cannot be, for there are an infinite number of symmetry algebras which have the observed symmetries as a subalgebra. There is however a second hypothesis which is widely believed.

Uniqueness: *There exists exactly one consistent unified theory of all the interactions and particles.*

If this hope is realized, then it suffices to find that one unified theory. The first fully consistent unified theory to be found will be the only one that can be found and it will thus have to be the true theory of nature. It has even been said that, because of this, physics no longer needs experimental input to progress. At the advent of string theory, this kind of talk was very common. The transition from physics as an experimental science to physics based on finding the single unified theory was even called the passage from modern to postmodern physics.

Given that, in a background-dependent theory, particles are classified by representations of the symmetries of the vacuum, it follows that the more unified a theory is the larger the symmetry of the background must be. This leads to the conjecture of

Maximal symmetry: *The unique unified theory will have the largest possible symmetry group consistent with the basic principles of physics, such as quantum theory and relativity.*

These three conjectures have driven much of the work in high-energy physics the last three decades. They led first to grand unified theories (large internal gauge groups), then to higher dimensional theories (which have larger symmetry groups) and also to supersymmetry.

While these conjectures come very naturally to anyone with training in elementary particle physics, it must be emphasized that they have arisen from a methodology which is thoroughly background dependent. The idea that the states of a theory are classified by representations of a symmetry group, however used to it we have become, makes no sense apart from a theory in which there is a fixed background, given by the spacetime geometry and the geometry of the spaces in which the fields live. Theories without a background, where the geometry is dynamical and time dependent, such as general relativity, have no symmetry groups which act on the space of their solutions. In general relativity symmetries only arise as accidental symmetries of particular solutions, they have no role in the formulation of the equations or space of solutions of the theory itself[26].

So the methodology of looking for theories with maximal symmetry only makes sense in a background-dependent context. Still, since an enormous amount of work has gone into pursuit of this idea we can ask how far it actually gets us.

The most important thing to know about how this programme turned out is that string theories at the background-dependent level did not turn out to be unique. There turn out to be five string theories in flat ten-dimensional spacetime background. Each of them becomes an infinite number of theories when the background is taken to be a static but curved spacetime. Many of these are spacetimes in which a certain number of dimensions remain flat while the others are compactified.

26 It must be emphasized that we are talking here of global symmetries, not gauge invariances. Spaces of states or of solutions do not transform under gauge transformations, they are left invariant.

To preserve the notion of a unique unification it was conjectured that there is nevertheless a unique unification of all the string theories, which has been called $\mathcal{M}$ theory. This was motivated by the discovery of evidence for conjectured duality transformations that relate states of the different string theories. This theory, which has so far not been constructed, is conjectured to include all the string theories in ten dimensions, plus one more theory, which is eleven-dimensional supergravity. This is the largest consistent possible supersymmetric gravity theory and, at least at a classical level, naturally incorporates many of the symmetries known or conjectured that act on states of the different string theories.

The search for $\mathcal{M}$ theory has mostly followed the methodology which follows the three principles we mentioned. One posits a maximal symmetry algebra $\mathcal{A}_{max}$ that contains at least the eleven-dimensional super-Poincaré algebra and then tries to construct a theory based on it. Candidates for this symmetry algebra include the infinite dimensional algebra E_{10} (Damour et al. 2002) and compact superalgebras which have the eleven-dimensional super-Poincaré algebra as a subalgebra such as $Osp(1, 32)$ and $Osp(1, 64)$.

However, not much work has been done on the problem of constructing $\mathcal{M}$ theory, despite it being apparently necessary for the completion of the programme of unification through symmetry. It is interesting to ask why this is.

If $\mathcal{M}$ theory is to be a unification of all the different background-dependent string theories, and hence treat them all on an equal footing, it cannot be formulated in terms of any single spacetime background. Hence, we expect that $\mathcal{M}$ theory must be a background-independent theory.

However, background-independent theories are very different from background-dependent theories, as we have seen already in this chapter. One reason for the lack of interest in background-independent approaches to $\mathcal{M}$ theory might be simply that it is difficult for someone schooled in background-dependent methods to make the transition to the study of background-independent theories.

But there is a better reason, which is that there is a built-in contradiction to the idea of $\mathcal{M}$ theory. As I just emphasized, the idea that $\mathcal{M}$ theory is based on the largest possible symmetry is one that only makes sense in a background-dependent context. But as we have also just seen, $\mathcal{M}$ theory must be background independent.

One way out is to posit that the symmetry of $\mathcal{M}$ theory, while acting formally like a background, will not be the symmetry of any classical metric geometry. Indeed this is true of the possibilities studied such as E_{11} and $Osp(1, 32)$. Still they privilege those geometries whose symmetries are subalgebras of the posited fundamental symmetry, such as the super-Poincaré algebra. It thus seems hard to avoid a situation in which the solutions which are background spacetimes of maximal symmetry will play a privileged role in the theory. This is unlike the case of general relativity in which the solutions of maximal symmetry may have special properties, like being the ground state with certain asymptotic conditions, but play no special role in the formulation of the dynamics of the theory.

If we are to formulate $\mathcal{M}$ theory as a truly background-independent theory, we need a new methodology, tailored to background-independent theories. In the next section we will begin a discussion of what a background-independent approach to unification might look like.

Before we do, there is one more issue about background-dependent theories that we should consider. *If the hypothesis of unification is correct, then what accounts for the fact that the observed particles and forces have the particular properties which distinguish them?*

The basic strategy of all modern theories of unification is to answer this question with the mechanism of:

Spontaneous symmetry breaking: *The distinctions between the observed particles and interactions result from a vacuum state of the theory not being invariant under all the symmetries of the dynamics.*

This means that the properties that distinguish the different particles and forces from each other are due precisely to their relationship with a choice of background, which is a vacuum state of the theory. If a theory can have different vacuum states, which preserve different subgroups of the symmetries of the dynamics, then the properties of the particles and forces will differ in each. Hence we see explicitly that in these theories the properties of particles are determined by their relationship to the background.

However, notice that something new is happening here, which is quite important for the relational/absolute debate. The point of *spontaneous* symmetry breaking is that the choice of background is a consequence of the dynamics and can also reflect the history of the system. Hence theories that incorporate spontaneous symmetry breaking take a step in the direction of relational theories in which the properties of elementary particles are determined by their relationships with a dynamically chosen vacuum state.

But if the choice of vacuum state is to be determined dynamically, the fundamental dynamics must be formulated in a way that is independent of a choice of background. That is, the more spontaneous symmetry breaking is used to explain distinctions between particles and interactions, the more the fundamental theory must be background independent.

In conventional quantum field theories this is realized to some extent. But the background of spacetime is generally not part of the dynamics. In string theory the choice of solution can involve the geometry and topology of space and time. Hence, we arrive again at the necessity to ground string theory on a background-independent theory.

7.6.1 The Challenge of the String Theory Landscape

Before we turn to see how to approach the problem of unification from a background-independent theory, we should try to draw some lessons from the status of the search based on background-dependent methods.

We can begin by asking, what tools have string theorists used to study the problem of unification?

A principal tool invoked in much recent work in string theory is effective field theory. An effective field theory is a semiclassical field theory which is constructed to represent the behaviour of the excitations of a vacuum state of a more fundamental theory below some specified energy scale. They have the great advantage that one can study a theory expanded around a particular solution, treating that solution as a fixed background. This lets us use many of the intuitions and tools developed in the study of background-dependent theories. But there are also disadvantages to the use of effective field theory. One is that the threshold of evidence required to establish as likely a string background is weakened. Whereas it was at first thought necessary to prove perturbative finiteness around the background to all orders, it is now thought sufficient to display a classical solution to an effective field theory, which is some version of supergravity coupled to branes.

But no effective field theory can stand on its own, for these are not consistent microscopic theories. The reliability of effective field theory must always be justified by an appeal to its derivation from that more fundamental theory. In the applications where it was first developed, effective field theory is derived as an approximation to a more fundamental theory. This is true in QCD (Quantum Chromodynamics) and the standard model, as well as in its applications to nuclear physics and condensed matter physics.

We can see this easily by considering cases in which we believe there is no good fundamental theory, such as interacting quantum field theories in five or more dimensions. We can construct effective field theories to our heart's content to describe the low-energy physics in such contexts. These may be approximations to cut-off quantum field theories, for example, based on lattices. But they are unlikely to be approximations to any Poincaré-invariant theories. This is because there is strong evidence that the only Poincaré-invariant quantum field theories in more than four spacetime dimensions are free.

However, in string theory, effective field theory is being used in a context where we do not know that there is a more fundamental theory. That more fundamental theory, if it exists, is the conjectured background-independent unification of the different string theories. But since we do not have this theory, in the form of a set of either equations or principles, we cannot be assured that it exists. Hence, by relying on effective field theory we may get ourselves in the situation in which we are studying semiclassical theories which are not approximations to any more fundamental theory.

But, nevertheless, if one insists on confining investigations to background-dependent methods, there is little alternative to reliance on effective field theory. In the absence of a derivation from a full quantum theory, one can still posit that the existence of a consistent effective field theory is sufficient to justify belief in a string background, and see where this takes us. One requires a weak form of consistency, which is that excitations of the solution, were they to exist, would be weakly coupled. Not surprisingly, perhaps, this approach leads to evidence for a landscape consisting of an infinite number of discrete string backgrounds (De Wolf et al. 2005). Even restricting the counting to backgrounds that have positive vacuum energy and broken supersymmetry leads to estimates of 10^{500} or more discrete vacua (Kachru et al. 2003).

It is interesting to note that the term 'landscape' implies the existence of a function, h, the height, such that the discrete vacua are at local minima of h. In the recent literature, the height h is a potential or free energy. While it is clear what is meant by this, it is perhaps worrying that the concept of energy is problematic in a cosmological or quantum gravity context. This is because, once the gravitational degrees of freedom are included, the energy of cosmological spacetimes is constrained to vanish. All cosmological solutions to diffeomorphism-invariant theories have the same energy; zero. There is in cosmology no ground state with zero energy; solutions with different potential energies are no more or less likely to exist, they just expand at different rates. Even if the background geometry is assumed fixed, energy and free energy are only defined on a background that has a timelike killing field.

But what could the height be, if not energy? The context which inspired the original use of the term landscape in string theory Smolin (1997a) was mathematical models of natural selection, in which the height h measures the fitness, which is the number of viable[27] progeny of a state. The term was introduced in (Smolin 1997a) to evoke the methodologies by which *fitness landscapes* are studied.

However, in the recent string theory literature on landscapes, the analogy to natural selection is not invoked. What then is the height? If it is energy, then that implies the existence of a fixed background, with a timelike killing field. But what is the background, when the space we are considering is a space of different vacua, with different geometries and topologies? There seems to be a confusion in which reference to a structure that depends on a fixed background is being invoked in the description of the space of possible backgrounds.

Another way to see that the notion of a fixed background is sneaking back into the theory is to consider the assumptions behind the probabilistic studies of the landscape.

There are, broadly speaking, two kinds of methods that might be brought to bear to the study of probability distributions on such landscapes of states. One may study distributions that are in *equilibrium, and hence static*, or one may study *non-equilibrium and hence time-dependent distributions.*

Almost all the recent work on probability distributions in the string theory landscape have taken the first kind of approach. Some, but not all, of this work evokes what we may call

The anthropic hope: *There are a vast number of unified theories, and a vast number of regions of the universe where they may act. Out of all of these, there will be a very small fraction where the laws of physics allow the existence of intelligent life. We find ourselves in one of these. Because the number of universes and theories is so vast, theory can make few predictions except those that follow from requiring our own existence.*

The reliance on the anthropic principle is unfortunate, because it can be shown that the use of the anthropic principle cannot lead to any falsifiable predictions. This is argued in detail in Smolin (2004a), to which the reader is referred. As a result, one has to suspect that a search for a unified theory of physics that in the end invokes the anthropic principle has reached a *reductio ad absurdum*. Somewhere along the line, in the search for a unified string theory, a wrong turn has been taken.

[27] Meaning they will have their own progeny.

It could be that the wrong turn is that string theory is based on physical hypotheses that have nothing to do with nature. But if this is not the case, some wrong direction must have been taken in the path that led from the conjecture of a unique unification within string theory to the present invocations of the anthropic principle.

We can see from the survey of the situation we just made that the dilemma we have arrived at seems to involve trying to use background-dependent notions, like energy, to do physics in a setting that must be background independent. For if there is a space of possible backgrounds, on which we are to do dynamics, it is obvious that the form of dynamics we employ cannot make reference to any given fixed background. Hence, it seems reasonable to suggest that the wrong turn is the failure to search for a background-independent foundation for the theory.

It is then interesting to note that the invocation of static probability distributions harks back to the absolute perspective. To see this we can ask, what is the time with respect to which the probability distribution is considered to be static? It cannot be the time within a given spacetime background, because the probability distribution lives on the space of possible backgrounds. Single universes may evolve, and may come and go, but there is hypothesized to be nevertheless a static and eternal distribution of universes with different properties. It is to this distribution, which exists absolutely and for all time, that we must go for an explanation of any properties of our universe. Thus, at the level of multiverses, static distributions on landscapes have more in common with Aristotle's way of thinking about cosmology than with general relativity.

Is there then an alternative methodology for treating the landscape, which would naturally arise from a background-independent theory? I would like to claim there is. The next sections are devoted to its motivation and description.

7.7 A RELATIONAL APPROACH TO THE PROBLEMS OF UNIFICATION AND DETERMINATION OF THE STANDARD MODEL PARAMETERS

Let us then assume we agree on the need to formulate a unified theory in a background-independent framework. Even without having a complete formulation of this kind in hand, it may be of interest to ask, what would a background-independent approach to the problem of unification look like? How would it address the problems raised in the last section? To approach these questions we return to the question of *how we are to explain the properties of the elementary particles*.

In a relational theory, as I explained earlier, the properties of the elementary entities can only have to do with relations they have to other elementary entities. Let us explore the implications of this.

The first implication is that any relational system with a large number of parts must be complex, in the sense of having no symmetries. The reason is Leibniz's principle of the identity of the indiscernible: if two entities have the same relations to the rest, they are to be identified. Each individual entity must then have a unique set of relations to the rest.

The elementary entities in general relativity are the events. An event is characterized by the information coming to it, from the past. We may call the information received by an event in spacetime, the *view* of that event. It literally consists of what an observer at that event would see looking out at their backwards light cone.

It follows that any two events in a spacetime must have different views. This implies that

1. *There are no symmetries.*
2. *The spacetime is not completely in thermal equilibrium.*

These are in fact true of our universe. The universe may be homogeneous above the enormous scale of 300*Mpc*, but on every smaller scale there is structure. Similarly, while the microwave background is in thermal equilibrium, numerous bodies and regions are out of equilibrium with each other.

Julian Barbour and I call a spacetime in which the view of each event is distinct a *Leibniz spacetime*. We note, with some wonder, that the fact that our universe is not completely in thermal equilibrium is due to the fact that gravitationally bound systems have negative specific heat, and therefore cannot evolve to unique equilibrium configurations. Furthermore, gravity causes small fluctuations to grow that would otherwise be damped. This is why the universe is filled with galaxies and stars. Thus, gravity, which as Einstein taught us is the force that necessarily exists due to the relational character of space and time, is at the same time the agent that keeps the world out of equilibrium and causes fluctuations to grow rather than to dissipate, which is a necessary condition for it to have a completely relational description.

There is a further consequence of taking the relational view seriously. In a relational theory, the relations that define the properties of elementary entities are not static they evolve in time according to some law. This means that the properties by which we characterize the interaction of an elementary particle with the rest of the universe are likely to include some which are not fixed a priori by the theory, but depend on solutions to dynamical equations. We can expect that this applies to all of the basic properties that characterize particles such as masses and charges.

7.8 RELATIONALISM AND NATURAL SELECTION

How far can we go to a relational explanation of the properties of the elementary particles in the standard model? While the anthropic principle itself is not explanatory, it is useful to go back to its starting point, which is an apparently true observation, which we may call

The anthropic observation: *Our universe is much more complex (in for example its astrophysics and chemistry) than most universes with the same laws but different values of the parameters of those laws (including masses, charges, etc.)*

This requires explanation. Unfortunately no principle has been found that explains the values of the physical parameters (which can be taken to be the parameters of the standard models of particle physics and cosmology). Given recent progress in string theory, there is no reason to expect such a principle to exist. Instead, as the relational argument suggests, those parameters are environmental, and can differ in different solutions of the fundamental theory. We then require a dynamical explanation for the anthropic observation. For it to be science, the explanation must make falsifiable predictions that are testable by real experiments.

There is only one mode of explanation I know of, developed by science, to explain why a system has parameters that lead to much more complexity than typical values of those parameters. This is natural selection.

It may be observed that natural selection is to some extent part of the movement from absolute to relational modes of explanation. There are several reasons to characterize it as such.

- *Natural selection follows the relational strategy.* Before it, properties that characterize species were believed to be eternal, and to have a priori explanations. These are replaced by a characterization of species that is relational and evolves in time as a result of interactions between it and other species.
- *The properties natural selection acts on, such as fitness, are relational quantities, in that they summarize consequences of relations between the properties of a species and other species.*
- *These properties are not fixed in advance, they evolve lawfully.*
- *A relational system requires a dynamical mechanism of individuation, leading to enough complexity that each element can be individuated by its relations to the rest.* Natural selection acts in this way, for example, it inhibits two species from occupying exactly the same niche. By doing so it increases the complexity, measured in terms of the relations between the different species.

This suggests the application of the mode of explanation of natural selection to cosmology.

This has been developed in Smolin (1997a), and it is successful in that it does lead to predictions that are falsifiable, but so far not falsified. The idea, briefly, is the following.

To apply natural selection to a population, there must be:

- A space of parameters for each entity, such as the genes or the phenotypes.
- A mechanism of reproduction.
- A mechanism for those parameters to change, but slightly, from parent to child.
- Differentiation, in that reproductive success strongly depends on the parameters.

By simple statistical reasoning, the population will evolve so that it occupies places in the parameter space leading to atypically large reproductive success, compared to typical parameter values. (Note that creatures with randomly chosen genes are dead.)

This can be applied to cosmology:

- The space of parameters is the space of parameters of the standard models of physics and cosmology. This is the analogue of phenotype. At a deeper level, this is to be explained by a space analogous to genotypes such as the space of possible string theories. This leads to the term the string theory landscape.

- The mechanism of reproduction is the formation of black holes. It is long conjectured that black hole singularities bounce, leading to the formation of new universes through new big bangs. There is increasing evidence that this is true in loop quantum gravity.
- We may conjecture that the low-energy parameters do change in such a bounce. There are a few calculations that support this (Smolin 2004a).
- The mechanism of differentiation is that universes with different parameters will have different numbers of black holes.

This leads to a simple prediction: our universe has many more black holes than universes with random values of the parameters. This implies that most ways to change the parameters of the standard models of particle physics and cosmology should have fewer black holes.

This leads to testable predictions. I'll mention one here: there can be no neutron stars with masses larger than 1.6 times the mass of the sun. I will not explain here how this prediction follows, but simply note that it is falsifiable[28]. So far all neutron stars observed have masses less than 1.45 solar masses, but new ones are discovered regularly.

7.9 WHAT ABOUT THE COSMOLOGICAL CONSTANT PROBLEM?

It is becoming clearer and clearer that the hardest problem faced by theoretical physics is the problem of accounting for the small value of the cosmological constant problem. The problem is so hard that it constitutes the strongest arguments yet given for an anthropic explanation, following an argument of Weinberg (2000a, 2000b).[29]

Given that background-dependent theories have failed to resolve it, it is important to ask whether background-independent approaches have done any better.

We mention several interesting results here:

- There is an argument for the relaxation of the cosmological constant in LQG, analogous to the Pecci–Quinn mechanism (Alexander 2005). This relies on a connection between the cosmological constant and parity breaking, which is natural within LQG.
- Volovik has argued, in a particular example, that if spacetime is emergent from more fundamental quantum degrees of freedom, then there is a dynamical mechanism which relaxes the ground state energy (Volovik 2005). This mechanism is missed if one formulates the theory in terms of an effective field theory that describes only the low-energy collective excitations on a fixed background.

28 Details of the argument can be found in Smolin (1997a, 2004a).

29 See Smolin (2004a) for a summary, references, and critique.

- Dreyer argues that the cosmological constant problem is in fact an artiefact of background-dependent approaches (Dreyer 2004). He proposes that the problem arises from the unphysical splitting of the degrees of freedom of a fundamental, background-dependent theory into a background, which has only classical dynamics, and quantum excitations of it. He presents an example from condensed matter physics in which exactly this occurs. In his model, one can calculate the ground state energy two ways: in terms of the fundamental Hamiltonian, which is a function of the elementary degrees of freedom, and in terms of an effective low-energy Hamiltonian which describes collective, emergent low-energy degrees of freedom. The zero point energy in the latter overestimates the ground state energy computed in the fundamental theory.
- The only approach to quantum gravity that predicted the correct magnitude of the observed cosmological constant is the causal set theory (Ahmed et al. 2004). There it naturally comes out that a universe with many events has a small cosmological constant. Whether the mechanism that works there extends to other background-independent approaches is an interesting open question.

While all these results are preliminary, what is remarkable is that new possibilities for resolving the cosmological constant problem appear when the problem is posed in a background-independent theory.

7.10 THE ISSUE OF EXTENDING QUANTUM THEORY TO COSMOLOGY

Let us now turn to the third issue raised in the introduction, whether a cosmological theory can be formulated in the same language as theories of small parts of the universe, or requires a new formulation. An aspect of this is the problem of quantum cosmology. In recent years new proposals to resolve this stubborn problem have been formulated in the context of background-independent approaches to quantum gravity.

7.10.1 Relational Approaches to Quantum Cosmology

In the last ten years several new proposals have been made concerning the foundational issues in quantum cosmology, which have gone under the name of relational quantum theories[30]. These have been inspired by the general philosophy of relationalism.

These approaches have been put forward, in slightly different ways, by Crane, Rovelli, and Markopoulou (Crane 1993a, 1993b, 1995; Rovelli 1996; Markopoulou 2000a, 2000b, 2000c; Hawkins et al. 2003). The mathematical apparatus needed to formalize this view has been studied by Butterfield and Isham (1997, 1998, 1999, 2000). While they differ as to details, they agree that a quantum theory of cosmology is not to be formulated in the language of ordinary quantum mechanics.

[30] For references for this section, see the corresponding discussion and references in Smolin (2004b).

One way to state the problem is to ask how we understand the quantum state: is it a complete and objective description of a physical system, in which case, how do we account for the measurement problem? Or is it a description of the information or knowledge that an observer has about a system they have isolated and studied? If this is the case, can we apply quantum theory to cosmology—or indeed to any system that contains observers?

There is a hint of relationalism in Bohr, who argued for a view something like the latter. Bohr always insisted that while there must be a line between the system and observer, that line is flexible; it may be drawn anywhere. This is frustrating for those who want to believe in a realist interpretation of the quantum state. A realist would argue that the observer and her instruments are physical systems. Consequently there must be a description in which they are included in the system being studied. Bohr replies there is no contradiction, because now we are speaking of the knowledge a second observer has of a system containing the first observer. According to Bohr then, each observer has a different wave function that describes the system they observe.

Relational approaches to quantum theory formalize this point of view. Rather than taking the Everett/many worlds view, and describing many universes in terms of a single quantum state, they posit that it requires many quantum states to describe a single universe. Each of these quantum states corresponds to a way of dividing the universe into two subsystems, such that one includes an observer.

A relational approach to quantum theory was proposed by Crane (1993a, 1993b, 1995), in a paper that anticipated some aspects of the holographic principle (t'Hooft 1993; Susskind 1995). In that paper, Crane proposed that there is no quantum state associated with the universe as a whole. Instead, there is a quantum state associated with every way of introducing an imaginary spatial boundary, splitting the universe into two. By analogy with topological field theory, he proposed that the Hilbert spaces on boundaries of $3 + 1$ dimensional spacetime should be built up out of state spaces of Chern–Simons theory. When fully developed, this proposal became the very fruitful isolated horizon approach to the quantum geometry and entropy of horizons.

Rovelli then developed a general framework for relational quantum theory (Rovelli 1996). The approaches of Rovelli, however, left open the precise structure that is to tie together the network of Hilbert spaces and algebras necessary to describe a whole universe. A template for such structure was given in the work of Butterfield and Isham, who showed how the consistent histories formulation could be interpreted in terms of a sheaf of Hilbert spaces (Butterfield and Isham 1997, 1998, 1999, 2000).

Markopoulou proposed that the structure tying together the different Hilbert spaces is the causal structure of spacetime (Markopoulou 2000a, 2000b, 2000c; Hawkins et al. 2003). In this formulation there is a Hilbert space for every event in a quantum spacetime. The state at each event is a density matrix that describes the quantum information available to an observer at that event. There are consistency conditions that prescribe how the flow of quantum information in a spacetime follows the causal structure of that spacetime. This is a generalization of quantum theory, for there need not be a quantum state associated with the whole system. (Indeed, it is related to a large class of such generalizations studied by Butterfield and Isham.)

This leads to a relational formulation of the holographic principle, sketched in (Markopoulou and Smolin (1999). The basic idea is that the events are associated with elements of surface. Each corresponds to a quantum channel, through which information flows from its causal past to its causal future. The area of such a channel is *defined* to be a measure of its channel capacity.

7.10.2 Relational Approaches to Going beyond Quantum Theory

Relational quantum theory gets us out of the paradoxes that arise from trying to describe the universe with a single quantum state. Still, there is, unfortunately, a problem with these approaches. This stems from the fact that the system of quantum states depends on the causal structure of spacetime being fixed. But in a quantum theory of gravity one is supposed to take a quantum sum over all possible histories of the universe, each with a different causal structure. This is to say that relational quantum theories appear to be as background dependent as ordinary quantum theory, it is just that they differ in how they are background dependent.

Can there be a fully background-independent approach to quantum theory? I believe that the answer is only if we are willing to go beyond quantum theory, to a hidden variables theory. I would like in closing then to briefly mention work in progress in this direction.

We know from the experimental disproof of the Bell inequalities that any viable hidden variables theory must be non-local. This suggests the possibility that the hidden variables are relational. That is, rather than giving a more detailed description of the state of an electron, relative to a background, the hidden variables may give a description of relations between that electron and the others in the universe.

The possibility of a relational hidden variables theory is suggested by a simple counting argument: in classical mechanics of N point particles, in three-dimensional space there are 6N phase space degrees of freedom. In quantum theory this is described by a complex function on the 3N dimensional configuration space—the wave function.

But a relational theory has in principle N^2 degrees of freedom, at least one for every pair of particles. Most of these are unobservable, by any local observer, because they involve relations between particles near to us and those very far away. Thus, any working out of a relational theory will have to treat them probabilistically. This will require a probability distribution, which is a real function on N^2 variables.

A real function on N^2 variables has much more information in it than a complex function on $3N$ variables. Thus, one can imagine deriving quantum mechanics for $3N$ variables from statistical mechanics for N^2 variables. Such a theory would be a non-local hidden variables theory.

This leads to a simple conjecture

Perhaps all the extra information, N^2 as compared to N, necessary for a completely relational theory, is the non-local hidden variables?

In the last few years two such relational hidden variables theories have been written down. Markopoulou and I have proposed one (Markopoulou and Smolin 2003), and

Stephen Adler (2002) proposes another. In our theory the non-local hidden variables are coded in a graph on N nodes, which is argued to arise from the low-energy limit of a relational theory like loop quantum gravity.

It is too soon to see if these theories will be successful. But they offer hope that taking relational ideas seriously may lead to a successful attack on all *five* of the problems mentioned in the introduction.

7.11 CONCLUSIONS

In this chapter I have described several partly relational, or background-independent, theories:

- General relativity.
- Relational approaches to quantum gravity, including loop quantum gravity, causal set models, causal dynamical triangulation models, and relational approaches to string/$\mathcal{M}$ theory.
- Relational approaches to extending quantum theory to cosmology.

Each is partly successful. Several are more successful than less relational alternatives. But none is completely successful and none is completely relational. They are not completely relational because each still has background structure, which is non-dynamical and must be specified in advance.

However, I believe we do learn something very important from these examples:

In several instances, the relational theory turns out to be more predictive and more falsifiable than background-dependent theories.

In particular, cosmological natural selection leads to falsifiable predictions, which anthropic approaches to the landscape so far do not. Furthermore, there is the very real possibility that the Planck scale will be probed in upcoming experiments, such as GLAST and AUGER (Smolin 2004b). Background-independent theories appear to give predictions for these experiments (Smolin 2005). String theory cannot, because it takes the symmetry of the background as input.

Why is this the case? I can only make some brief remarks here. The difference between relational and non-relational theories is between:

1. Explanations that refer ultimately to a network of relationships amongst equally physical entities, which evolve dynamically.

versus

2. Explanations that refer to relationships between dynamical entities and an a priori, non-dynamical, background.

The former are more constrained, hence harder to construct. More of what is observed is subject to law, as there is no background to be freely chosen. Hence, it appears that relational, background-independent theories are more testable, and more explanatory.

This is the reason for my provocative hypothesis. If it is true that *the reason that string theory finds itself in the situation described in the introduction is that no background-dependent theory could successfully solve the five key problems mentioned there. If this is true, then the only thing to do is to go back and work on the less studied road of relational theories.*

At the same time, I have tried here to explain the key problems still faced by the relational road. Some of these have to do with the problem of time. Others have to do with the inverse problem. We saw it in the discussion of causal set models, which are the only purely relational theories I discussed. The inverse problem is that there are many more discrete relational structures than those that approximate local, continuous structures such as classical spacetimes. So a purely relational theory that explains the fact that the world, at least on scales larger than the Planck scale, appears to be continuous and low-dimensional, must explain why those local and low dimensional structures dominate in an ensemble of histories most of which don't remotely resemble local, low-dimensional structures.

Let me close by recalling the extent to which the last three decades of theoretical physics are anomalous, compared with the previous history of physics. Many ideas have been studied, but few have been subject to the only kind of test that really matters, which is experiment. The hope behind this chapter and the work it represents is that by following the relational strategy we may be led to invent theories that are more falsifiable, whose study will lead us back to the normal practice of science where theory and experiment evolve hand in hand.

ACKNOWLEDGMENTS

My awareness of the centrality of the problem of background independence and its roots in the history of physics is due primarily to Julian Barbour, who has served as a philosophical guru to a whole generation of workers in quantum gravity. My understanding of these issues has been deepened through conversations and collaborations with many people including Abhay Ashtekar, John Baez, Louis Crane, Chris Isham, Fotini Markopoulou, Roger Penrose, Carlo Rovelli, and John Stachel. The work described in the closing section is joint work with Fotini Markopoulou and is largely based on her ideas. Finally, many thanks to Freddie Cachazo, David Rideout, and Simon Saunders for very helpful suggestions which improved the manuscript.

REFERENCES

Adler, S. (2002) *Statistical Dynamics of Global Unitary Invariant Matrix Models as Pre-Quantum Mechanics*. ArXiv:hep-th/0206120.

Agishtein, M. E., and A. A. Migdal (1992) "Simulations of Four-Dimensional Simplicial Quantum Gravity". *Mod. Phys. Lett. A* 7: 1039.

Ahmed, M., S. Dodelson, P. B. Greene, and R. Sorkin (2004) "Everpresent Lambda". *Phys. Rev. D,* **69**: 103523.

Alexander, H. G. (ed.) (1956) *The Leibniz–Clarke Correspondence.* Mancheste: Manchester University Press. (For an annotated selection, see **www.bun.kyoto-u.ac.jp/~suchii/leibniz-clarke.html**).

Alexander, S. (2005) "A Quantum Gravitational Relaxation of the Cosmological Constant". ArXiv:hep-th/0503146.
Ambjorn, J. (1995) "Quantum Gravity Represented as Dynamical Triangulations." *Class. Quant. Grav.* 12: 2079.
____ and R. Loll (1998) *Nucl. Phys.* B536: 407. ArXiv:hep-th/9805108.
____ Z. Burda, J. Jurkiewicz, and C. F. Kristjansen (1992) "Quantum Gravity Represented as Dynamical Triangulations". *Acta Phys. Polon. B* 23: 99.
____ K. N. Anagnostopoulos, and R. Loll (2000a) "Crossing the c=1 Barrier in 2d Lorentzian Quantum Gravity". ArXiv:hep-lat/9909129, *Phys.Rev, D* 6: 1 044010.
____ J. Jurkiewicz, and R. Loll (2000b) "Nonperturbative Lorentzian Path Integral for Gravity". *Phys. Rev. Lett.* 85: 924–7. ArXiv:hep-th/0002050.
____ (2001a) "Dynamically Triangulating Lorentzian Quantum Gravity". *Nucl. Phys.* B610: 347–8. ArXiv:hep-th/0105267.
____ (2001b) "Nonperturbative 3D Lorentzian Quantum Gravity". *Phys. Rev. D* 64: 044011. ArXiv:hep-th/0011276.
____ and G. Vernizzi (2001c) "Lorentzian 3d Gravity with Wormholes via Matrix Models". *JHEP* **0109**: 022. ArXiv:hep-th/0106082.
____ A. Dasgupta, J. Jurkiewiczcy, and R. Loll (2002) "A Lorentzian Cure for Euclidean Troubles". ArXiv:hep-th/0201104.
____ J. Jurkiewicz, and R. Loll (2004) "Emergence of a 4D World from Causal Quantum Gravity". ArXiv:hep-th/0404156.
____ (2005a) "Reconstructing the Universe". ArXiv:hep-th/0505154.
____ (2005b) "Spectral Dimension of the Universe." ArXiv:hep-th/0505113.
____ (2005c) "Semiclassical Universe from First Principles." *Phys. Rev. Lett.* **B607**: 205–13.
Arnsdorf, M., and L. Smolin (2001) "The Maldacena Conjecture and Rehren Duality". ArXiv:hep-th/0106073.
Ashtekar, A. (1988) *New Perspectives in Canonical Gravity.* Naples: Bibliopolis.
____ (1991) *Lectures on Non-Perturbative Canonical Gravity.* Advanced Series in Astrophysics and Cosmology 6. Singapore: World Scientific.
____ and J. Lewandowski (2004) "Background Independent Quantum Gravity: A Status Report'. ArXiv:gr-qc/0404018.
Azuma, T. (2001) "Investigation of Matrix Theory via Super Lie Algebra". ArXiv:hep-th/0103003.
____ and M. Bagnoud (2002) "Curved-Space Classical Solutions of a Massive Supermatrix Model". ArXiv:hep-th/0209057.
____ S. Iso, H. Kawai, and Y. Ohwashi (2001) "Supermatrix Models". *Nucl. Phys. B* 610: 251–79.
Baez, J. (1998) "Spin Foam Models". *Class. Quant. Grav.* 15: 1827–58. ArXiv:gr-qc/9709052.
____ (2000) "An Introduction to Spin Foam Models of Quantum Gravity and BF Theory". *Lecture Notes in Physics*, 543: 25–94.
Bagnoud, M., L. Carlevaro, and A. Bilal (2002) "Supermatrix Models for M-Theory Based on osp(1—32,R)". *Nucl. Phys. B* 641: 61–92.
Barbour, J. (1984) "Leibnizian Time, Machian Dynamics, and Quantum Gravity". in R. Penrose and C. Isham (eds.), *Quantum Concepts In Space and Time.* Oxford: Clarendon Press (pp. 236–46).
____ (1989) *Relative or Absolute Motion: The Discovery of Dynamics.* Cambridge: Cambridge University Press.
____ (2000) *The End of Time.* Oxford: Oxford University Press.

—— (2001) *The Discovery of Dynamics: A Study from a Machian Point of View of the Discovery and the Structure of Dynamical Theories.* Oxford: Oxford University Press.
—— and H. Pfister (eds.) (1996) *Mach's Principle: From Newton's Bucket to Quantum Gravity.* Berlin: Berkhäuser Verlag.
Berkovits, N. (2004) "Covariant Multiloop Superstring Amplitudes". ArXiv:hep-th/0410079.
Bombelli, L., J. H. Lee, D. Meyer, and R. Sorkin (1987) "Space-Time as a Causal Set". *Phys. Rev. Lett.* 59: 52.
Butterfield, J., and C. J. Isham (1997) "Topos Theory and Consistent Histories: The Internal Logic of the Set of all Consistent Sets". *Int. J. of Theor. Phys.* 36: 785–814.
—— (1998) "A Topos Perspective on the Kochen–Specker Theorem: I. Quantum States as Generalized Valuations". ArXiv:quant-ph/9803055.
—— (1999) "A Topos Perspective on the Kochen–Specker Theorem: II. Conceptual Aspects and Classical Analogues". *Int. J. Theor. Phys.* 38: 827–59.
—— (2000) "Some Possible Roles for Topos Theory in Quantum Theory and Quantum Gravity". *Found. Phys.* 30: 1707–35.
Connes, A., and H. Moscovici (2005) "Background Independent Geometry and Hopf Cyclic Cohomology". ArXiv:math.QA/0505475.
Crane, L (1993a) "Categorical Physics". ArXiv:hep-th/9301061.
—— (1993b) "Topological Field Theory as the Key to Quantum Gravity". In J. Baez (ed.), *Knot Theory and Quantum Gravity.* Oxford: Oxford University Press.
—— (1995) "Clocks and Categories, is Quantum Gravity Algebraic?" *J. Math. Phys.* 36: 6180–93.
Damour, T., M. Henneaux, and H. Nicolai (2002) "E10 and a 'Small Tension Expansion' of M Theory". *Phys. Rev. Lett.* 89: 221601.
Dasgupta, A., and R. Loll (2001) *Nucl. Phys.* B606: 357. ArXiv:hep-th/0103186.
DeWolfe, O, A. Giryavets, S. Kachru, and W. Taylor (2005) "Type IIA Moduli Stabilization". ArXiv:hep-th/0505160.
D'Hoker, E., and D. H. Phong (2002a) "Two Loop Superstrings, 1. Main Formulas". *Phys. Lett. B* 529: 241.
—— (2002b) "Lectures on Two-Loop Superstrings". ArXiv:hep-th/0211111.
Dittrich, B., and R. Loll (2002) "A Hexagon Model for 3D Lorentzian Quantum Cosmology". ArXiv:hep-th/0204210.
Dreyer, O. (2004) "Background Independent Quantum Field Theory and the Cosmological Constant Problem". ArXiv:hep-th/0409048.
Earman, J. (1989) *World Enough and Spacetime: Absolute vs. Relational Theories of Space and Time.* Cambridge, MA: MIT Press.
Fleischhack, C. (2004) "Representations of the Weyl Algebra in Quantum Geometry". ArXiv:math-ph/0407006
Francks, R. and R. S. Woolhouse (ed.) (1999) *Leibniz.* Oxford Philosophical Texts. Oxford: Oxford University Press.
Freidel, L., and E. R. Livine (2005) "Ponzano–Regge Model Revisited II: Feynman Diagrams and Effective Field". ArXiv:hep-th/0502106.
Gambini, R., and J. Pullin (1996) *Loops, Knots, Gauge Theories and Quantum Gravity.* Cambridge: Cambridge University Press.
Hawkins, E., F. Markopoulou, and H. Sahlmann (2003) "Evolution in Quantum Causal Histories". ArXiv:hep-th/0302111.
Hertog, T., G. T. Horowitz, and K. Maeda (2003) "Negative Energy Density in Calabi-Yau Compactifications". *JHEP* 0305: 060.

Isham, C. J., R. Penrose, and D. W. Sciama (eds.) (1975) *Quantum Gravity: An Oxford Symposium.* Oxford: Clarendon Press.

Kachru, S., R. Kallosh, A. Linde, and S. P. Trivedi (2003) "de Sitter Vacua in String Theory". ArXiv:hep-th/0301240.

Kuchař, K. (1976) "Dynamics of Tensor Fields in Hyperspace: III". *J. Math. Phys.* 17: 792.

____ (1982) "Conditional Symmetries in Parametrized Field Theories". *J. Math. Phys.* 23: 1647.

Leibniz, G. F. (1698) *The Monadology.* Translated by Robert Latta, available at **http://oregonstate.edu/instruct/phl302/texts/leibniz/monadology.html.**

Lewandowski, J., A. Okolow, H. Sahlmann, and T. Thiemann (2005) "Uniqueness of Diffeomorphism Invariant States on Holonomy-Flux Algebras". ArXiv:gr-qc/0504147.

Ling, Y., and L. Smolin (2001) "Eleven Dimensional Supergravity as a Constrained Topological Field Theory". *Nucl. Phys. B* 601: 191.

Loll, R. (2001) *Nucl. Phys. B* (Proc. Suppl.) 94 96. ArXiv:hep-th/0011194.

Mach, E. (1893) *The Science of Mechanics.* La Salle, IL: Open Court.

Malament, D. (forthcoming) "Classical Relativity Theory". ArXiv:gr-qc/0506065. To appear in J. Butterfield and J. Earman (eds.), *Handbook of the Philosophy of Physics.* Amsterdam: Elsevier.

Markopoulou, F. (1997) "Dual Formulation of Spin Network Evolution". ArXiv:gr-qc/9704013.

____ (2000a) "Quantum Causal Histories". *Class. Quant. Grav.* 17: 2059–72.

____ (2000b) "The Internal Description of a Causal Set: What the Universe Looks Like from the Inside". *Commun. Math. Phys.* 211: 559–83.

____ (2000c) "An Insider's Guide to Quantum Causal Histories". *Nucl. Phys. Proc. Suppl.* 88: 308–13.

____ (2002) "Planck-Scale Models of the Universe". ArXiv:gr-qc/0210086.

____ and L. Smolin (1999) "Holography in a Quantum Spacetime". ArXiv:hep-th/9910146.

____ (2003) "Quantum Theory from Quantum Gravity". ArXiv:gr-qc/0311059.

____ (2004) "Gauge Fixing in Causal Dynamical Triangulations". ArXiv:hep-th/0409057.

____ (forthcoming) "The Role of Non-Locality in Quantum Gravity". In preparation.

Martin, X., D. O'Connor, D. P. Rideout, and R. D. Sorkin (2001) "On the "Renormalization" Transformations Induced by Cycles of Expansion and Contraction in Causal Set Cosmology". *Phys. Rev. D* 63: 084026. ArXiv:gr-qc/0009063.

Norton, J. D. (1987) "Einstein, the Hole Argument and the Reality of Space". In J. Forge (ed.), *Measurement, Realism and Objectivity.* Boston: D. Reidel (pp. 153–88).

Okolow, A., and J. Lewandowski (2003) "Diffeomorphism Covariant Representations of the Holonomy Flux *-Algebra". *Class. Quant. Grav.* 20: 3543–3568.

Perez, A. (2003) "Spin Foam Models for Quantum Gravity". Topical Review in *Class. Quant. Grav.* 20 R43. ArXiv:gr-qc/0301113.

Rideout, D. P. (2002) "Dynamics of Causal Sets". ArXiv:gr-qc/0212064.

____ and R. D. Sorkin (2000) "A Classical Sequential Growth Dynamics for Causal Sets". *Phys. Rev. D* 61 024002. ArXiv:gr-qc/9904062.

____ (2001) "Evidence for a Continuum Limit in Causal Set Dynamics". *Phys. Rev. D,* 63: 104011. ArXiv:gr-qc/0003117.

Rovelli, C. (1991) "What is Observable in Classical and Quantum Gravity?" *Class. Quant. Grav.* 8: 297.

____ (1996) "Relational Quantum Mechanics". *Int. J. of Theor. Phys.* 35: 1637.

____ (1998) "Loop Quantum Gravity." **www.livingreviews.org/Articles/Volume1/1998-1rovelli.**

—— (2000) "Notes for a Brief History of Quantum Gravity". ArXiv:gr-qc/0006061.
—— (2004) *Quantum Gravity*. Cambridge: Cambridge University Press.
Sahlmann, H. (2002a) "Some Comments on the Representation Theory of the Algebra Underlying Loop Quantum Gravity". ArXiv:gr-qc/0207111.
—— (2002b) "When do Measures on the Space of Connections Support the Triad Operators of Loop Quantum Gravity?" ArXiv:gr-qc/0207112.
—— and T. Thiemann (2003a) "On the Superselection Theory of the Weyl Algebra for Diffeomorphism Invariant Quantum Gauge Theories". ArXiv:gr-qc/0302090.
—— and —— (2003b) "Irreducibility of the Ashtekar–Isham–Lewandowski Representation". ArXiv:gr-qc/0303074.
Saunders, S. (2003) "Indiscernibles, General Covariance, and Other Symmetries: The Case for Non-reductive Relationalism". In A. Ashtekar, D. Howard, J. Renn, S. Sarkar, and A. Shimony (eds.), *Revisiting the Foundations of Relativistic Physics: Festschrift in Honour of John Stachel*. Dardrecht.
Smolin, L. (1992) 'Recent Developments in Nonperturbative Quantum Gravity'. In J. Perez-Mercader et al. (ed.), *Quantum Gravity and Cosmology*. Singapore: World Scientific.
—— (1997a) *Life of the Cosmos*. New York: Oxford University Press and London: Wiedenfeld & Nicolson.
—— (1997b) "The Future of Spin Networks". ArXiv:gr-qc/9702030 in the Penrose Festschrift.
—— (1997c) "Chern–Simons Theory in 11 Dimensions as a Non-perturbative Phase of $\mathcal{M}$ Theory". ArXiv:hep-th/9703174.
—— (2000a) "Strings as Perturbations of Evolving Spin-Networks". *Nucl. Phys. Proc. Suppl.* 88: 103–13.
—— (2000b) "The Strong and the Weak Holographic Principles". ArXiv:hep-th/0003056.
—— (2000c) "$\mathcal{M}$ Theory as a Matrix Extension of Chern-Simons Theory". *Nucl. Phys. B* 591: 227–42.
—— (2000d) "The Cubic Matrix Model and a Duality between Strings and Loops". ArXiv:hep-th/0006137.
—— (2001a) "The Present Moment in Quantum Cosmology: Challenges to the Arguments for the Elimination of Time". ArXiv:gr-qc/0104097.
—— (2001b) "The Exceptional Jordan Algebra and the Matrix String". ArXiv:hep-th/0104050.
—— (2002) "Quantum Gravity with a Positive Cosmological Constant". ArXiv:hep-th/0209079.
—— (2003) "How Far are we from the Quantum Theory of Gravity?" ArXiv:hep-th/0303185.
—— (2004a) "Scientific Alternatives to the Anthropic Principle". ArXiv:hep-th/0407123.
—— (2004b) "An Invitation to Loop Quantum Gravity". ArXiv:hep-th/0408048.
—— (2005) "Falsifiable Predictions from Semiclassical Quantum Gravity". ArXiv:hep-th/0501091.
—— (forthcoming) "Essay on the Methodology and Ethics of Science". In preparation.
—— (forthcoming) "A Quantization of Topological $\mathcal{M}$ Theory".
Stachel, J. (1989) "Einstein's Search for General Covariance, 1912–15". In D. Howard and J. Stachel (eds.), *Einstein and the History of General Relativity*. Vol. i of *Einstein Studies*. Boston: Birkhäuser (pp. 63–100).
Susskind, L. (1995) "The World as a Hologram". *J. Math. Phys.* 36: 6377.
Thiemann, T. (2001) *Introduction to Modern Canonical Quantum General Relativity*. Cambridge: Cambridge University Press (to appear). ArXiv:gr-qc/0110034.

t'Hooft, G. (1993) "Dimensional Reduction in Quantum Gravity". In A. Alo, J. Ellis, and S. Randjbar-Daemi (eds.), *Salanfestschrift*. Singapore: World Scientific.
Volovik, G. E. (2005) "Emergent Physics on Vacuum Energy and Cosmological Constant". ArXiv:cond-mat/0507454.
Weinberg, S. (2000a) "*A Priori* Probability Distribution of the Cosmological Constant". *Phys. Rev. D* 61: 103505.
____ (2000b) "The Cosmological Constant Problems". ArXiv:astro-ph/0005265.

8

Quantum Quandaries: A Category-Theoretic Perspective

John Baez

ABSTRACT

General relativity may seem very different from quantum theory, but work on quantum gravity has revealed a deep analogy between the two. General relativity makes heavy use of the category nCob, whose objects are $(n-1)$-dimensional manifolds representing 'space' and whose morphisms are n-dimensional cobordisms representing 'spacetime'. Quantum theory makes heavy use of the category Hilb, whose objects are Hilbert spaces used to describe 'states', and whose morphisms are bounded linear operators used to describe 'processes'. Moreover, the categories nCob and Hilb resemble each other far more than either resembles Set, the category whose objects are sets and whose morphisms are functions. In particular, both Hilb and nCob but not Set are $*$-categories with a nonCartesian monoidal structure. We show how this accounts for many of the famously puzzling features of quantum theory: the failure of local realism, the impossibility of duplicating quantum information, and so on. We argue that these features only seem puzzling when we try to treat Hilb as analogous to Set rather than nCob, so that quantum theory will make more sense when regarded as part of a theory of spacetime.

8.1 INTRODUCTION

Faced with the great challenge of reconciling general relativity and quantum theory, it is difficult to know just how deeply we need to rethink basic concepts. By now it is almost a truism that the project of quantizing gravity may force us to modify our ideas about spacetime. Could it also force us to modify our ideas about the quantum? So far this thought has appealed mainly to those who feel uneasy about quantum theory and hope to replace it by something that makes more sense. The problem is that the success and elegance of quantum theory make it hard to imagine promising replacements. Here I would like to propose another possibility, namely that *quantum theory will make more sense when regarded as part of a theory of spacetime.* Furthermore, I claim that *we can only see this from a category-theoretic perspective*—in particular, one that de-emphasizes the primary role of the category of sets and functions.

Part of the difficulty of combining general relativity and quantum theory is that they use different sorts of mathematics: one is based on objects such as manifolds, the other on objects such as Hilbert spaces. As 'sets equipped with extra structure', these look like very different things, so combining them in a single theory has always seemed a bit like trying to mix oil and water. However, work on topological quantum field theory has uncovered a deep analogy between the two. Moreover, this analogy operates at the level of categories.

We shall focus on two categories in this chapter. One is the category Hilb whose objects are Hilbert spaces and whose morphisms are linear operators between these. This plays an important role in quantum theory. The other is the category nCob whose objects are $(n-1)$-dimensional manifolds and whose morphisms are n-dimensional manifolds going between these. This plays an important role in relativistic theories where spacetime is assumed to be n-dimensional: in these theories the objects of nCob represent possible choices of 'space', while the morphisms—called 'cobordisms'—represent possible choices of 'spacetime'.

While an individual manifold is not very much like a Hilbert space, the *category* nCob turns out to have many structural similarities to the *category* Hilb. The goal of this chapter is to explain these similarities and show that the most puzzling features of quantum theory all arise from ways in which Hilb resembles nCob more than the category Set, whose objects are sets and whose morphisms are functions.

Since sets and functions capture many basic intuitions about macroscopic objects, and the rules governing them have been incorporated into the foundations of mathematics, we naturally tend to focus on the fact that any quantum system has a *set* of states. From a Hilbert space we can indeed extract a set of states, namely the set of unit vectors modulo phase. However, this is often more misleading than productive, because this process does not define a well-behaved map—or more precisely, a functor—from Hilb to Set. In some sense the gap between Hilb and Set is too great to be usefully bridged by this trick. However, many of the ways in which Hilb differs from Set are ways in which it resembles nCob! This suggests that the interpretation of quantum theory will become easier, not harder, when we finally succeed in merging it with general relativity.

In particular, it is easy to draw pictures of the objects and morphisms of nCob, at least for low n. Doing so lets us *visualize* many features of quantum theory. This is not really a new discovery: it is implicit in the theory of Feynman diagrams. Whenever one uses Feynman diagrams in quantum field theory, one is secretly working in some category where the morphisms are graphs with labelled edges and vertices, as shown in Figure 8.1.

The precise details of the category depend on the quantum field theory in question: the labels for edges correspond to the various *particles* of the theory, while the labels for vertices correspond to the *interactions* of the theory. Regardless of the details, categories of this sort share many of the structural features of both nCob and Hilb. Their resemblance to nCob, namely their topological nature, makes them a powerful tool for visualization. On the other hand, their relation to Hilb makes them useful in calculations.

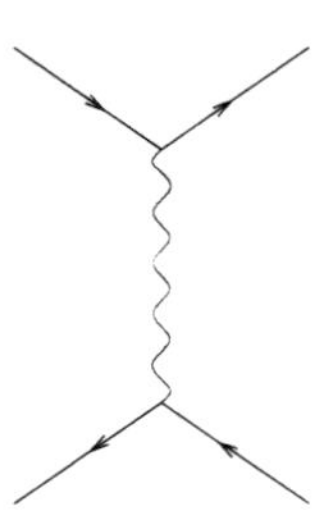

Figure 8.1. A Feynman diagram

Though Feynman diagrams are far from new, the fact that they are morphisms in a category only became appreciated in work on quantum gravity, especially string theory and loop quantum gravity. Both these approaches stretch the Feynman diagram concept in interesting new directions. In string theory, Feynman diagrams are replaced by 'string worldsheets': two-dimensional cobordisms mapped into an ambient spacetime, as shown in Figure 8.2. Since these cobordisms no longer have definite edges and vertices, there are no labels anymore. This is one sense in which the various particles and interactions are all unified in string theory. The realization that processes in string theory could be described as morphisms in a category was crystallized by Segal's definition of 'conformal field theory' (Segal 2004).

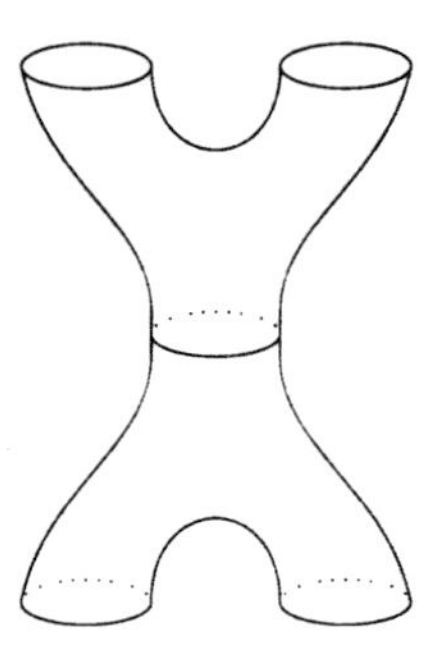

Figure 8.2. A string worldsheet

Loop quantum gravity is moving towards a similar picture, though with some important differences. In this approach processes are described by 'spin foams'. These are a two-dimensional generalization of Feynman diagrams built from vertices, edges, and faces, as shown in Figure 8.3. They are not mapped into an ambient spacetime: in this approach spacetime is nothing but the spin foam itself—or more precisely, a linear combination of spin foams. Particles and interactions are not 'unified' in these models, so there are labels on the vertices, edges, and faces, which depend on the details of the model in question. The category-theoretic underpinnings of spin foam models were explicit from the very beginning (Baez 1998), since they were developed after Segal's work on conformal field theory, and also after Atiyah's

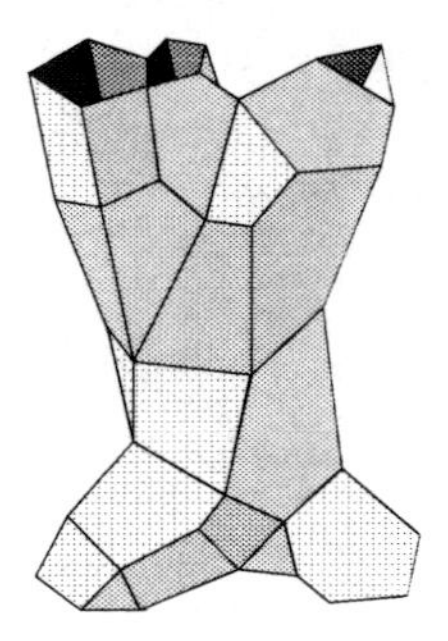

Figure 8.3. A spin foam

work on topological quantum field theory (Atiyah 1989), which exhibits the analogy between nCob and Hilb in its simplest form.

There is not one whit of experimental evidence for either string theory or loop quantum gravity, and both theories have some serious problems, so it might seem premature for philosophers to consider their implications. It indeed makes little sense for philosophers to spend time chasing every short-lived fad in these fast-moving subjects. Instead, what is worthy of reflection is that these two approaches to quantum gravity, while disagreeing heatedly on so many issues (Smolin 2001; Vaas 2004), have so much in common. It suggests that in our attempts to reconcile the quantum-theoretic notions of *state* and *process* with the relativistic notions of *space* and *spacetime*, we have a limited supply of promising ideas. It is an open question whether these ideas will be up to the task of describing nature. But this actually makes it more urgent, not less, for philosophers to clarify and question these ideas and the implicit assumptions upon which they rest.

Before plunging ahead, let us briefly sketch the contents of this paper. In §8.2 we explain the analogy between nCob and Hilb by recalling Atiyah's definition of 'topological quantum field theory', or 'TQFT' for short. In §8.3, we begin by noting that unlike many familiar categories, neither Hilb nor nCob is best regarded as a category whose objects are sets equipped with extra structures and properties, and whose morphisms are functions preserving these extra structures. In particular, operators between Hilbert spaces are not required to preserve the inner product. This raises the question of precisely what role the inner product plays in the category Hilb. Of course the inner product is crucial in quantum theory, since we use it to compute transition amplitudes between states—but how does it manifest itself mathematically in the structure of Hilb? One answer is that it gives a way to 'reverse' an operator $T: H \to H'$, obtaining an operator $T^*: H' \to H$ called the 'adjoint' of T such that

$$\langle T^*\phi, \psi\rangle = \langle \phi, T\psi\rangle$$

for all $\psi \in H$ and $\phi \in H'$. This makes Hilb into something called a '$*$-category': a category where there is a built-in way to reverse any process. As we shall see, it is easy to compute transition amplitudes using the $*$-category structure of Hilb. The category nCob is also a $*$-category, where the adjoint of a spacetime is obtained simply by switching the roles of future and past. On the other hand, Set cannot be made into a $*$-category. All this suggests that both quantum theory and general relativity will be best understood in terms of categories quite different from the category of sets and functions.

In §8.4 we tackle some of the most puzzling features of quantum theory, namely those concerning joint systems: physical systems composed of two parts. It is in the study of joint systems that one sees the 'failure of local realism' that worried Einstein so terribly (Einstein et al. 1935), and was brought into clearer focus by Bell (1964). Here is also where one discovers that one 'cannot clone a quantum state'—a result due to Wooters and Zurek (1982) which serves as the basis of quantum cryptography. As explained in §8.4, both these phenomena follow from the failure of the tensor product to be 'Cartesian' in a certain sense made precise by category theory. In Set, the usual product of sets is Cartesian, and this encapsulates many of our usual intuitions about ordered pairs, like our ability to pick out the components a and b of any pair (a, b), and our ability to 'duplicate' any element a to obtain a pair (a, a). The fact that we cannot do these things in Hilb is responsible for the failure of local realism and the impossibility of duplicating a quantum state. Here again the category Hilb resembles nCob more than Set. Like Hilb, the category nCob has a non-Cartesian tensor product, given by the disjoint union of manifolds. Some of the mystery surrounding joint systems in quantum theory dissipates when one focuses on the analogy to nCob and stops trying to analogize the tensor product of Hilbert spaces to the Cartesian product of sets.

This chapter is best read as a follow-up to my paper 'Higher-Dimensional Algebra and Planck-Scale Physics' (2001), since it expands on some of the ideas already on touched upon there.

8.2 LESSONS FROM TOPOLOGICAL QUANTUM FIELD THEORY

Thanks to the influence of general relativity, there is a large body of theoretical physics that does not presume a fixed topology for space or spacetime. The idea is that after having assumed that spacetime is n-dimensional, we are in principle free to choose any $(n - 1)$-dimensional manifold to represent space at a given time. Moreover, given two such manifolds, say S and S', we are free to choose any n-dimensional manifold-with-boundary, say M, to represent the portion of spacetime between them, so long as the boundary of M is the union of S and S'. In this situation we write $M: S \to S'$, even though M is not a function from S to S', because we may think of M as the process of time passing from the moment S to the moment S'. Mathematicians call M a **cobordism** from S to S'. For example, in Figure 8.4 we depict a two-dimensional manifold M going from a one-dimensional manifold S (a pair of circles) to a one-dimensional manifold S' (a single circle). Physically, this cobordism represents a process in which two separate spaces collide to form a single one! This is an example of 'topology change'.

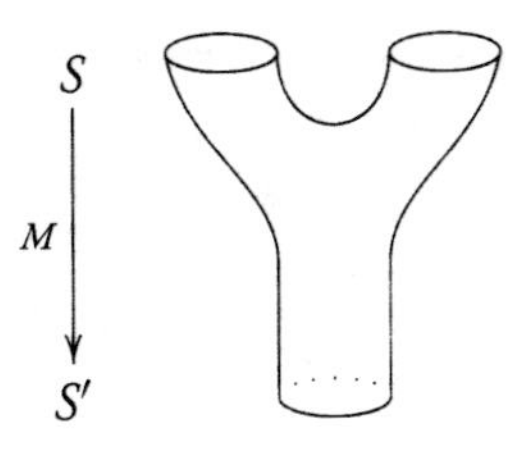

Figure 8.4. A cobordism

All this has a close analogue in quantum theory. First, just as we can use any $(n-1)$-manifold to represent space, we can use any Hilbert space to describe the states of some quantum system. Second, just as we can use any cobordism to represent a spacetime going from one space to another, we can use any operator to describe a process taking states of one system to states of another. More precisely, given systems whose states are described using the Hilbert spaces H and H', respectively, any bounded linear operator $T: H \to H'$ describes a process that carries states of the first system to states of the second. We are most comfortable with this idea when the operator T is unitary, or at least an isometry. After all, given a state described as a unit vector $\psi \in H$, we can only be sure $T\psi$ is a unit vector in H' if T is an isometry. So, only in this case does T define a *function* from the *set* of states of the first system to the *set* of states of the second. However, the interpretation of linear operators as processes makes sense more generally. One way to make this interpretation precise is as follows: given a unit vector $\psi \in H$ and an orthonormal basis ϕ_i of H', we declare that the **relative probability** for a system prepared in the state ψ to be observed in the state ϕ_i after undergoing the process T is $|\langle\phi_i, T\psi\rangle|^2$. By this, we mean that the probability of observing the system in the ith state divided by the probability of observing it in the jth state is

$$\frac{|\langle\phi_i, T\psi\rangle|^2}{|\langle\phi_j, T\psi\rangle|^2}.$$

The use of non-unitary operators to describe quantum processes is not new. For example, projection operators have long been used to describe processes like sending a photon through a polarizing filter. However, these examples traditionally arise when we treat part of the system (e.g. the measuring apparatus) classically. It is often assumed that at a fundamental level, the laws of nature in quantum theory describe time evolution using unitary operators. But as we shall see in §8.3, this assumption should be dropped in theories where the topology of space can change. In such theories we should let *all* the morphisms in Hilb qualify as 'processes', just as we let *all* morphisms in nCob qualify as spacetimes.

Having clarified this delicate point, we are now in a position to clearly see a structural analogy between general relativity and quantum theory, in which $(n-1)$-dimensional manifolds representing space are analogous to Hilbert spaces, while cobordisms describing spacetime are analogous to operators. Indulging in some lofty

Table 8.1. Analogy between general relativity and quantum theory

General relativity	Quantum theory
$(n-1)$-dimensional manifold	Hilbert space
(space)	(states)
cobordism between $(n-1)$-dimensional manifolds	operator between Hilbert spaces
(spacetime)	(process)
composition of cobordisms	composition of operators
identity cobordism	identity operator

rhetoric, we might say that *space* and *state* are aspects of *being*, while *spacetime* and *process* are aspects of *becoming*. We summarize this analogy in Table 8.1.

This analogy becomes more than mere rhetoric when applied to topological quantum field theory (Baez 2001). In quantum field theory on curved spacetime, space and spacetime are not just manifolds: they come with fixed 'background metrics' that allow us to measure distances and times. In this context, S and S' are Riemannian manifolds, and $M: S \to S'$ is a **Lorentzian cobordism** from S to S': that is, a Lorentzian manifold with boundary whose metric restricts at the boundary to the metrics on S and S'. However, topological quantum field theories are an attempt to do background-free physics, so in this context we drop the background metrics: we merely assume that space is an $(n-1)$-dimensional manifold and spacetime is a cobordism between such manifolds. A topological quantum field theory then consists of a map Z assigning a Hilbert space of states $Z(S)$ to any $(n-1)$-manifold S and a linear operator $Z(M): Z(S) \to Z(S')$ to any cobordism between such manifolds. This map cannot be arbitrary, though: for starters, it must be a *functor* from the category of n-dimensional cobordisms to the category of Hilbert spaces. This is a great example of how every sufficiently good analogy is yearning to become a functor.

However, we are getting a bit ahead of ourselves. Before we can talk about functors, we must talk about categories. What is the category of n-dimensional cobordisms, and what is the category of Hilbert spaces? The answers to these questions will allow us to make the analogy in Table 8.1 much more precise.

First, recall that a **category** consists of a collection of objects, a collection of morphisms $f: A \to B$ from any object A to any object B, a rule for composing morphisms $f: A \to B$ and $g: B \to C$ to obtain a morphism $gf: A \to C$, and for each object A an identity morphism $1_A : A \to A$. These must satisfy the associative law $f(gh) = (fg)h$ and the left and right unit laws $1_A f = f$ and $f 1_A = f$ whenever these composites are defined. In many cases, the objects of a category are best thought of as *sets equipped with extra structure*, while the morphisms are *functions preserving the extra structure*. However, this is true neither for the category of Hilbert spaces nor for the category of cobordisms.

In the category Hilb we take the objects to be Hilbert spaces and the morphisms to be bounded linear operators. Composition and identity operators are defined as usual. Hilbert spaces are indeed sets equipped with extra structure, but bounded linear operators do not preserve all this extra structure: in particular, they need not preserve the inner product. This may seem like a fine point, but it is important, and we shall explore its significance in detail in §8.3.

In the category nCob we take the objects to be $(n-1)$-dimensional manifolds and the morphisms to be cobordisms between these. (For technical reasons mathematicians usually assume both to be compact and oriented.) Here the morphisms are not functions at all! Nonetheless we can 'compose' two cobordisms $M\colon S \to S'$ and $M'\colon S' \to S''$, obtaining a cobordism $M'M\colon S \to S''$, as in Figure 8.5. The idea here is that the passage of time corresponding to M followed by the passage of time corresponding to M' equals the passage of time corresponding to $M'M$. This is analogous to the familiar idea that waiting t seconds followed by waiting t' seconds is the same as waiting $t' + t$ seconds. The big difference is that in topological quantum field theory we cannot measure time in seconds, because there is no background metric available to let us count the passage of time. We can only keep track of topology change. Just as ordinary addition is associative, composition of cobordisms satisfies the associative law:

$$(M''M')M = M''(M'M).$$

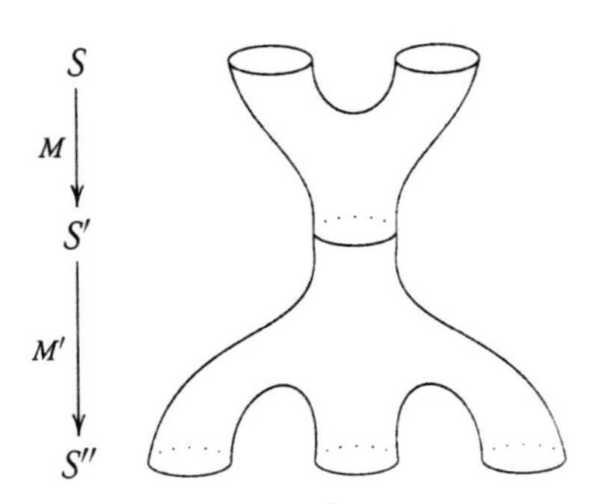

Figure 8.5. Composition of cobordisms

Furthermore, for any $(n-1)$-dimensional manifold S representing space, there is a cobordism $1_S\colon S \to S$ called the 'identity' cobordism, which represents a passage of time during which the topology of space stays constant. For example, when S is a circle, the identity cobordism 1_S is a cylinder, as shown in Figure 8.6. In general, the identity cobordism 1_S has the property that

$$1_S M = M$$

and

$$M 1_S = M$$

whenever these composites are defined. These properties say that an identity cobordism is analogous to waiting 0 seconds: if you wait 0 seconds and then wait t more seconds, or wait t seconds and then wait 0 more seconds, this is the same as waiting t seconds.

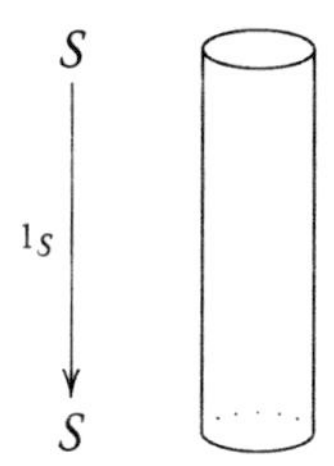

Figure 8.6. An identity cobordism

A **functor** between categories is a map sending objects to objects and morphisms to morphisms, preserving composition and identities. Thus, in saying that a topological quantum field theory is a functor

$$Z\colon n\mathrm{Cob} \to \mathrm{Hilb},$$

we merely mean that it assigns a Hilbert space of states $Z(S)$ to any $(n-1)$-dimensional manifold S and a linear operator $Z(M)\colon Z(S) \to Z(S')$ to any n-dimensional cobordism $M\colon S \to S'$ in such a way that:

- For any n-dimensional cobordisms $M\colon S \to S'$ and $M'\colon S' \to S''$,

$$Z(M'M) = Z(M')Z(M).$$

- For any $(n-1)$-dimensional manifold S,

$$Z(1_S) = 1_{Z(S)}.$$

Both these axioms make sense if one ponders them a bit. The first says that the passage of time corresponding to the cobordism M followed by the passage of time corresponding to M' has the same effect on a state as the combined passage of time corresponding to $M'M$. The second says that a passage of time in which no topology change occurs has no effect at all on the state of the universe. This seems paradoxical at first, since it seems we regularly observe things happening even in the absence of topology change. However, this paradox is easily resolved: a topological quantum field theory describes a world without local degrees of freedom. In such a world, nothing local happens, so the state of the universe can only change when the topology of space itself changes.

Unless elementary particles are wormhole ends or some other sort of topological phenomenon, it seems our own world is quite unlike this. Thus, we hasten to reassure the reader that this peculiarity of topological quantum field theory is *not* crucial to our overall point, which is the analogy between categories describing space and spacetime and those describing quantum states and processes. If we were doing quantum field theory on curved spacetime, we would replace nCob with a category where the objects are n-dimensional Riemannian manifolds and most of the morphisms are Lorentzian cobordisms between these. In this case a cobordism $M\colon S \to S'$ has not just a topology but also a geometry, so we can use cylinder-shaped cobordisms of different 'lengths' to describe time evolution for different amounts of time. The identity morphism is then described by a cylinder of 'length zero'. This degenerate cylinder is not really a Lorentzian cobordism, which leads to some technical complications. However, Segal showed how to get around these in his axioms for a conformal field theory (Segal 2004). There are some further technical complications arising from the fact that, except in low dimensions, we need to use the C*-algebraic approach to quantum theory, instead of the Hilbert space approach (Arageorgis et al. 2002). Here the category Hilb should be replaced by one where the objects are C*-algebras and the morphisms are completely positive maps between their duals (Hawkins et al. 2003).

Setting aside these nuances, our main point is that treating a TQFT as a functor from nCob to Hilb is a way of making very precise some of the analogies between general relativity and quantum theory. However, we can go further! A TQFT is more than just a functor. It must also be compatible with the 'monoidal category' structure of nCob and Hilb, and to be physically well behaved it must also be compatible with their '$*$-category' structure. We examine these extra structures in the next two sections.

8.3 THE $*$-CATEGORY OF HILBERT SPACES

What is the category of Hilbert spaces? While we have already given an answer, this is actually a tricky question, one that makes many category theorists acutely uncomfortable.

To understand this, we must start by recalling that one use of categories is to organize discourse about various sorts of 'mathematical objects': groups, rings, vector spaces, topological spaces, manifolds, and so on. Quite commonly these mathematical objects are sets equipped with extra structure and properties, so let us restrict attention to this case. Here by **structure** we mean operations and relations defined on the set in question, while by **properties** we mean axioms that these operations and relations are required to satisfy. The division into structure and properties is evident from the standard form of mathematical definitions such as 'a widget is a set equipped with ... such that ...' Here the structures are listed in the first blank, while the properties are listed in the second.

To build a category of this sort of mathematical object, we must also define morphisms between these objects. When the objects are *sets equipped with extra structure and properties*, the morphisms are typically taken to be *functions that preserve the extra structure*. At the expense of a long digression we could make this completely precise—and also more general, since we can also build categories by equipping not sets but objects of other categories with extra structure and properties. However, we prefer to illustrate the idea with an example. We take an example closely related to but subtly different from the category of Hilbert spaces: the category of complex vector spaces.

A **complex vector space** is a set V equipped with extra structure consisting of operations called addition

$$+ : V \times V \to V$$

and scalar multiplication

$$\cdot : \mathbb{C} \times V \to V,$$

which in turn must have certain extra properties: commutativity and associativity together with the existence of an identity and inverses for addition, associativity and the unit law for scalar multiplication, and distributivity of scalar multiplication over addition. Given vector spaces V and V', a **linear operator** $T\colon V \to V'$ can be defined as a function preserving all the extra structure. This means that we require

$$T(\psi + \phi) = T(\psi) + T(\phi)$$

and

$$T(c\psi) = cT(\psi)$$

for all $\psi, \phi \in V$, and $c \in \mathbb{C}$. Note that the properties do not enter here. Mathematicians define the category Vect to have complex vector spaces as its objects and linear operators between them as its morphisms.

Now compare the case of Hilbert spaces. A Hilbert space H is a set equipped with all the structure of a complex vector space but also some more, namely an inner product

$$\langle \cdot, \cdot \rangle \colon H \times H \to \mathbb{C}.$$

Similarly, it has all the properties of a complex vector spaces but also some more: for all $\phi, \psi, \psi' \in H$ and $c \in \mathbb{C}$ we have the equations

$$\langle \phi, \psi + \psi' \rangle = \langle \phi, \psi \rangle + \langle \phi, \psi' \rangle,$$
$$\langle \phi, c\psi \rangle = c\langle \phi, \psi \rangle,$$
$$\langle \phi, \psi \rangle = \overline{\langle \psi, \phi \rangle},$$

together with the inequality

$$\langle \psi, \psi \rangle \geq 0$$

where equality holds only if $\psi = 0$; furthermore, the norm defined by the inner product must be complete. Given Hilbert spaces H and H', a function $T\colon H \to H'$ that preserves all the structure is thus a linear operator that preserves the inner product:

$$\langle T\phi, T\psi \rangle = \langle \phi, \psi \rangle$$

for all $\phi, \psi \in H$. Such an operator is called an **isometry**.

If we followed the pattern that works for vector spaces and many other mathematical objects, we would thus define the category Hilb to have Hilbert spaces as objects and isometries as morphisms. However, this category seems too constricted to suit what physicists actually do with Hilbert spaces: they frequently need operators that aren't isometries! Unitary operators are always isometries, but self-adjoint operators, for example, are not.

The alternative we adopt in this chapter is to work with the category Hilb whose objects are Hilbert spaces and the morphisms are bounded linear operators. However, this leads to a curious puzzle. In a precise technical sense, the category of finite-dimensional Hilbert spaces and linear operators between these is *equivalent* to the category of finite-dimensional complex vector spaces and linear operators. So, in defining this category, we might as well ignore the inner product entirely! The puzzle is thus what role, if any, the inner product plays in this category.

The case of general, possibly infinite-dimensional Hilbert spaces is subtler, but the puzzle persists. The category of all Hilbert spaces and bounded linear operators between them is *not* equivalent to the category of all complex vector spaces and linear operators. However, it *is* equivalent to the category of 'Hilbertizable' vector spaces—that is, vector spaces equipped with a topology coming from *some* Hilbert space structure—and continuous linear operators between these. So, in defining this category, what matters is not the inner product but merely the topology it gives rise to. The point is that bounded linear operators don't preserve the inner product, just the topology, and a structure that is not preserved might as well be ignored, as far as the category is concerned.

My resolution of this puzzle is simple but a bit upsetting to most category theorists. I admit that the inner product is inessential in defining the category of Hilbert spaces and bounded linear operators. However, I insist that it plays a crucial role in making this category into a $*$-category!

What is a $*$-category? It is a category C equipped with a map sending each morphism $f: X \to Y$ to a morphism $f^*: Y \to X$, satisfying

$$1_X^* = 1_X,$$
$$(fg)^* = g^* f^*,$$

and

$$f^{**} = f.$$

To make Hilb into a $*$-category we define T^* for any bounded linear operator $T: H \to H'$ to be the **adjoint** operator $T^* H' \to H$, given by

$$\langle T^* \psi, \phi \rangle = \langle \psi, T\phi \rangle.$$

We see by this formula that the inner products on both H and H' are required to define the adjoint of T.

In fact, we can completely recover the inner product on every Hilbert space from the $*$-category structure of Hilb. Given a Hilbert space H and a vector $\psi \in H$, there is a unique operator $T_\psi: \mathbb{C} \to H$ with $T_\psi(1) = \psi$. Conversely, any operator from $\mathbb{C}$ to H determines a unique vector in H this way. So, we can think of elements of a Hilbert space as morphisms from $\mathbb{C}$ to this Hilbert space. Using this trick, an easy calculation shows that

$$\langle \phi, \psi \rangle = T_\phi^* T_\psi.$$

The right-hand side is really a linear operator from $\mathbb{C}$ to $\mathbb{C}$, but there is a canonical way to identify such a thing with a complex number. So, everything about inner products is encoded in the $*$-category structure of Hilb. Moreover, this way of thinking about the inner product formalizes an old idea of Dirac. The operator T_ψ is really just Dirac's 'ket' $|\psi\rangle$, while T_ϕ^* is the 'bra' $\langle\phi|$. Composing a ket with a bra, we get the inner product.

This shows how adjoints are closely tied to the inner product structure on Hilbert space. But what is the physical significance of the adjoint of an operator, or more generally the $*$ operation in any $*$-category? Most fundamentally, the $*$ operation gives us a way to 'reverse' a morphism even when it is not invertible. If we think of inner products as giving transition amplitudes between states in quantum theory, the equation $\langle T^*\phi, \psi\rangle = \langle \phi, T\psi\rangle$ says that T^* is the unique operation we can perform on any state ϕ so that the transition amplitude from $T\psi$ to ϕ is the same as that from ψ to $T^*\phi$. So, in a suggestive but loose way, we can say that the process described by T^* is some sort of 'time-reversed' version of the process described by T. If T is unitary, T^* is just the inverse of T. But, T^* makes sense even when T has no inverse!

This suggestive but loose relation between $*$ operations and time reversal becomes more precise in the case of nCob. Here the $*$ operation really *is* time reversal. More precisely, given an n-dimensional cobordism $M: S \to S'$, we let the **adjoint** cobordism $M^*: S' \to S$ to be the same manifold, but with the 'past' and 'future' parts of its boundary switched, as in Figure 8.7. It is easy to check that this makes nCob into a $*$-category.

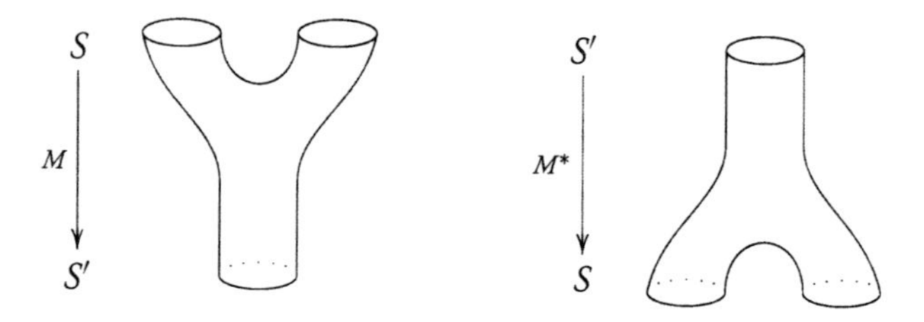

Figure 8.7. A cobordism and its adjoint

In a so-called **unitary** topological quantum field theory (the terminology is a bit unfortunate), we demand that the functor $Z: n\mathrm{Cob} \to \mathrm{Hilb}$ preserve the $*$-category structure in the following sense:

$$Z(M^*) = Z(M)^*.$$

All the TQFTs of interest in physics have this property, and a similar property holds for conformal field theories and other quantum field theories on curved spacetime. This means that in the analogy between general relativity and quantum theory, *the analogue of time reversal is taking the adjoint of an operator between Hilbert spaces*. To 'reverse' a spacetime $M: S \to S'$ we formally switch the notions of future and past, while to 'reverse' a process $T: H \to H'$ we take its adjoint.

Taking this analogy seriously leads us in some interesting directions. First, since the $*$ operation in nCob is given by time reversal, while $*$ operation in Hilb is defined using the inner product, there should be some relation between time reversal and the inner product in quantum theory! The details remain obscure, at least to me, but we can make a little progress by pondering the following equation, which we originally introduced as a 'trick' for expressing inner products in terms of adjoint operators:

$$\langle \phi, \psi \rangle = T_\phi^* T_\psi.$$

An equation this important should not be a mere trick! To try to interpret it, suppose that in some sense the operator T_ψ describes 'the process of preparing the system to be in the state ψ', while T_ϕ^* describes the process of 'observing the system to be in the state ϕ'. Given this, $T_\phi^* T_\psi$ should describe the process of first preparing the system to be in the state ψ and then observing it to be in the state ϕ. The above equation then relates this composite process to the transition amplitude $\langle \phi, \psi \rangle$. Moreover, we see that 'observation' is like a time-reversed version of 'preparation'. All this makes a rough intuitive kind of sense. However, these ideas could use a great deal of elaboration and clarification. I mention them here mainly to open an avenue for further thought.

Second, and less speculatively, the equation $Z(M^*) = Z(M)^*$ sheds some light on the relation between topology change and the failure of unitarity, mentioned already in §8.2. In any $*$-category, we may define a morphism $f: x \to y$ to be **unitary** if $f^*f = 1_x$ and $ff^* = 1_y$. For a morphism in Hilb this reduces to the usual definition of unitarity for a linear operator. One can show that a morphism M in nCob is unitary if M involves no topology change, or more precisely, if M is diffeomorphic to the Cartesian product of an interval and some $(n-1)$-dimensional manifold. (The converse is true in dimensions $n \leq 3$, but it fails in higher dimensions.) A TQFT satisfying $Z(M^*) = Z(M)^*$ maps unitary morphisms in nCob to unitary morphisms in Hilb, so for TQFTs of this sort, *absence of topology change implies unitary time evolution.* This fact reinforces a point already well known from quantum field theory on curved spacetime, namely that unitary time evolution is not a built-in feature of quantum theory but rather the consequence of specific assumptions about the nature of spacetime (Arageorgis et al. 2002).

To conclude, it is interesting to contrast nCob and Hilb with the more familiar category Set, whose objects are sets and whose morphisms are functions. There is no way to make Set into a $*$-category, since there is no way to 'reverse' the map from the empty set to the one-element set. So, our intuitions about sets and functions help us very little in understanding $*$-categories. The problem is that the concept of function is based on an intuitive notion of process that is asymmetrical with respect to past and future: a function $f: S \to S'$ is a relation such that each element of S is related to exactly one element of S', but not necessarily vice versa. For better or worse, this built-in 'arrow of time' has no place in the basic concepts of quantum theory.

Pondering this, it soon becomes apparent that if we want an easy example of a $*$-category other than Hilb to help build our intuitions about $*$-categories, we should use not Set but Rel, the category of sets and *relations*. In fact, quantum theory can be seen as a modified version of the theory of relations in which Boolean algebra has been replaced by the algebra of complex numbers! To see this, note that a linear operator between two Hilbert spaces can be described using a matrix of complex numbers as soon as we pick an orthonormal basis for each. Similarly, a relation R

between sets S and S' can be described by a matrix of truth values, namely the truth values of the propositions yRx where $x \in S$ and $y \in S'$. Composition of relations can be defined as matrix multiplication with 'or' and 'and' playing the roles of 'plus' and 'times'. It is easy to check that this is associative and has an identity morphism for each set, so we obtain a category Rel with sets as objects and relations as morphisms. Furthermore, Rel becomes a $*$-category if we define the relation R^* by saying that xR^*y if and only if yRx. Just as the matrix for the linear operator T^* is the conjugate transpose of the matrix for T, the matrix for the relation R^* is the transpose of the matrix for R.

So, the category of Hilbert spaces closely resembles the category of relations. The main difference is that binary truth values describing whether or not a transition is possible are replaced by complex numbers describing the amplitude with which a transition occurs. Comparisons between Hilb and Rel are a fruitful source of intuitions not only about $*$-categories in general but also about the meaning of 'matrix mechanics'. For some further explorations along these lines, see the work of Abramsky and Coecke (2004).

8.4 THE MONOIDAL CATEGORY OF HILBERT SPACES

An important goal of the enterprise of physics is to describe, not just one physical system at a time, but also how a large complicated system can be built out of smaller simpler ones. The simplest case is a so-called 'joint system': a system built out of two separate parts. Our experience with the everyday world leads us to believe that to specify the state of a joint system, it is necessary and sufficient to specify states of its two parts. (Here and in what follows, by 'states' we always mean what physicists call 'pure states'.) In other words, a state of the joint system is just an ordered pair of states of its parts. So, if the first part has S as its set of states, and the second part has T as its set of states, the joint system has the Cartesian product $S \times T$ as its set of states.

One of the more shocking discoveries of the twentieth century is that this is *wrong*. In both classical and quantum physics, given states of each part we get a state of the joint system. But only in classical physics is every state of the joint system of this form! In quantum physics are also 'entangled' states, which can only be described as *superpositions* of states of this form. The reason is that in quantum theory, the states of a system are no longer described by a set, but by a Hilbert space. Moreover—and this is really an extra assumption—the states of a joint system are described not by the Cartesian product of Hilbert spaces, but by their tensor product.

Quite generally, we can imagine using objects in any category to describe physical systems, and morphisms between these to describe processes. In order to handle joint systems, this category will need to have some sort of 'tensor product' that gives an object $A \otimes B$ for any pair of objects A and B. As we shall explain, categories of this sort are called 'monoidal'. The category Set is an example where the tensor product is just the usual Cartesian product of sets. Similarly, the category Hilb is a monoidal category where the tensor product is the usual tensor product of Hilbert spaces. However, these two examples are very different, because the product in Set is 'Cartesian' in a certain technical sense, while the product in Hilb is not. This turns out to explain a lot about why joint systems behave so counter-intuively in quantum physics. Moreover, it is yet another way in which Hilb resembles nCob more than Set.

To see this in detail, it pays to go back to the beginning and think about Cartesian products. Given two sets S and T, we define $S \times T$ to be the set of all ordered pairs (s, t) with $s \in S$ and $t \in T$. But what is an ordered pair? This depends on our approach to set theory. We can use axioms in which ordered pairs are a primitive construction, or we can define them in terms of other concepts. For example, in 1914, Wiener defined the ordered pair (s, t) to be the set $\{\{\{s\}, \emptyset\}, \{t\}\}$. In 1922, Kuratowski gave the simpler definition $(s, t) = \{\{s\}, \{s, t\}\}$. We can use the still simpler definition $(s, t) = \{s, \{s, t\}\}$ if our axioms exclude the possibility of sets that contain themselves. Various other definitions have also been tried. In traditional set theory we arbitrarily choose one approach to ordered pairs and then stick with it. Apart from issues of convenience or elegance, it does not matter which we choose, so long as it 'gets the job done'. In other words, all these approaches are all just technical tricks for implementing our goal, which is to make sure that $(s, t) = (s', t')$ if and only if $s = s'$ and $t = t'$.

It is a bit annoying that the definition of ordered pair cannot get straight to the point and capture the concept without recourse to an arbitrary trick. It is natural to seek an approach that focuses more on the *structural role* of ordered pairs in mathematics and less on their *implementation*. This is what category theory provides.

The reason traditional set theory arbitrarily chooses a specific implementation of the ordered pair concept is that it seems difficult to speak precisely about 'some thing (s, t)—I don't care what it is—with the property that $(s, t) = (s', t')$ iff $s = s'$ and $t = t'$'. So, the first move in category theory is to stop focusing on ordered pairs and instead focus on Cartesian products of sets. What properties should the Cartesian product $S \times T$ have? To make our answer applicable not just to sets but to objects of other categories, it should not refer to elements of $S \times T$. So, the second move in category theory is to describe the Cartesian product $S \times T$ in terms of functions to and from this set.

The Cartesian product $S \times T$ has functions called 'projections' to the sets S and T:

$$p_1 \colon S \times T \to S, \qquad p_2 \colon S \times T \to T.$$

Secretly we know that these pick out the first or second component of any ordered pair in $S \times T$:

$$p_1(s, t) = s, \qquad p_2(s, t) = t.$$

But, our goal is to characterize the product by means of these projections without explicit reference to ordered pairs. For this, the key property of the projections is that given any element $s \in S$ and any element $t \in T$, there exists a unique element $x \in S \times T$ such that $p_1(x) = s$ and $p_2(x) = T$. Furthermore, as a substitute for elements of the sets S and T, we can use functions from an arbitrary set to these sets.

Thus, given two sets S and T, we define their **Cartesian product** to be any set $S \times T$ equipped with functions $p_1 : S \times T \to S$, $p_2 : S \times T \to T$ such that for any set X and functions $f_1 : X \to S$, $f_2 : X \to T$, there exists a unique function $f : X \to S \times T$ with

$$f_1 = p_1 f, \qquad f_2 = p_2 f.$$

Note that with this definition, the Cartesian product is not unique! Wiener's definition of ordered pairs gives a Cartesian product of the sets S and T, but so does Kuratowski's, and so does any other definition that 'gets the job done'. However, this does not lead to any confusion, since one can easily show that any two choices of Cartesian product are isomorphic in a canonical way. For a proof of this and other facts about Cartesian products, see for example the textbook by McLarty (1995).

All this generalizes painlessly to an arbitrary category. Given two objects A and B in some category, we define their Cartesian product (or simply **product**) to be any object $A \times B$ equipped with morphisms

$$p_1 : A \times B \to A, \qquad p_2 : A \times B \to B,$$

called **projections**, such that for any object X and morphisms $f_1 : X \to A, f_2 : X \to B$, there is a unique morphism $f : X \to A \times B$ with $f_1 = p_1 f$ and $f_2 = p_2 f$. The product may not exist, and it may not be unique, but it is unique up to a canonical isomorphism. Category theorists therefore feel free to speak of 'the' product when it exists.

We say a category has **binary products** if every pair of objects has a product. One can also talk about n-ary products for other values of n, but a category with binary products has n-ary products for all $n \geq 1$, since we can construct these as iterated binary products. The case $n = 1$ is trivial, since the product of one object is just that object itself (up to canonical isomorphism). The only remaining case is $n = 0$. This is surprisingly important. A 0-ary product is usually called a **terminal object** and denoted 1: it is an object such that that for any object X there exists a unique morphism from X to 1. Terminal objects are unique up to canonical isomorphism, so we feel free to speak of 'the' terminal object in a category when one exists. The reason we denote the terminal object by 1 is that in Set, any set with one element is a terminal object. If a category has a terminal object and binary products, it has n-ary products for all n, so we say it has **finite products**.

It turns out that these concepts capture much of our intuition about joint systems in classical physics. In the most stripped-down version of classical physics, the states of a system are described as elements of a mere *set*. In more elaborate versions, the states of a system form an object in some fancier category, such as the category of *topological spaces* or *manifolds*. But, just like Set, these fancier categories have finite products—and we use this fact when describing the states of a joint system.

To sketch how this works in general, suppose we have any category with finite products. To do physics with this, we think of any of the objects of this category as describing some physical system. It sounds a bit vague to say that a physical system is 'described by' some object A, but we can make this more precise by saying that states of this system are morphisms $f: 1 \to A$. When our category is Set, a morphism of this sort simply picks out an element of the set A. In the category of topological spaces, a morphism of this sort picks out a point in the topological space A—and similarly for the category of manifolds, and so on. For this reason, category theorists call a morphism $f: 1 \to A$ an **element** of the object A.

Next, we think of any morphism $g: A \to B$ as a 'process' carrying states of the system described by A to states of the system described by B. This works as follows: given a state of the first system, say $f: 1 \to A$, we can compose it with g to get a state of the second system, $gf: 1 \to B$.

Then, given two systems that are described by the objects A and B, respectively, we decree that the joint system built from these is described by the object $A \times B$. The projection $p_1: A \times B \to A$ can be thought of as a process that takes a state of the joint system and discards all information about the second part, retaining only the state of the first part. Similarly, the projection p_2 retains only information about the second part.

Calling these projections 'processes' may strike the reader as strange, since 'discarding information' sounds like a subjective change of our *description* of the system, rather than an objective physical process like time evolution. However, it is worth noting that in special relativity, time evolution corresponds to a change of coordinates $t \mapsto t + c$, which can also be thought of as change of our description of the system. The novelty in thinking of a projection as a physical process really comes, not from the fact that it is 'subjective', but from the fact that it is not invertible.

With this groundwork laid, we can use the definition of 'product' to show that a state of a joint system is just an ordered pair of states of each part. First suppose we have states of each part, say $f_1: 1 \to A$ and $f_2: 1 \to B$. Then there is a unique state of the joint system, say $f: 1 \to A \times B$, which reduces to the given state of each part when we discard information about the other part: $p_1 f = f_1$ and $p_2 f = f_2$. Conversely, every state of the joint system arises this way, since given $f: 1 \to A \times B$ we can recover f_1 and f_2 using these equations.

However, the situation changes drastically when we switch to quantum theory! The states of a quantum system can still be thought of as forming a set. However, we do not take the product of these sets to be the set of states for a joint quantum system. Instead, we describe states of a system as unit vectors in a Hilbert space, modulo phase. We define the Hilbert space for a joint system to be the *tensor product* of the Hilbert spaces for its parts.

The tensor product of Hilbert spaces is not a Cartesian product in the sense defined above, since given Hilbert spaces H and K there are no linear operators $p_1: H \otimes K \to H$ and $p_2: H \otimes K \to K$ with the required properties. This means that from a (pure) state of a joint quantum system we cannot extract (pure) states of its parts. This is the key to Bell's 'failure of local realism'. Indeed, under quite general conditions one can derive Bell's inequality from the assumption that pure states of a joint system determine pure states of its parts, so violations of Bell's inequality should be seen as an indication that this assumption fails.

The Wooters–Zurek argument that 'one cannot clone a quantum state' (1982) is also based on the fact that the tensor product of Hilbert spaces is not Cartesian. To get some sense of this, note that whenever A is an object in some category for which the product $A \times A$ exists, there is a unique morphism

$$\Delta : A \to A \times A$$

such that $p_1 \Delta = 1_A$ and $p_2 \Delta = 1_A$. This morphism is called the **diagonal** of A, since in the category of sets it is the map given by $\Delta(a) = (a, a)$ for all $a \in A$, whose graph is a diagonal line when A is the set of real numbers. Conceptually, the role of a diagonal morphism is to *duplicate* information, just as the projections *discard* information. In applications to physics, the equations $p_1 \Delta = 1_A$ and $p_2 \Delta = 1_A$ says that if we duplicate a state in A and then discard one of the two resulting copies, we are left with a copy identical to the original.

In Hilb, however, since the tensor product is not a product in the category-theoretic sense, it makes no sense to speak of a diagonal morphism $\Delta : H \to H \otimes H$. In fact, a stronger statement is true: there is no natural (i.e. basis-independent) way to choose a linear operator from H to $H \otimes H$ other than the zero operator. So, there is no way to duplicate information in quantum theory.

Since the tensor product is not a Cartesian product in the sense explained above, what exactly is it? To answer this, we need the definition of a 'monoidal category'. Monoidal categories were introduced by Mac Lane (1963) in the early 1960s, precisely in order to capture those features common to all categories equipped with a well-behaved but not necessarily Cartesian product. Since the definition is a bit long, let us first present it and then discuss it:

Definition. *A* **monoidal category** *consists of:*

(i) *a category* $\mathcal{M}$,
(ii) *a functor* $\otimes : \mathcal{M} \times \mathcal{M} \to \mathcal{M}$,
(iii) *a* **unit object** $I \in \mathcal{M}$,
(iv) *natural isomorphisms called the* **associator***:*

$$a_{A,B,C} : (A \otimes B) \otimes C \to A \otimes (B \otimes C),$$

the **left unit law***:*

$$\ell_A : I \otimes A \to A,$$

and the **right unit law***:*

$$r_A : A \otimes I \to A,$$

such that the following diagrams commute for all objects $A, B, C, D \in \mathcal{M}$*:*

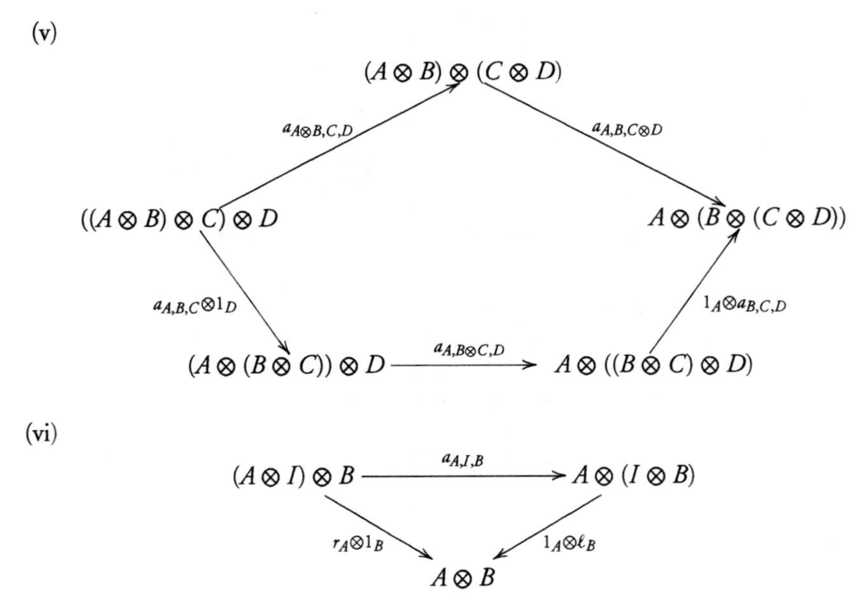

This obviously requires some explanation! First, it makes use of some notions we have not explained yet, ruining our otherwise admirably self-contained treatment of category theory. For example, what is $\mathcal{M} \times \mathcal{M}$ in clause (ii) of the definition? This is just the category whose objects are pairs of objects in $\mathcal{M}$, and whose morphisms are pairs of morphisms in $\mathcal{M}$, with composition of morphisms done componentwise. So, when we say that the tensor product is a functor $\otimes \colon \mathcal{M} \times \mathcal{M} \to \mathcal{M}$, this implies that for any pair of objects $x, y \in \mathcal{M}$ there is an object $x \otimes y \in \mathcal{M}$, while for any pair of morphisms $f \colon x \to x', g \colon y \to y'$ in $\mathcal{M}$ there is a morphism $f \otimes g \colon x \otimes y \to x' \otimes y'$ in $\mathcal{M}$. Morphisms are just as important as objects! For example, in Hilb, not only can we take the tensor product of Hilbert spaces, but also we can take the tensor product of bounded linear operators $S \colon H \to H'$ and $T \colon K \to K'$, obtaining a bounded linear operator

$$S \otimes T \colon H \otimes K \to H' \otimes K'.$$

In physics, we think of $S \otimes T$ as a joint process built from the processes S and T 'running in parallel'. For example, if we have a joint quantum system whose two parts evolve in time without interacting, any time evolution operator for the whole system is given by the tensor product of time evolution operators for the two parts.

Similarly, in nCob the tensor product is given by disjoint union, both for objects and for morphisms. In Figure 8.8 we show two spacetimes M and M' and their tensor product $M \otimes M'$. This as a way of letting two spacetimes 'run in parallel', like independently evolving separate universes. The resemblance to the tensor product of morphisms in Hilb should be clear. Just as in Hilb, the tensor product in nCob is not a Cartesian product: there are no projections with the required properties. There is also no natural choice of a cobordism from S to $S \otimes S$. This means that the very nature of topology prevents us from finding spacetimes that 'discard' part

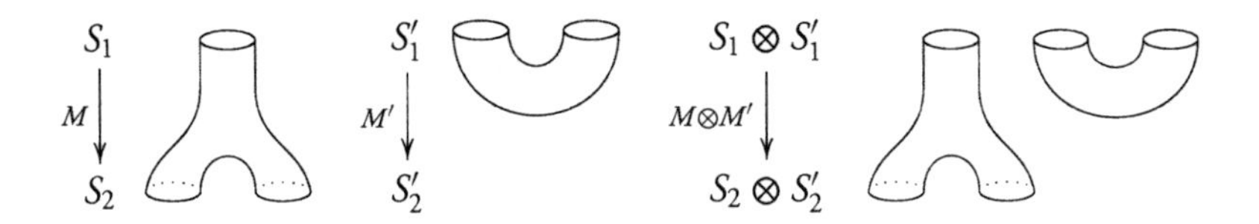

Figure 8.8. Two cobordisms and their tensor product

of space, or 'duplicate' space. Seen in this light, the fact that we cannot discard or duplicate information in quantum theory is not a flaw or peculiarity of this theory. *It is a further reflection of the deep structural analogy between quantum theory and the conception of spacetime embodied in general relativity.*

Turning to clause (iii) in the definition, we see that a monoidal category needs to have a 'unit object' I. This serves as the multiplicative identity for the tensor product, at least up to isomorphism: as we shall see in the next clause, $I \otimes A \cong A$ and $A \otimes I \cong A$ for every object $A \in \mathcal{M}$. In Hilb the unit object is $\mathbb{C}$ regarded as a Hilbert space, while in nCob it is the empty set regarded as an $(n-1)$-dimensional manifold. Any category with finite products gives a monoidal category in which the unit object is the terminal object 1.

This raises an interesting point of comparison. In classical physics we describe systems using objects in a category with finite products, and a state of the system corresponding to the object A is just a morphism $f\colon 1 \to A$. In quantum physics we describe systems using Hilbert spaces. Is a state of the system corresponding to the Hilbert space H the same as a bounded linear operator $T\colon \mathbb{C} \to H$? Almost, but not quite! As we saw in §8.3, such operators are in one-to-one correspondence with vectors in H: any vector $\psi \in H$ corresponds to an operator $T_\psi\colon \mathbb{C} \to H$ with $T_\psi(1) = \psi$. States, on the other hand, are the same as unit vectors modulo phase. Any non-zero vector in H gives a state after we normalize it, but different vectors can give the same state, and the zero vector does not give a state at all. So, quantum physics is really different from classical physics in this way: we cannot define states as morphisms from the unit object. Nonetheless, we have seen that the morphisms $T\colon \mathbb{C} \to H$ play a fundamental role in quantum theory: they are just Dirac's 'kets'.

Next, let us ponder clause (iv) of the definition of monoidal category. Here we see that the tensor product is associative, but only *up to a specified isomorphism*, called the 'associator'. For example, in Hilb we do not have $(H \otimes K) \otimes L = H \otimes (K \otimes L)$, but there is an obvious isomorphism

$$a_{H,K,L}\colon (H \otimes K) \otimes L \to H \otimes (K \otimes L)$$

given by

$$a_{H,K,L}((\psi \otimes \phi) \otimes \eta) = \psi \otimes (\phi \otimes \eta).$$

Similarly, we do not have $\mathbb{C} \otimes H = H$ and $H \otimes \mathbb{C} = H$, but there are obvious isomorphisms

$$\ell_H\colon \mathbb{C} \otimes H \to H, \qquad r_H\colon H \otimes \mathbb{C} \to H.$$

Moreover, all these isomorphisms are 'natural' in a precise sense. For example, when we say the associator is natural, we mean that for any bounded linear operators $S\colon H \to H'$, $T\colon K \to K'$, $U\colon L \to L'$ the following square diagram commutes:

$$\begin{array}{ccc}
(H \otimes K) \otimes L & \xrightarrow{a_{H,K,L}} & H \otimes (K \otimes L) \\
\Big\downarrow{\scriptstyle (S\otimes T)\otimes U} & & \Big\downarrow{\scriptstyle S\otimes(T\otimes U)} \\
(H' \otimes K') \otimes L' & \xrightarrow{a_{H',K',L'}} & H' \otimes (K' \otimes L')
\end{array}$$

In other words, composing the top morphism with the right-hand one gives the same result as composing the left-hand one with the bottom one. This compatibility condition expresses the fact that no arbitrary choices are required to define the associator: in particular, it is defined in a basis-independent manner. Similar but simpler 'naturality squares' must commute for the left and right unit laws.

Finally, what about clauses (v) and (vi) in the definition of monoidal category? These are so-called 'coherence laws', which let us manipulate isomorphisms with the same ease as if they were equations. Repeated use of the associator lets us construct an isomorphism from any parenthesization of a tensor product of objects to any other parenthesization—for example, from $((A \otimes B) \otimes C) \otimes D$ to $A \otimes (B \otimes (C \otimes D))$. However, we can actually construct *many* such isomorphisms—and in this example, the pentagonal diagram in clause (v) shows two. We would like to be sure that all such isomorphisms from one parenthesization to another are equal. In his fundamental paper on monoidal categories, Mac Lane (1963) showed that the commuting pentagon in clause (v) guarantees this, not just for a tensor product of four objects, but for arbitrarily many. He also showed that clause (vi) gives a similar guarantee for isomorphisms constructed using the left and right unit laws.

8.5 CONCLUSIONS

Our basic intuitions about mathematics are to some extent abstracted from our dealings with the everyday physical world (Lakoff and Núñez 2000). The concept of a *set*, for example, formalizes some of our intuitions about piles of pebbles, herds of sheep, and the like. These things are all pretty well described by classical physics, at least in their gross features. For this reason, it may seem amazing that mathematics based on set theory can successfully describe the microworld, where quantum physics reigns supreme. However, beyond the overall 'surprising effectiveness of mathematics', this should not really come as a shock. After all, set theory is sufficiently flexible that any sort of effective algorithm for making predictions can be encoded in the language of set theory: even Peano arithmetic would suffice.

But, we should not be lulled into accepting the primacy of the category of sets and functions just because of its flexibility. The mere fact that we *can* use set theory as a framework for studying quantum phenomena does not imply that this is the most enlightening approach. Indeed, the famously counter-intuitive behaviour of the microworld suggests that not only set theory but even classical logic is not optimized for understanding quantum systems. While there are no real paradoxes, and one can compute everything to one's heart's content, one often feels that one is grasping these systems 'indirectly', like a nuclear power plant operator handling radioactive material behind a plate glass window with robot arms. This sense of distance is reflected in the endless literature on 'interpretations of quantum mechanics', and also in the constant invocation of the split between 'observer' and 'system'. It is as if classical logic continued to apply to us, while the mysterious rules of quantum theory apply only to the physical systems we are studying. But of course this is not true: we are part of the world being studied.

To the category theorist, this raises the possibility that quantum theory might make more sense when viewed, not from the category of sets and functions, but *within some other category*: for example Hilb, the category of Hilbert spaces and bounded linear operators. Of course it is most convenient to define this category and study it with the help of set theory. However, as we have seen, the fact that Hilbert spaces are sets equipped with extra structure and properties is almost a distraction when trying to understand Hilb, because its morphisms are not functions that preserve this extra structure. So, we can gain a new understanding of quantum theory by trying to accept Hilb on its own terms, unfettered by preconceptions taken from the category Set. As Corfield (2003) writes: 'Category theory allows you to work on structures without the need first to pulverise them into set theoretic dust. To give an example from the field of architecture, when studying Notre Dame cathedral in Paris, you try to understand how the building relates to other cathedrals of the day, and then to earlier and later cathedrals, and other kinds of ecclesiastical building. What you don't do is begin by imagining it reduced to a pile of mineral fragments.'

In this chapter, we have tried to say quite precisely how some intuitions taken from Set fail in Hilb. Namely: unlike Set, Hilb is a $*$-category, and a monoidal category where the tensor product is non-Cartesian. But, what makes this really interesting is that these ways in which Hilb differs from Set are precisely the ways it resembles nCob, the category of $(n-1)$-dimensional manifolds and n-dimensional cobordisms going between these manifolds. In general relativity these cobordisms represent 'spacetimes'. Thus, from the category-theoretic perspective, a bounded linear operator between Hilbert spaces acts more like a *spacetime* than a *function*. This not only sheds a new light on some classic quantum quandaries, it also bodes well for the main task of quantum gravity, namely to reconcile quantum theory with general relativity.

At best, we have only succeeded in sketching a few aspects of the analogy between Hilb and nCob. In a more detailed treatment we would explain how both Hilb and nCob are 'symmetric monoidal categories with duals'—a notion which subsumes being a monoidal category and a $*$-category. Moreover, we would explain how unitary topological quantum field theories exploit this fact to the hilt. However, a discussion of this can be found elsewhere (Baez and Dolan 1995), and it necessarily leads us into deeper mathematical waters which are not of such immediate philosophical interest. So, instead, I would like to conclude by saying a bit about the progress people have made in learning to think within categories other than Set.

It has been known for quite some time in category theory that each category has its own 'internal logic', and that while we can reason externally about a category using classical logic, we can also reason *within it* using its internal logic—which gives a very different perspective. For example, our best understanding of intuitionistic logic has long come from the study of categories called 'topoi', for which the internal logic differs from classical logic mainly in its renunciation of the principle of excluded middle (Boileau and Joyal 1981; Coste 1972; Mitchell 1972). Other classes of categories have their own forms of internal logic. For example, ever since the work of Lambek and Scott (1986), the typed lambda-calculus, so beloved by theoretical computer scientists has been understood to arise as the internal logic of 'Cartesian closed' categories. More generally, Lawvere's algebraic semantics allows us to see any 'algebraic theory' as the internal logic of a category with finite products (Lawvere 1963).

By now there are many textbook treatments of these ideas and their ramifications, ranging from introductions that do not assume prior knowledge of category theory (Crole 1993; McLarty 1995), to more advanced texts that do (Barr and Wells 1983; Johnstone 2002; Lambek and Scott 1986; MacLane and Moerdijk 1992). Lawvere has also described how to do classical physics in a topos (Lawvere 1979; Lawvere and Schanuel 1986). All this suggests that the time is ripe to try thinking about quantum physics using the internal logic of Hilb, or *n*Cob, or related categories. However, the textbook treatments and even most of the research literature on category-theoretic logic focus on categories where the monoidal structure is Cartesian. The study of logic within more general monoidal categories is just beginning. More precisely, while generalizations of 'algebraic theories' to categories of this sort have been studied for many years in topology and physics (Loday et al. 1997; Markl et al. 2002), it is hard to find work that explicitly recognizes the relation of such theories to the traditional concerns of logic, or even of quantum logic. For some heartening counter-examples, see the work of Abramsky and Coecke (2004), and also of Mauri (internet resource). So, we can only hope that in the future, more interaction between mathematics, physics, logic, and philosophy will lead to new ways of thinking about quantum theory—and quantum gravity—that take advantage of the internal logic of categories like Hilb and *n*Cob.

ACKNOWLEDGMENTS

I thank Aaron Lauda for preparing most of the figures in this chapter, and thank Bob Coecke, David Corfield, Daniel Ruberman, Issar Stubbe, and especially James Dolan for helpful conversations. I would also like to thank the Physics Department of UC Santa Barbara for inviting me to speak on this subject, and Fernando Souza for inviting me to speak on this subject at the American Mathematical Society meeting at San Francisco State University, both in October 2000.

REFERENCES

Abramsky, S., and B. Coecke (2004) "A Categorical Semantics of Quantum Protocols". ArXiv:quant-ph/0402130.

Arageorgis, A., J. Earman, and L. Ruetsche (2002) "Weyling the Time away: The Non-Unitary Implementability of Quantum Field Dynamics on Curved Spacetime". *Studies in the History and Philosophy of Modern Physics*, 33: 151–84.
Atiyah, M. F. (1989) "Topological Quantum Field Theories". *Publ. Math. IHES Paris*, 68: 175–86.
—— (1990) *The Geometry and Physics of Knots*. Cambridge: Cambridge University Press.
Baez, J. (1987) "Bell's Inequality for C*-Algebras". *Lett. Math. Phys.* 13: 135–6.
—— (1998) "Spin Foam Models." *Class. Quantum Grav.* 15: 1827–58. ArXiv:gr-qc/9709052.
—— (1999) "An Introduction to Spin Foam Models of Quantum Gravity and *BF* Theory". In H. Gausterer and H. Grosse (eds.), *Geometry and Quantum Physics*. Berlin: Springer-Verlag (pp. 25–93). ArXiv:gr-qc/9905087.
—— (2001) "Higher-Dimensional Algebra and Planck-Scale Physics". In C. Callender and N. Huggett (eds.), *Physics Meets Philosophy at the Planck Length*. Cambridge: Cambridge University Press (pp. 177–95). ArXiv:gr-qc/9902017.
—— and J. Dolan (1995) "Higher-Dimensional Algebra and Topological Quantum Field Theory". *Jour. Math. Phys.* 36: 6073–105. ArXiv:q-alg/9503002.
Barr, M., and C. Wells (1983) *Toposes, Triples and Theories*. Berlin: Springer-Verlag. Revised and corrected version available at **www.cwru.edu/artsci/math/wells/pub/ttt.html**.
Bell, J. S. (1964) "On the Einstein–Podolsky–Rosen Paradox". *Physics*, 1: 195–200.
Boileau, A., and A. Joyal (1981) "La Logique des topos". *J. Symb. Logic*, 46: 6–16.
Corfield, D. (2003) *Towards a Philosophy of Real Mathematics*. Cambridge: Cambridge University Press.
Coste, M. (1972) *Langage interne d'un topos*. Paris: Seminaire Bénabou, Université Paris-Nord.
Crole, R. L. (1993) *Categories for Types*. Cambridge: Cambridge University Press.
Einstein, A., B. Podolsky, and N. Rosen (1935) "Can Quantum-Mechanical Description of Physical Reality be Considered Complete?" *Phys. Rev.* 47: 77.
Hawkins, E., F. Markopoulou, and H. Sahlmann (2003) "Evolution in Quantum Causal Histories". *Class. Quant. Grav.* 20: 3839–54. ArXiv:hep-th/0302111.
Johnstone, P. (2002) "Toposes as Theories". In *Sketches of an Elephant: A Topos Theory Compendium*, vol. ii. Oxford: Oxford University Press (pp. 805–1088).
Kanamori, A. (2003) "The Empty Set, the Singleton, and the Ordered Pair". *Bull. Symb. Logic*, 9: 273–98. Also available at **www.math.ucla.edu/~asl/bsl/0903/0903-001.ps**.
Lakoff, G., and R. Núñez (2000) *Where Mathematics Comes From: How the Embodied Mind Brings Mathematics into Being*. New York: Basic Books.
Lambek, J. (1980) "From λ-Calculus to Cartesian Closed Categories". In J. P. Seldin and J. R. Hindley (eds.), *To H. B. Curry: Essays on Combinatory Logic, Lambda Calculus and Formalism*. New York: Academic Press (pp. 375–402).
—— and P. J. Scott (1986) *Introduction to Higher-Order Categorical Logic*. Cambridge: Cambridge University Press.
Lawvere, F. W. (1963) 'Functorial Semantics of Algebraic Theories'. Ph.D. dissertation, Columbia University.
—— (1979) "Categorical Dynamics". In *Proceedings of Aarhus May 1978 Open House on Topos Theoretic Methods in Geometry*. Aarhus: Aarhus University Press.
—— (1980) "Toward the Description in a Smooth Topos of the Dynamically Possible Motions and Deformations of a Continuous Body". *Cahiers de Top. et Géom. Diff. Cat.* 21: 337–92.
—— and S. Schanuel (eds.) (1986) *Categories in Continuum Physics*. Berlin: Springer-Verlag.
Loday, J.-L., J. Stasheff, and A. Voronov (eds.) (1997) *Operads: Proceedings of Renaissance Conferences*. Providence, RI: American Mathematical Society.

Mac Lane, S. (1963) "Natural Associativity and Commutativity". *Rice Univ. Stud.* 49: 28–46.
——and I. Moerdijk (1992) *Sheaves in Geometry and Logic: A First Introduction to Topos Theory*. Berlin: Springer-Verlag.
McLarty, C. (1995) *Elementary Categories, Elementary Toposes*. Oxford: Clarendon Press.
Markl, M., S. Shnider, and J. Stasheff (2002) *Operads in Algebra, Topology and Physics*. Providence, RI: American Mathematical Society.
Mauri, L. (internet resource) "Algebraic Theories in Monoidal Categories". Available at **www.math.rutgers.edu/~mauri.**
Mitchell, W. (1972) "Boolean Topoi and the Theory of Sets". *J. Pure Appl. Alg.* 2: 261–74.
Segal, G. (2004) "The Definition of a Conformal Field Theory". In U. L. Tillmann (ed.), *Topology, Geometry and Quantum Field Theory: Proceedings of the 2002 Oxford Symposium in Honour of the 60th Birthday of Graeme Segal.* Cambridge: Cambridge University Press.
Smolin, L. (2001) *Three Roads to Quantum Gravity*. New York: Basic Books.
——(2003) "How Far are We from the Theory of Quantum Gravity?" ArXiv:hep-th/0303185.
Vaas, R. (2004) "The Duel: Strings Versus Loops." Trans. M. Bojowald and A. Sen. ArXiv:physics/0403112.
Wootters, W. K., and W. H. Zurek (1982) "A Single Quantum Cannot Be Cloned". *Nature*, 299: 802–3.

Index